AF739669

Laboratory Manual for Electrical Machinery Control Fundamentals

RIVER PUBLISHERS SERIES IN ENERGY ENGINEERING AND SYSTEMS

Series Editor

BOBBY RAUF
Semtrain LLC, USA

The "River Publishers Series in Energy Engineering and Systems" is a series of comprehensive academic and professional books focussing on the theory and applications behind various energy-related technologies and control systems. The series features handbooks for related technology, as well as manuals on the fundamentals and theoretical aspects of energy engineering projects.

The main aim of this series is to be a reference for academics, researchers, managers, engineers, and other professionals in related matters with energy engineering and control systems.

Topics covered in the series include, but are not limited to:

- Energy engineering;
- Machinery;
- Testing;
- HVAC;
- Electricity and electronics;
- Water systems;
- Industrial control systems;
- Web based systems;
- Automation;
- Lighting controls.

For a list of other books in this series, visit www.riverpublishers.com

Laboratory Manual for Electrical Machinery Control Fundamentals

Dale R. Patrick
USA

Stephen W. Fardo
Department of Applied Engineering and Technology,
Eastern Kentucky University, USA

Vigyan (Vigs) Chandra
Eastern Kentucky University, USA

Brian W. Fardo
Department of Applied Engineering and Technology and
the Noel Studio for Academic Creativity,
Eastern Kentucky University, USA

River Publishers

Routledge
Taylor & Francis Group
NEW YORK AND LONDON

Published 2026 by River Publishers
River Publishers
Broagervej 10, 9260 Gistrup, Denmark
www.riverpublishers.com

Distributed exclusively by Routledge
605 Third Avenue, New York, NY 10017, USA
4 Park Square, Milton Park, Abingdon, Oxon OX14 4RN

Laboratory Manual for Electrical Machinery Control Fundamentals / Dale R. Patrick, Stephen W. Fardo, Vigyan (Vigs) Chandra and Brian W. Fardo.

Routledge is an imprint of the Taylor & Francis Group, an informa business

ISBN 978-87-438-1016-2 (paperback)
ISBN 978-87-438-1013-1 (online)
ISBN 978-8-743-81012-4 (ebook master)
ISBN 978-87-438-1015-5 (epub)

DOI: 10.1201/9788743810124

Contents

Preface

This Laboratory/Activities manual is designed for use with the **Electrical Machinery Control Fundamentals** textbook. The activities in this manual are closely tied to the textbook. The activities have been divided into seven units in parallel with the text.

Unit I – Electrical Basics

Unit II – Electrical Distribution Basics

Unit III – Electric Motor Basics

Unit IV – Electrical Machinery Control Basics

Unit V – Electronic Control Basics

Unit VI – Industrial Electronic Control Systems

Unit VII – Digital Control Basics and Applications

Each laboratory activity reinforces the important concepts in the corresponding unit from the textbook. Laboratory activities begin with the learning objective and a list of equipment needed to complete the activity. Following the list of equipment, step-by-step instructions are included to take you through the laboratory. In these laboratories, you are expected to set up electrical circuits, test motors, collect data, and answer questions. Charts are provided in many of the activities for your data.

**Important Note:

Working with electricity and machinery can be dangerous. Always use safe laboratory practices. These laboratory activities are designed to be used in an engineering or technical laboratory with proper **supervision**. If you have any questions about **safe laboratory practices**, discuss them with your instructor, supervisor, or other skilled professional.

Safety is always a primary factor for laboratory activities. Some lab activities should only be completed with supervision. You may complete certain steps of the "Procedure" and the "Analysis" for better understanding without applying voltage.

Hands-on learning is very important. These laboratory activities are designed to increase your understanding of electrical machinery control. Take your time while doing these activities. Your knowledge and skills will be enriched.

Dale R. Patrick
Stephen W. Fardo
Vigyan (Vigs) Chandra
Brian W. Fardo

List of Figures

List of Abbreviations

AWG	American Wire Gage
BCD	Binary coded decimal
BCO	Binary-coded octal
EMF	Electromotive force
GF	Gauge factor
LASCR	Light-activated silicon controlled rectifier
LDR	Light-dependent resistor
LED	Light-emitting diode
LVDT	Linear variable differential transformer
MSI	Medium-scale integration
NC	Normally closed
NEC	National Electrical Code
NO	Normally open
SCR	Silicon-controlled rectifier
SPST	Single-pole, single-throw

UNIT I

Electrical Basics

Activity 1

Electrical Conductors

Objectives

In this activity, you will measure the diameter of several electrical conductors using a wire gauge.

Equipment and Materials

- Wire gauge
- Electrical conductors of various sizes

Procedure

1. Obtain a wire gauge (see Figure 1.1) and several conductors. The wire gauge will give conductor readings in AWG (American Wire Gage).
2. Measure the diameter of several conductors with the gauge.
3. Record the values.
4. Along with the AWG value, record whether the conductor is copper or aluminum.
5. If an insulation type is marked on the wire, record the insulation type also.

DOI: 10.1201/9788743810124-2

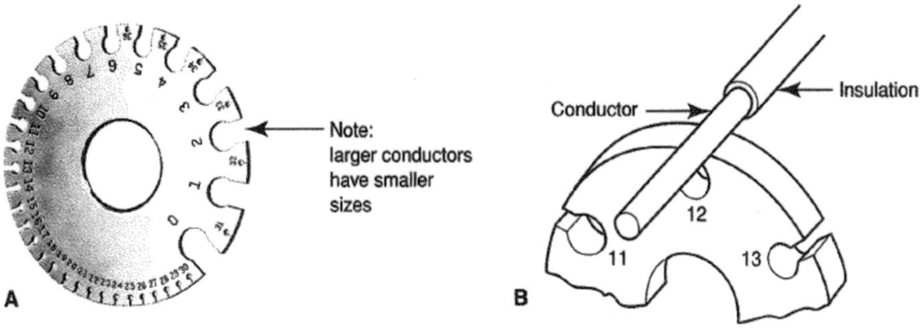

Figure 1.1 (A) American Wire Gage is used to measure conductor size. (B) Using an AWG.

Activity 2

Electrical Tool Identification

Objectives

In this activity, you will learn to correctly identify the types of tools used by electrical technicians. It is important to know the correct names of the tools used by technically skilled individuals.

Equipment and Materials

- Electrical tools used in your lab or shop

Procedure

1. Study the illustrations of electrical tools in Chapter 2 of your textbook.
2. Learn the **correct** names of these tools.
3. Learn the uses of the tools as you study them.
4. Several common tools are shown in Figure 2.1. Provide the correct names of the tools shown in Figure 2.1.
5. The instructor/supervisor can now give you a test to see how many of the electrical tools used in your laboratory or shop you can identify.

DOI: 10.1201/9788743810124-3

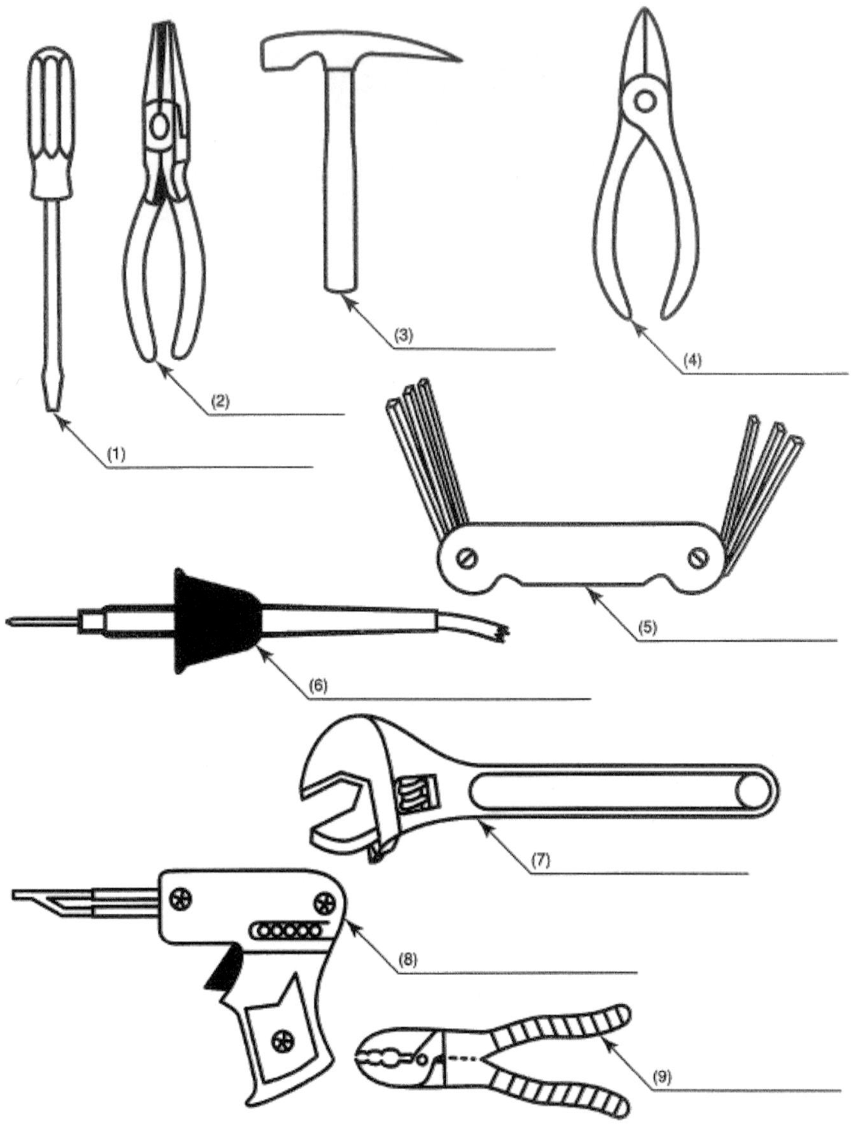

Figure 2.1 Electrical hand tools. Correctly fill in the blanks with the name of each tool.

Activity 3

Soldering and Terminal Connections

Objectives

In this activity, you will demonstrate your skill in three important exercises: (1) soldering, (2) desoldering, and (3) terminal connection. An understanding of the proper soldering and terminal connection techniques is needed by technicians who design, service, or repair equipment.

Equipment and Materials

- Soldering iron – 25 to 50 W
- Desoldering tool
- Solder
- No. 22 insulated wire – length
- Printed circuit board
- Component-mounting strip
- Resistor (any value)
- Wire stripper
- Side-cutting pliers
- Needle-nose pliers
- “Solderless” connector and wire
- Crimping tool

DOI: 10.1201/9788743810124-4

Safety

Exercise caution while soldering. Use protective eyewear and clothing during soldering operations.

Procedure

Section A: Soldering iron preparation and component mounting

1. Obtain a soldering iron and some solder.
2. Allow the soldering iron to heat to its operating temperature.
3. Apply a small amount of solder to tin the tip of the iron.
4. If available, obtain a component-mounting strip and resistor.
5. Using the proper soldering techniques found in Chapter 2 of your textbook, solder the resistor onto the component-mounting strip.
6. Show the component-mounting strip to the instructor/supervisor for approval.

Approval: ______________________________

Section B: Printed circuit board soldering

1. Obtain a printed circuit board, a length of No. 22 insulated wire, and a pair of wire strippers.
2. Place the wire near the holes in the printed circuit board where you intend to solder it to determine the length of wire required.
3. Strip 1/4" to 3/8" of insulation from one end of the wire with the wire stripper.
4. Cut off any excess wire and strip the insulation off the other end of the wire. The wire should lie flat on the printed circuit board when soldered.
5. Using proper soldering techniques, solder both ends of the wire to the printed circuit board terminals.
6. Show the printed circuit board to the instructor/supervisor for approval.

Approval:______________________________

Section C: Printed circuit board desoldering

1. Obtain a desoldering tool.
2. Allow the desoldering tool to heat to its operating temperature.
3. Using the proper desoldering techniques discussed in Chapter 2 of your textbook, remove the piece of wire from the printed circuit board without damaging the conductive strips on the board.
4. Show the printed circuit board to the instructor/supervisor for approval.

Approval:____

Section D: Terminal connection installation

1. Obtain a "solderless" terminal connector crimping tool and a length of insulated wire from your instructor.
2. Strip enough insulation from the wire so that the terminal connector fits properly.
3. Use the crimping tool to fasten the terminal connector to the wire.
4. Show the terminal connector and wire to the instructor/supervisor for approval.

Approval:____

Activity 4

Basic Electrical Symbols

Objectives

Working with control systems requires the ability to read and recognize electrical schematic symbols. Electrical schematics are used to maintain, adjust, and repair electronic systems and controllers. Many of these symbols will be used in the activities that follow.

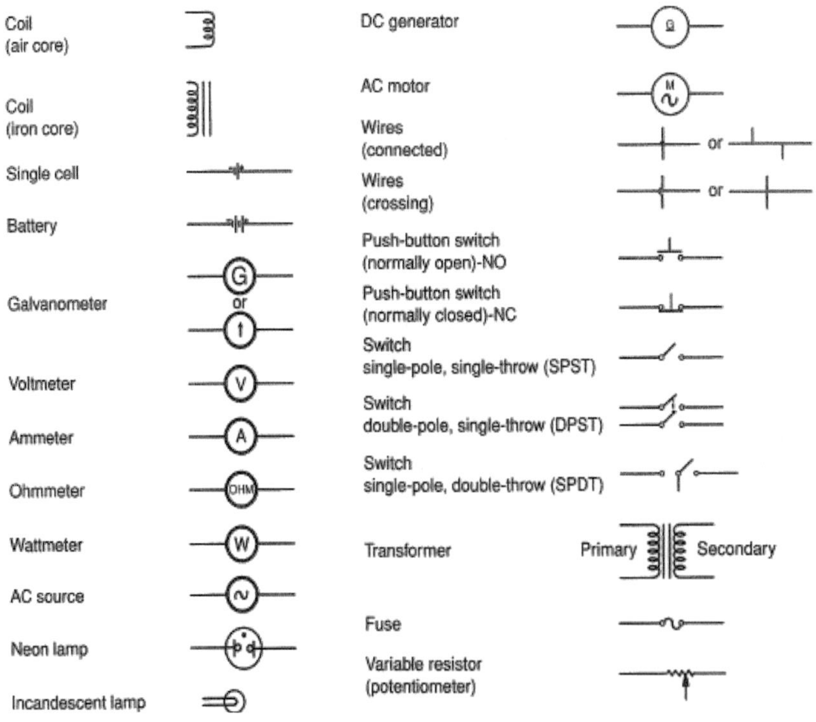

Figure 4.1 Electrical schematic symbols.

DOI: 10.1201/9788743810124-5

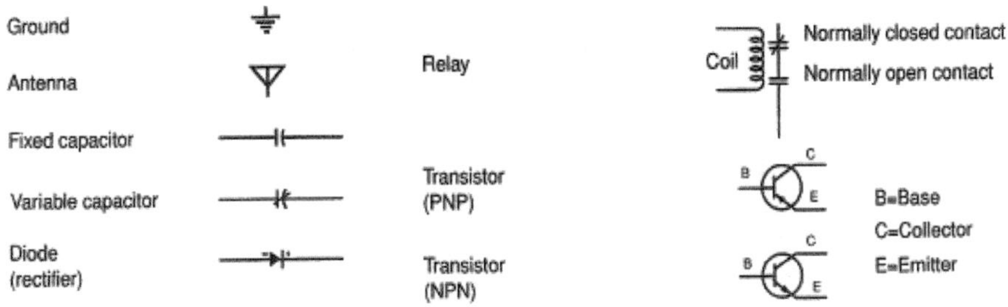

Figure 4.1 Continued

In this activity, you should study and learn to recognize the basic symbols shown in Figure 4.1. You can demonstrate your understanding by completing the self-test on electrical symbols.

Analysis

Without looking at the symbols in Figure 4.1, identify each of the following symbols. Provide the correct response for each.

1. ______

2. 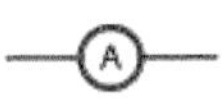 ______

3.  ______

4. 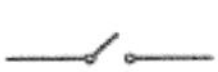 ______

5.  ______

6.  ______

7.  ______

8. 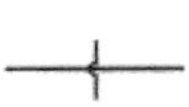 ______

9. 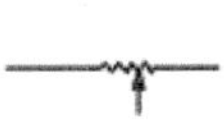 ______

10. ______

11. 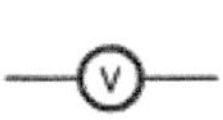 ______

12. ______

Activity 5

Electrical Components, Equipment, and Symbols

Objectives

Basic to the study of any technical subject is the understanding of the language and symbols used. The study of control technology is dependent on the graphic language of electrical diagrams. Various components, equipment, and symbols are used. You should become familiar with the graphic symbols used in electrical diagrams and learn to recognize the symbols used for common electrical devices.

Equipment and Materials

- Multi meter
- Soldering iron – 25 to 50 W
- SPST switch
- Potentiometer (any value)
- Connecting wires for the circuit board
- Battery – 6 volt
- Lamp with socket – 6 volt

Safety

Exercise caution while soldering. Use protective eyewear and clothing during soldering operations.

DOI: 10.1201/9788743810124-6

Procedure

1. The wires you will use in this activity are paths for the movement of electrons. Wires can be connected to each other at almost any angle and can cross each other without being connected. The graphic symbols representing these wires, called conductors, are illustrated in Figure 5.1.

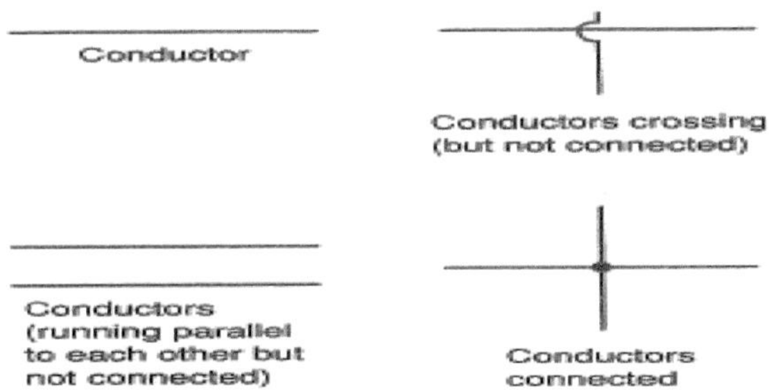

Figure 5.1 Different schematic symbols for conductors.

2. In the illustration of Figure 5.2, how many times do conductors cross each other without connecting, and how many times are conductors connected to each other?

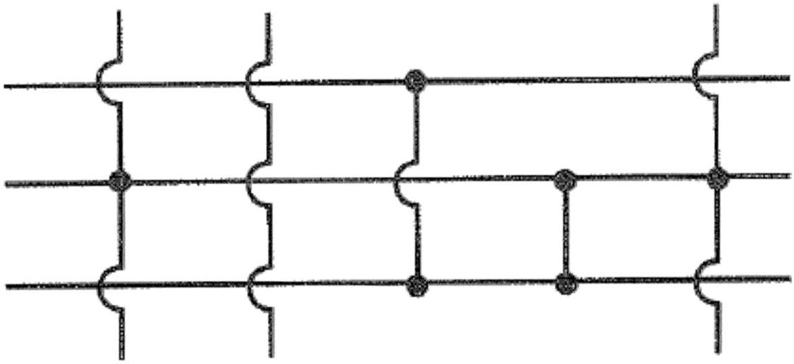

Figure 5.2 Various conductors in a schematic.

Conductors cross each other_______ times.

Conductors are connected to each other___________ times.

3. A single-pole, single-throw (SPST) switch is a device used to allow current to flow when closed, or on, and interrupt the flow of current when open, or off. A switch is used to turn the light in a room on and off. The symbols representing the SPST switch in its two conditions, on and off, are illustrated in Figure 5.3.

Figure 5.3 Schematic symbols for a single-pole, single-throw switch.

4. Fixed resistors are used frequently to control current. The symbol for a fixed resistor is illustrated in Figure 5.4.

Figure 5.4 Fixed resistor.

5. A light bulb or lamp is a common component used in flashlights and other devices to produce light. The symbol of an incandescent lamp is shown in Figure 5.5.

Figure 5.5 Incandescent lamp.

6. A potentiometer, or pot, is a resistive component that can be adjusted to control current. The illustrations of Figure 5.6 represent the schematic symbols for a potentiometer. Notice that the center connection of the potentiometer is designated by an arrow. It represents the adjustable portion of the device.

Figure 5.6 Schematic symbols for a potentiometer.

7. The battery is a chemical source of electrical energy that causes current to flow through conductors, resistors, and other components. The symbol in Figure 5.7 is used for a battery. Note that one of the battery connectors is labeled with a + sign and the other is labeled with a – sign. It is always important to observe how a battery is connected to a circuit.

Figure 5.7 Battery.

8. Use your circuit board or trainer unit to set up the circuit as shown in Figure 5.8. Two types of circuit boards that can be used for your experiment are shown in Figure 5.9. You should study the layout of the board you are using. Be sure that you connect this circuit and the others you will construct properly. You should become very familiar with the uses of a circuit board. For most circuit boards, it is necessary to cut wire of the proper diameter to various sizes and strip about ¼ inch of insulation from both ends. The wires are then used to make circuit connections. The switch and potentiometer should have wires of about 2 inches long soldered to their terminals for connecting to the circuit board. They will be used for other experiments.

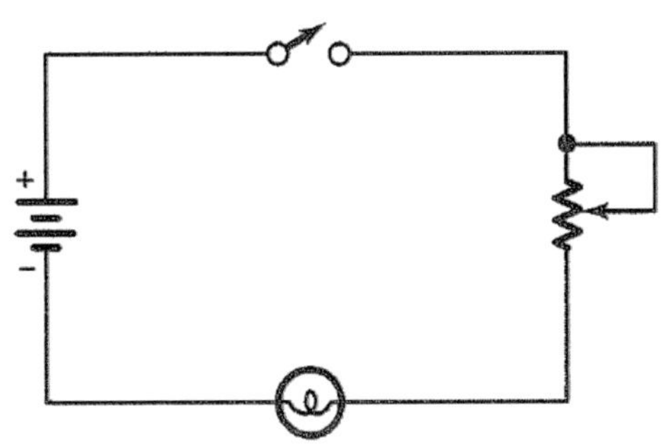

Figure 5.8 Series circuit with potentiometer.

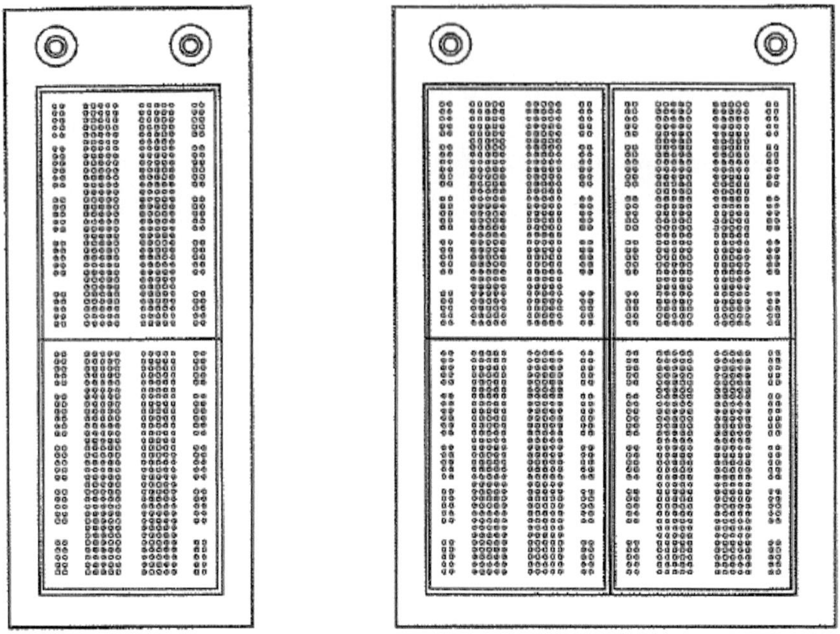

Figure 5.9 Circuit trainer board.

9. Complete the circuit connection using your board. Close the SPST switch and adjust the potentiometer from its maximum counterclockwise position to its maximum clockwise position. What happens to the lamp? If the circuit is constructed properly, the lamp should get brighter and then dimmer as the potentiometer is adjusted.

10. Adjust the potentiometer until the lamp is at its brightest and open the SPST switch. What happened to the lamp?

Analysis

1. Draw the symbols for the indicated components in the space that follows.

   Fixed resistor

   Potentiometer

   Open SPST switch

2. Draw the symbols for conductors in the space below.

   Conductors crossing (not connected)

   Conductors connected

3. Draw the symbol for a battery, placing the positive sign and the negative sign at the proper sides of the symbol.

4. Draw the symbol for a lamp.

Activity 6

Resistor Color Code

Objectives

Resistors are one of the most often used electrical control components. In this activity, you will review the resistor color code and show your knowledge by identifying the values of several resistors on a self-test. Study the resistor color codes shown below, then complete the self-test.

Resistor Color Code

Color	1st Significant Figure	2nd Significant Figure	Multiplier	Tolerance
Black	0	0	1	
Brown	1	1	10	
Red	2	2	100	
Orange	3	3	1000	
Yellow	4	4	10,000	
Green	5	5	100,000	
Blue	6	6	1,000,000	
Violet	7	7	10,000,000	
Gray	8	8		
White	9	9		
Gold*	—	—	0.1	5%
Silver*	—	—	0.01	10%
No color	—	—		20%

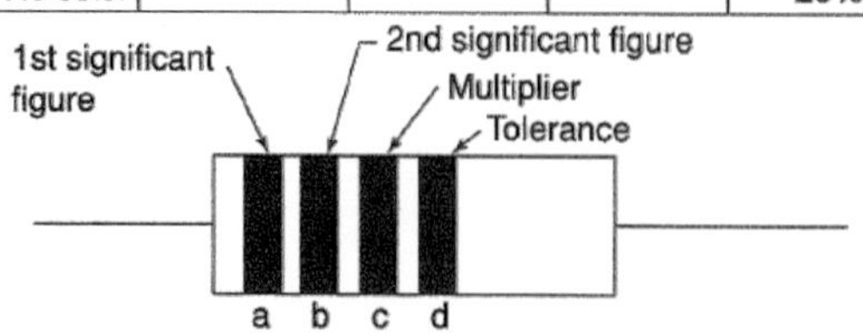

Resistor

DOI: 10.1201/9788743810124-7

- When resistors have a value of less than 10 ohms, the third color band is a decimal multiplier. The two colors used are gold and silver.

Resistor Color Code – Self Test

For each of the following color codes, identify the resistance value and tolerance of the resistor. Place your answer in the proper blank space.

1. Answer: ________ ohms ________, tolerance_______.
   (a) red
   (b) red
   (c) blue

2. Answer: ________ ohms ________ , tolerance______.
   (a) black
   (b) brown
   (c) brown
   (d) gold

3. Answer: ________ ohms ________, tolerance_______.
   (a) gray
   (b) red
   (c) black
   (d) silver

4. Answer: ________ ohms ________ , tolerance______.
   (a) white
   (b) brown
   (c) orange
   (d) gold

5. Answer: ________ ohms ________, tolerance_______.
   (a) yellow
   (b) violet
   (c) orange

6. Answer: ________ ohms ________, tolerance_______.
   (a) brown
   (b) green
   (c) orange

7. Answer: ________ ohms ________, tolerance_______.
   (a) blue
   (b) gray
   (c) black
   (d) gold

8. Answer: ________ ohms ________, tolerance______.
   (a) green
   (b) blue
   (c) silver
   (d) gold

9. Answer: ________ ohms ________, tolerance______.
   (a) brown
   (b) black
   (c) black

10. Answer: ________ ohms ________, tolerance______.
    (a) orange
    (b) white
    (c) black
    (d) gold

11. Answer: ________ ohms ________, tolerance______.
    (a) red
    (b) violet
    (c) gold
    (d) gold

12. Answer: ________ ohms ________, tolerance______.
    (a) brown
    (b) brown
    (c) black
    (d) gold

13. Answer: ________ ohms ________, tolerance______.
    (a) violet
    (b) green
    (c) orange
    (d) gold

14. Answer: ________ ohms ________, tolerance_______.
    (a) yellow
    (b) violet
    (c) green

15. Answer: ________ ohms ________, tolerance_______.
    (a) green
    (b) blue
    (c) orange

16. Answer: ________ ohms ________, tolerance_______.
    (a) blue
    (b) gray
    (c) orange

17. Answer: ________ ohms ________, tolerance_______.
    (a) yellow
    (b) orange
    (c) brown
    (d) gold

18. Answer: ________ ohms ________, tolerance_______.
    (a) orange
    (b) orange
    (c) red
    (d) gold

19. Answer: ________ ohms ________, tolerance_______.
    (a) brown
    (b) black
    (c) green

20. Answer: ________ ohms ________, tolerance_______.
    (a) red
    (b) red
    (c) red

Activity 7

Measuring Resistance

Objectives

Resistance is the term used to describe the opposition encountered by an electrical current. Resistance is measured in ohms.

In the previous activity, you learned how to determine the value of a resistor by observing its color bands. However, the resistance of many components cannot be determined by observation and therefore must be measured. In this activity, you will learn how to measure resistance with the ohmmeter portion of a multimeter. **If you are using a digital meter, utilize this activity to learn to read an analog scale to measure resistance. You may use the non-lab "Procedure Steps" and complete the "Analysis" items at the end.**

Equipment and Materials

- Multimeter
- Resistors
- Potentiometer

Procedure

1. The multimeter is used for basic electronic measurement. It can be used as an ammeter, a voltmeter, or an ohmmeter. Simply adjust the function selection switch for the desired function. Figure 7.1 shows the controls, including the function-select switch, of an analog multimeter.

DOI: 10.1201/9788743810124-8

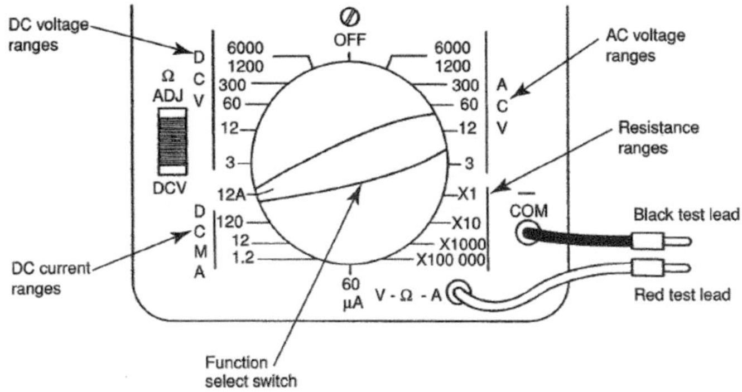

Figure 7.1 Controls of a typical multimeter.

You will notice that the function-select switch is in the center of some circular divisions and acts as a dial. You will also notice that a portion of the circular divisions is designated as the ohms function. Further, notice that the area of the divisions located within the ohms-function space is divided into four positions: X 1, X 100, X 1000 and X 100,000. The meter that you are using will be different from the one illustrated. Nevertheless, it will exhibit the same basic characteristics as the multimeter shown. Adjust the function-select switch on your meter to the ohms function. List below the different positions within the ohms-function area on your meter.

____

____

2. Now that you have adjusted the meter to the ohms function, connect the test leads to the meter. Notice that the test leads used with the meter are (ordinarily) red and black. The colors are used in order to help you distinguish between the positive and negative polarities of the meter. For our purposes, red indicates the positive polarity (+), and black indicates the negative (–) polarity. Insert the red test lead in the hole or jack marked with a (+) and the black test lead in the hole or jack marked with a (–) (note: the meter you are using may be different from the one illustrated in Figure 7.1. If this is so, plug the red and black test leads into the appropriate jacks and proceed). Touch the test leads together and describe what happens to the needle located on the scale of the meter.

____

____

3. You have now adjusted the function selection switch to the ohms function of the meter, inserted the proper test leads into the proper jacks, and seen that when the test leads are touched together, or shorted, the needle of the meter moves from the extreme left-hand side of the meter scale to the extreme right-hand side. If you are using a digital meter, the display will indicate 0. When the leads are not touched together, the digit 1 will appear on the left display in most cases.

____

____

4. You should now become familiar with the scale of the meter. Figure 7.2 is a scale of a multimeter. If you are using a digital meter, use this section as a review of analog scales.

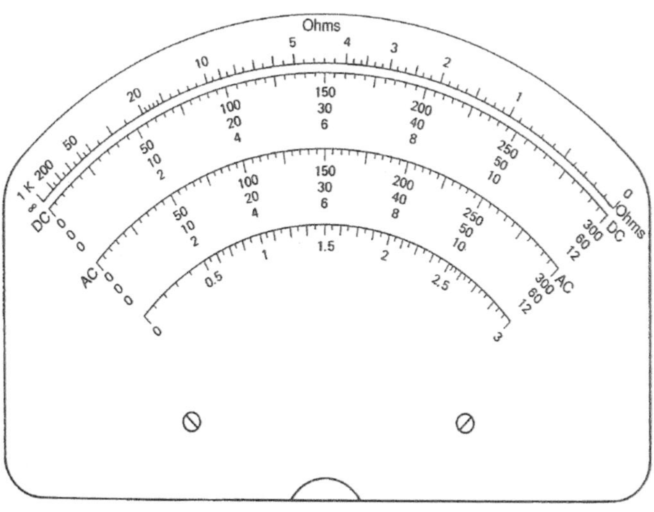

Figure 7.2 Multimeter scale.

You will notice that the top scale (from 0 to infinite, labeled **ohms**) is used for measuring ohms only. On most meters, the top scale is designated as the ohms scale. Record the location and color of the ohms scale used on your meter.

____

Location Color

5. To measure any resistance in ohms, you must select the proper position or range. On the meter illustrated in Figure 7.1, there are four ranges: _____ , _____, _____, and _____. The ohmmeter must be properly **zeroed** before attempting to accurately measure resistance. To zero the ohmmeter, connect the two test leads together. This should cause the needle to move from infinity on the left to zero (0) on the right. Infinity represents a very **high** resistance, whereas zero represents a very **low** resistance. If the needle does not reach zero or if it goes past zero when the test leads are shorted, then the control marked **ohms adjust** must be used to adjust the needle to rest on zero when the test leads are touched together. There should be a similar adjustment on your meter.

   Adjust the meter to each of its ranges and practice zeroing the meter for each of these ranges. The ohmmeter must be zeroed prior to each resistance measurement and after each range change. Otherwise, your measurements will be incorrect.

6. Notice that the ohms scale on the meter is **nonlinear**. The divisions on the right side of the scale are farther apart than those on the left side. This provides a more accurate measurement of resistance when the needle of the meter deflects and stops somewhere between the center of the ohms scale and zero. Choosing the proper range adjustment will control where the needle deflects. Adjust the function-select switch to the range of the ohms function. Zero the meter and then connect it across a resistor. Record the precise resistance (in ohms) indicated by the needle on the ohms scale. Adjust the function-select switch to the range of the ohms function. Zero the meter and then connect it across a potentiometer as shown below. Record the precise resistance (in ohms) indicated by the needle on the ohms scale. The range selected for this operation was _______, which means that the number to which the needle points will be multiplied by _________.

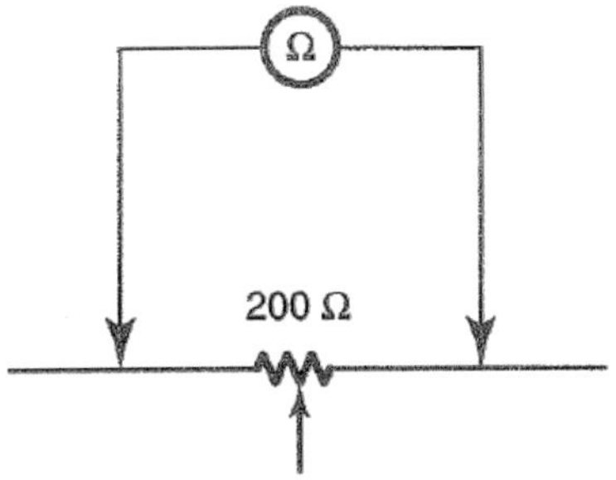

________________ Ω

7. Adjust the control of the potentiometer while the ohmmeter is connected. Describe how this action affects the measured resistance of the potentiometer.

8. Alter the meter connections to the potentiometer as illustrated below.

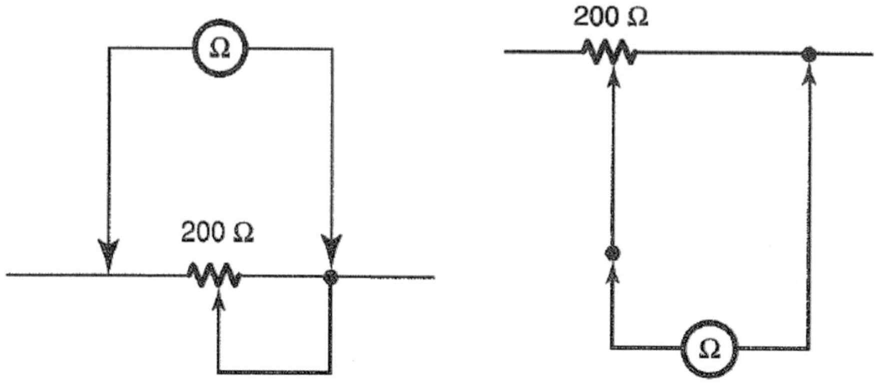

9. Adjust the potentiometer both clockwise and counterclockwise and describe how this action affects its measured resistance.

Analysis

1. Why is the ohms scale of the multimeter considered to be nonlinear?

2. Where on the ohms scale are the most accurate measurements found?

3. What is meant by the range on the ohmmeter?

4. What is meant by zeroing the ohmmeter?

5. Why is it necessary to zero the ohmmeter?

6. If the range of the ohmmeter was set to X 100,000 and the needle pointed to 0.6 on the ohms scale of Figure 7.2, what would be the value of the resistance being measured?

Activity 8

Measuring Voltage

Objectives

Voltage is the force that causes electrical current to flow through conductors or paths. This force, sometimes called electromotive force (EMF), is measured in volts. It is essential that anyone involved in electronic work be able to use the necessary equipment to measure voltage accurately. In this activity, you will:

1. Learn to measure voltage with a voltmeter.
2. Construct basic electrical circuits.

If you are using a digital meter, utilize this activity to learn to read an analog scale to measure voltage. You may use the non-lab "Procedure Steps" and complete the "Analysis" items at the end.

Equipment and Materials

- Multimeter
- Resistors
- Potentiometer
- Battery or power supply – 6 volt
- Lamp with socket – 6 volt
- Connecting wires

DOI: 10.1201/9788743810124-9

Procedure

1. You learned in the previous activity that a multimeter can be used to measure resistance. Likewise, a multimeter can be used to measure voltage. You will also recall from the previous activity that the function-select switch can be adjusted to cause the meter to perform many measurement functions using several different ranges. When the function-select switch is adjusted to 3 V on the DC volts range, the meter is capable of measuring a **maximum** of 3 V. The same is true for the remaining ranges within the DC volts function. The numerical value of the chosen range indicates the maximum value of voltage that can be measured on the range. The meter you are using will differ from the multimeter that is illustrated in Figure 8.1. Nevertheless, it will exhibit the same basic characteristics as the one described.

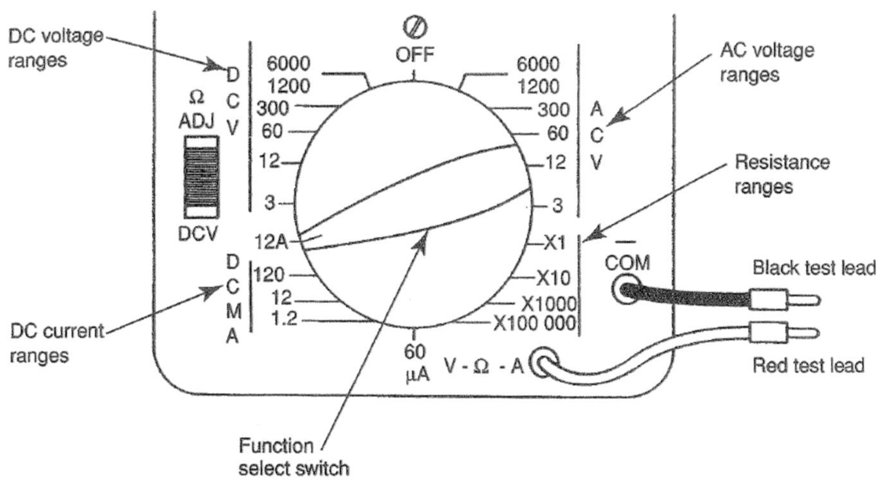

Figure 8.1 Sketch of multimeter settings.

2. Adjust the function-select switch on your meter to the lowest range of the DC volts function. List the different ranges within the DC volts function on your meter.

_______________, _______________, _______________, _______________

3. Connect the red and black test leads to the meter by inserting the appropriate ends into the proper jacks on the face of the meter. The red test lead should be inserted into the jack labeled + and the black test lead should be inserted into the jack labeled – COM (note: the meter you are using may be different. If this is the case, plug the red and black test leads into the appropriate jacks and proceed).

4. You now have the multimeter properly equipped and adjusted to measure DC volts. You should now become familiar with that portion of the scale of the meter used for measuring DC volts. **Part A** of Figure 8.2 shows the scale of a very commonly used multimeter. Notice that the scale immediately under the ohms scale is the DC volts scale. You will also notice that there are three DC volts scales: 0-12 V, 0-60 V and 0-300 All DC voltages are measured using one of these scales. The question becomes, what scales are to be used with what meter ranges? Notice that each of the DC voltage ranges corresponds to a number on the right side of the meter scale or a multiple or divisor of that number.

5. You will notice that when you choose the 12V, 60V or 300V range, you read the scale directly. On these ranges, the number to which the needle points is the actual value of the voltage being measured. When you select the 3 V range, the number to which the needle points must be divided by 100. For example, if the needle of the meter points to the number 50 while the meter is adjusted to the 3 V range, the measured voltage is 0.5 V (note: the meter you are using will differ from the one just described. Generally, the meter you are using will have several scales. Some of these scales can be read directly, whereas others will require the use of a multiplier or divisor. Identify the scales which can be read directly and the ones which must be used with multipliers or divisors).

6. Complete **Part B** of Figure 8.2 by indicating the proper voltage scale and voltage value. Also show the multiplier or divisor for the voltage range for the meter scale shown. This will give you practice using a meter scale.

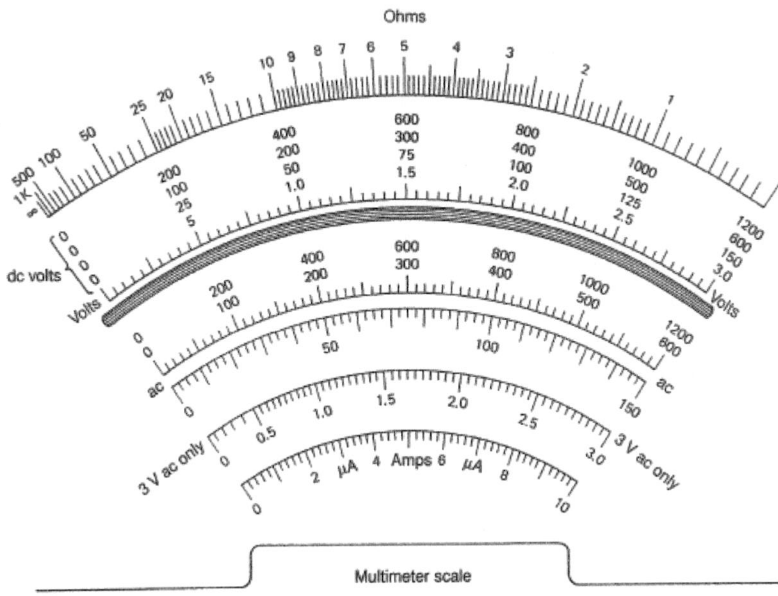

A

Multimeter range	Proper scale to read	Number needle points to	Voltage value	Multiplier	Divisor
Example: 3 V	0 to 3	1.5	1.5	None	None
3 V		2.0			
15 V		100			
60 V		300			
150 V		75			
600 V		200			
1200 V		800			
15 V		125			
60 V		350			
150 V		125			
60 V		125			
15 V		75			

B

Figure 8.2 Multimeter scale and range settings.

7. You are now almost ready to measure DC voltage. Before actually doing so you should become familiar with some facts about choosing the DC voltage range. You will recall from Step 1 that the numerical value of the chosen range indicates the **maximum** value of voltage that can be measured on that range. When the range selected is 12 V, the maximum voltage that the meter can measure is 12 V. Any voltage above 12 V

could damage the meter on this range. This is true for most voltage scales on most multimeters. If you are trying to measure a voltage that is totally unknown (no indications as to its approximate value), you should start the measuring procedure by choosing the **highest** range on your meter and slowly adjusting the range downward until a voltage reading is indicated on the upper or right-hand half of the meter scale. Complete Figure 8.3 by showing the proper meter range to be selected, the proper meter scale to be read, and the proper multiplier and divisor to be used with each voltage to be measured. Use the scale of your laboratory meter and complete the figure.

Voltage to be measured	Range to select	Scale to read	Multiplier	Divisor
Example: 2.0 V	3V	0 to 3.0	None	None
3.5 V				
0.5 V				
6 V				
14 V				
17 V				
30 V				
65 V				
100 V				
250 V				
1100 V				
Unknown				

Figure 8.3 Complete the chart.

8. You also need to consider meter **polarity** when measuring DC voltage. Correct matching of meter polarity to the voltage polarity is vitally important. If proper care is not exercised, the meter will deflect backwards, causing damage to its internal components. Meter polarity is simple to determine. The positive red test is connected to the positive side of the DC voltage to be measured. The negative black test lead is connected to the negative side of the DC voltage to be measured.

9. Measure and record the actual voltage of a battery. The voltmeter is always connected **across** the voltage, or in parallel with the voltage to be measured. Actual voltage of battery ________ V.

10. Construct the circuit illustrated. Measure and record the voltage supplied by the battery and the voltage across the lamp.

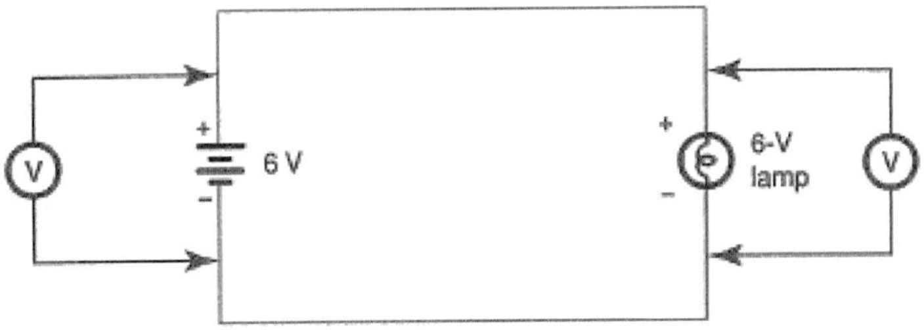

Battery voltage = ________________V

Lamp voltage = ________________V

11. A certain amount of voltage is always necessary to cause current to flow. This voltage is called the **voltage drop** and is found across any component through which current flows. The polarity of this voltage drop is determined by the direction of the movement of electrons from negative to positive. Thus, if electrons were moving through a resistor in the direction shown in the left side of the resistor would be negative (–) and the right side would be positive (+). The black test lead of the meter would be connected to the left side of the resistor, and the red test lead would be connected to the right side. This procedure would measure the voltage drop developed across this resistor. The illustrations

indicate the proper meter polarity for measuring the voltage drop across each resistor.

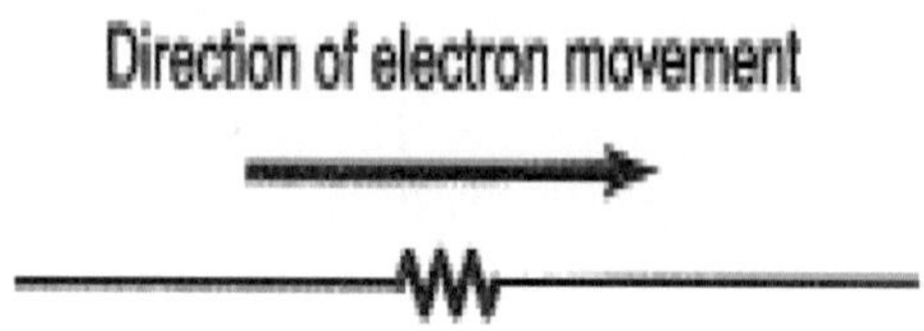

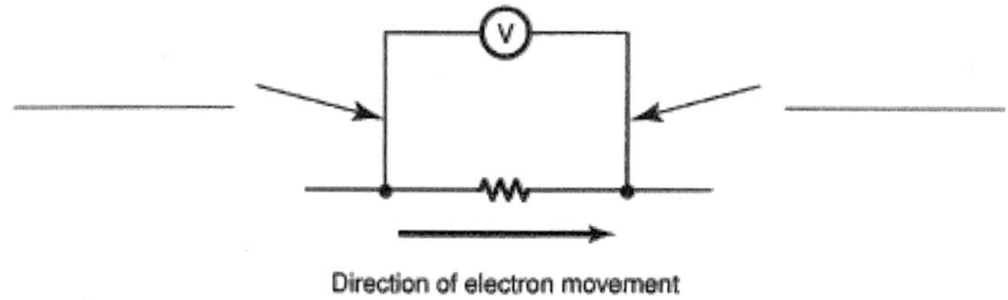

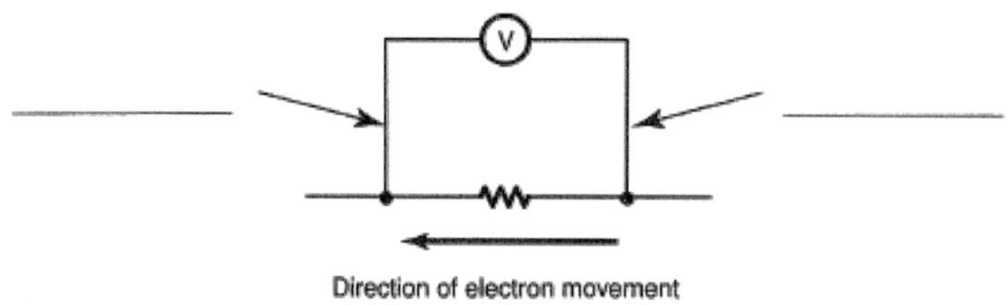

12. Construct the circuit. Measure and record the battery voltage and the voltage drops, and record below. Remember that electrons move from negative to positive.

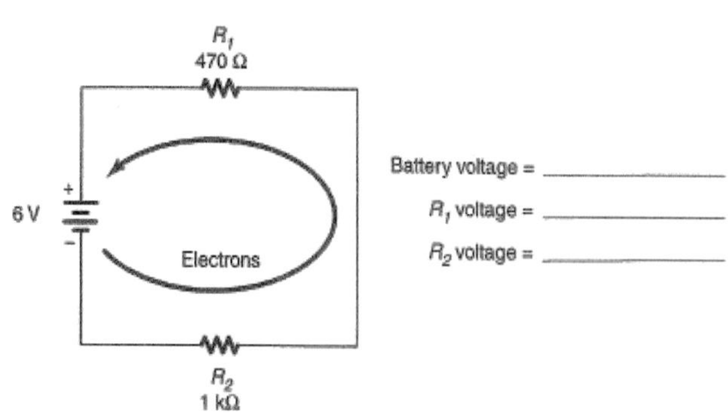

13. Construct the circuit. Measure and record the voltage drops across and each resistor.

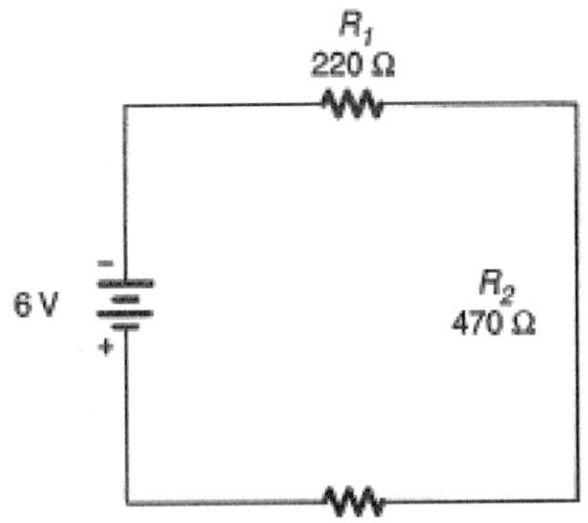

R_1 voltage = ______________

R_2 voltage = ______________

14. Construct the circuit shown.

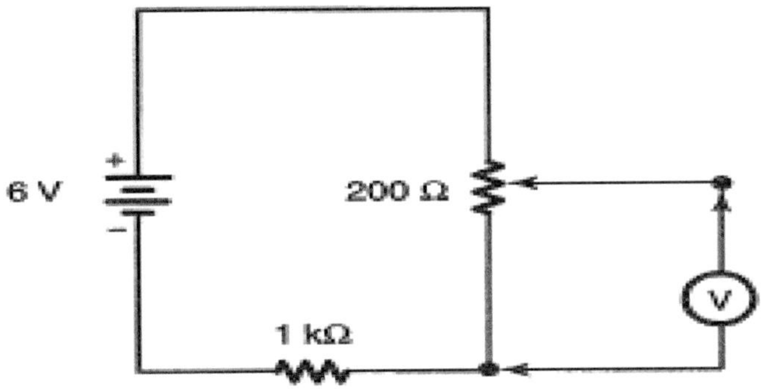

15. Adjust the potentiometer both counterclockwise and clockwise and record the voltages.

Counterclockwise voltage = __________ V.

Clockwise voltage = __________ V.

Analysis

1. What is voltage?

2. If you were measuring an unknown voltage, what meter range would you choose? Why?

3. What determines the polarity of a voltage drop developed by a resistor through which electrons are moving?

4. What causes the voltage drop developed across any component through which electrons are moving?

5. When is it necessary to use a multiplier or divisor with a scale on a multimeter?

6. How is the proper range of a multimeter selected?

7. Could 4 V be measured on the 3 V range of the multimeter? Why?

Activity 9

Measuring Current

Objectives

Current is the movement of electrons from one location to another through conductors or paths. Current is measured in amperes (the number of electrons moving past a given point in a circuit per second), milliamperes (0.001 A), or microamperes. It is essential that anyone involved in electronic work learn to measure current values accurately. In this activity, you will:

1. Learn to measure current with an ammeter.
2. Construct basic electrical currents.
3. Convert amperes to milliamperes or microamperes and vice versa.

If you are using a digital meter, utilize this activity to learn to read an analog scale to measure current. You may use the non-lab "Procedure Steps" and complete the "Analysis" items at the end.

Equipment and Materials

- Multimeter
- Resistors
- 6 V battery or power supply
- SPST switch
- Connecting wires

DOI: 10.1201/9788743810124-10

Procedure

1. You have learned from previous activities that a multimeter can be used to measure resistance and voltage. In addition to these electrical quantities, a multimeter can be used to measure current. Figures 9.1 and 9.2 show the controls and the scale of a common multimeter. The function selection switch controls the measurement function, as well as the range of the meter. Figure 9.1 indicates that the function-select switch can be adjusted to one of five ranges within the DC function. Thus, when the function select switch is placed in the position within the DC function, the meter is capable of measuring a maximum of 120 mA. Likewise, the other DC current ranges of the multimeter are similarly arranged. The numerical value of the chosen range shows the maximum value of current that can be measured on that range.

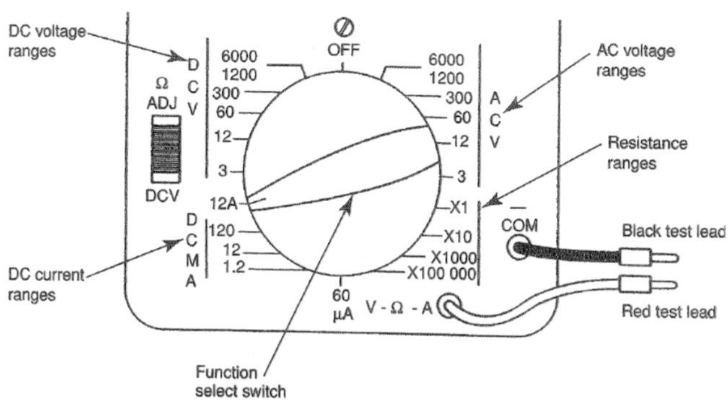

Figure 9.1 Controls of a typical multimeter.

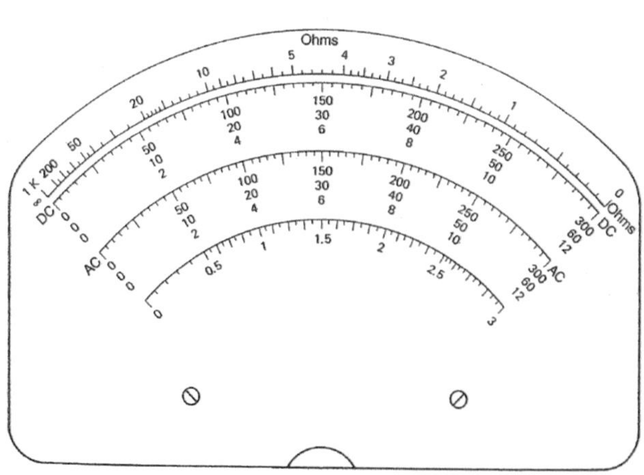

Figure 9.2 Multimeter scale.

2. The meter you are using may differ somewhat from the illustrated multimeter. However, it should exhibit the same basic characteristics as the one described. Adjust the function selection switch on your meter to the lowest range of the direct current function. List below the different ranges for measuring direct current on your meter.

____________, __________________, __________________, ____________

3. Plug the black and red test leads into the jacks labeled (–) and (+), respectively (note: if your meter is different from the one illustrated, plug the test leads into the appropriate jacks and proceed).

4. You are now ready to become familiar with the portion of the scale of the meter used for measuring direct current. Figure 9.2 illustrates the scale of a multimeter. Notice that there are three scales used for measuring DC. They are the same as those used to measure DC voltage. There are five DC ranges that can be used with this meter. All current measurements will be read on these scales, using the proper multiplier or divisor. Shows the proper multiplier or divisor to be used with the de scales, for any of the five ranges (note: your meter may have different scales and multipliers from those in Figure 9.3. You should identify them and proceed).

Multimeter current range	Multimeter scale	Multiplier	Divisor
60 μA	0-60	None	None
1.2 mA	0-12	—	10
12 mA	0-12	None	None
120 mA	0-12	10	—
12 A	0-12	None	None

5. Observe that each range of de has a specific multiplier or divisor or is read directly on the scale. For example, if the needle of the meter points to the number 6 on the DC scale, and the meter is adjusted to the 12 mA range, the measured current is 6 mA. If the needle points to 6 while the meter is adjusted to the 120 mA range, the measured current is 60 mA ($6 \times 10 = 60$, or $6 + 0.1 = 60$).

6. Complete Figure 9.3 by determining the proper current value when the needle points to a specific number. Also, determine the correct multiplier and divisor to be used with the indicated current range. You should use the equivalent ranges, scales, multipliers, and divisors of the meter of Figures 9.1 and 9.2.

Equivalent range	Number needle points to	Current value	Multiplier	Divisor
Example: 12 mA	8	8 mA	None	None
120 mA	2			
120 mA	7			
1.2 mA	4			
12 mA	9			
60 µA	5			
120 mA	7.5			
1.2 mA	3.2			
60 µA	6			
12 A	3			

Figure 9.3 Fill in the values.

7. Most multimeters are capable of measuring currents above 1 A Up to 12 A of current can be measured with the meter illustrated. You will recall that the numerical value of the chosen range indicates the maximum value of current that can be measured on the range. For example, when the range selected is 12 mA, the meter is capable of measuring a maximum of 12 mA. Any current above 12 mA could damage the meter in this range. This is true for most current scales of most multimeters. If you are trying to measure an unknown current (no indication of its approximate value), you should start the measuring procedure by choosing the highest range on your meter and slowly adjusting the range downward until a current reading is indicated on the scale. Complete Figure 9.5 by indicating the proper meter range to be selected and the proper multiplier and divisor to be used with each value of current to be measured (note: always select the smallest range that will allow you to measure the current). Use your meter's ranges and scales for completing Figure 9.4.

Current to be measured	Range to select	Multiplier	Divisor
Example: 0.9 mA	1 mA	0.1	10
2 A			
900 mA			
25 μA			
2 mA			
85 mA			
6 mA			
0.5 mA			
16 mA			
9 mA			
Unknown current			

Figure 9.4 Fill in the values.

8. **Polarity** of the meter is important when measuring current. The multimeter must always be connected in such a way as to allow the current to flow through the meter in the proper direction. Otherwise, the needle will deflect backwards, possibly causing internal damage to the meter. If the current is to flow through the meter in the proper direction, consideration must be given to the direction of the current. Current flows from negative to positive; thus, if the meter is connected correctly into a circuit, current will flow through the black test lead (–), through the meter, and then through the red test lead (+). The illustration of Figure 9.6 should be helpful.

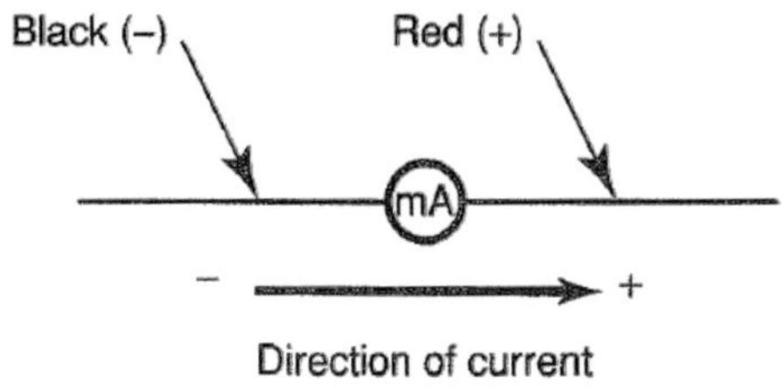

9. The illustrations of show which leads should be red (+) and which should be black (–) for the current meters indicated.

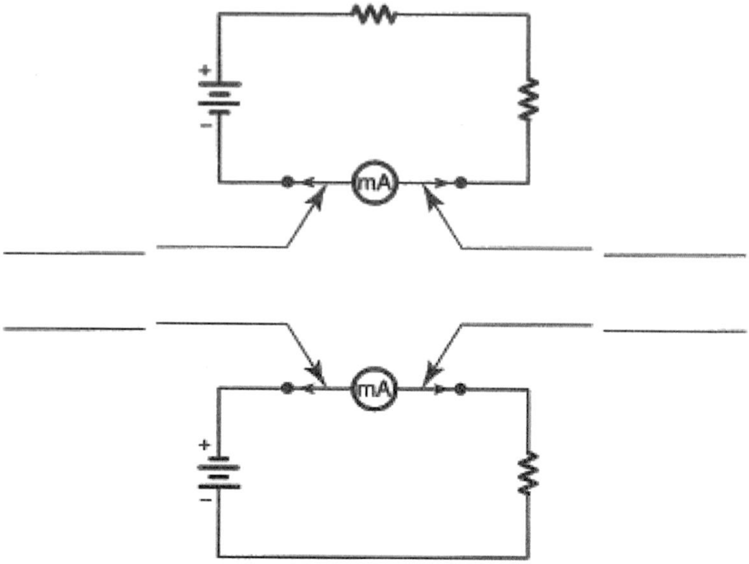

10. Adjust the meter to the de function, 120 mA range (or equivalent), and connect it into the circuit, as illustrated. Be sure to observe the proper meter polarity.

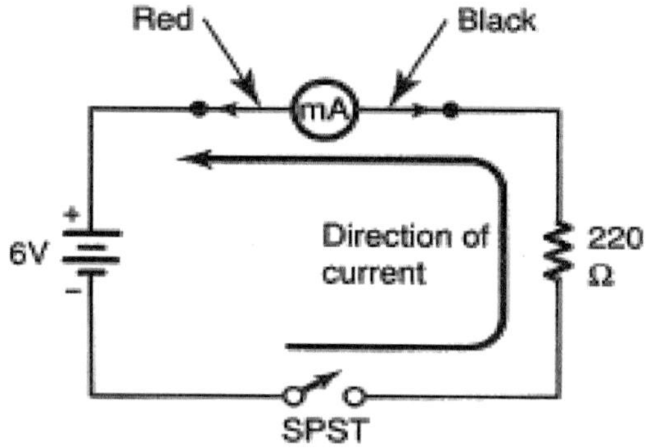

11. Close the SPST switch and record the current indicated by the meter.

_________mA, or _______A.

12. Open the SPST switch and add the additional resistor illustrated.

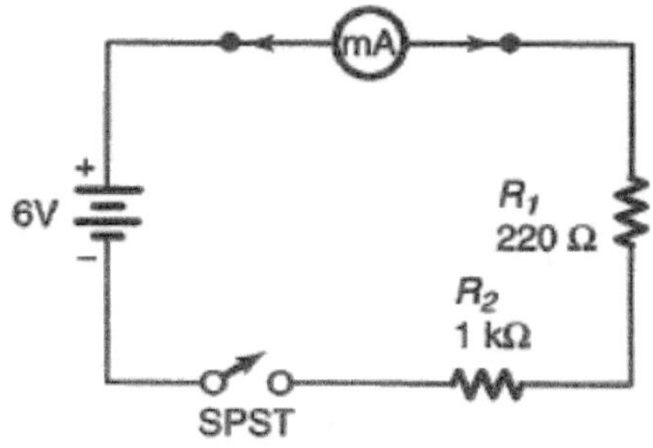

13. Change the range of the multimeter to 12 mA (or equivalent) and close the switch. Record the current indicated.

    ________ mA , or _________ A.

14. Disconnect the circuit and connect the circuit.

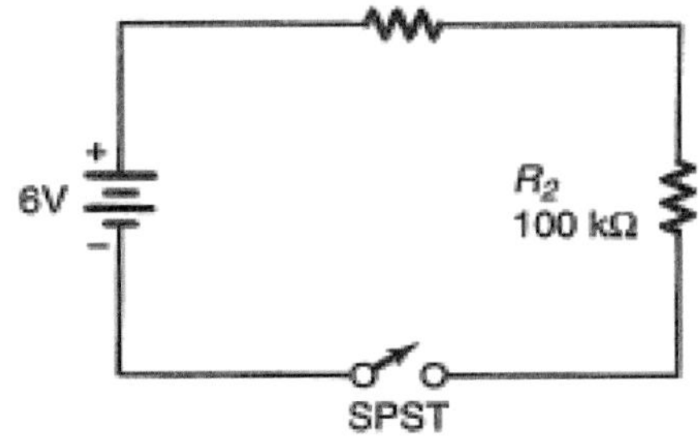

15. Set the meter to the range (or equivalent), and connect the meter to the circuit, observing proper polarity. Close the switch and record the current. _____ mA , or __________ A.

16. Disconnect the circuit illustrated.

17. Construct the circuit illustrated.

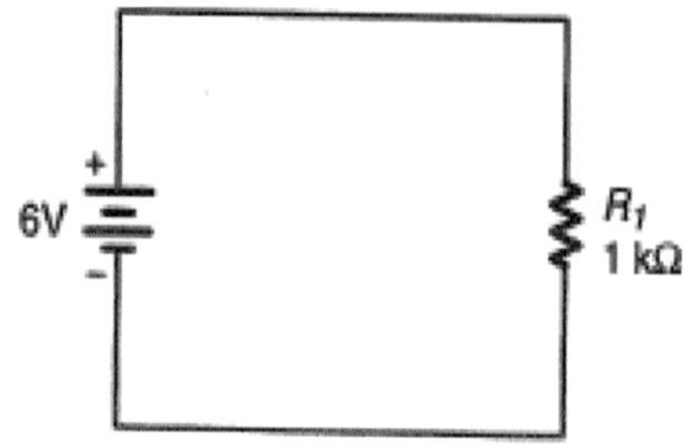

18. Adjust the meter to the range (or equivalent) and connect it into the circuit illustrated, observing proper polarity. Record the current. _____ mA , or ___________ A.

Analysis

1. What is current?

2. What do each of the DC ranges on the multimeter indicate?

3. Describe how the multimeter is properly connected to a circuit to measure current.

4. Why is proper polarity important to observe when one is measuring current?

5. What meter ranges would be selected on your meter to measure the following currents?

Current	Range
0.009 A	_____
0.11 A	_____
0.8 A	_____
0.5 mA	_____
0.000040 A	_____

6. What is the proper procedure to be followed when attempting to measure an unknown current?

7. What is the proper procedure to be followed when attempting to measure a current greater than 2 A?

Activity 10

Basic Electrical Problem Solving

Objectives

Basic electrical problems are often encountered in any area that involves electrical control systems. The most basic problems in electrical systems involve Ohm's law. Ohm's law is a mathematical formula that explains the relationship between voltage, current, and resistance. This relationship must be understood before electrical concepts are meaningful. In this activity, you will complete some practical problems by applying Ohm's law. Refer to Figure 10.1 for assistance.

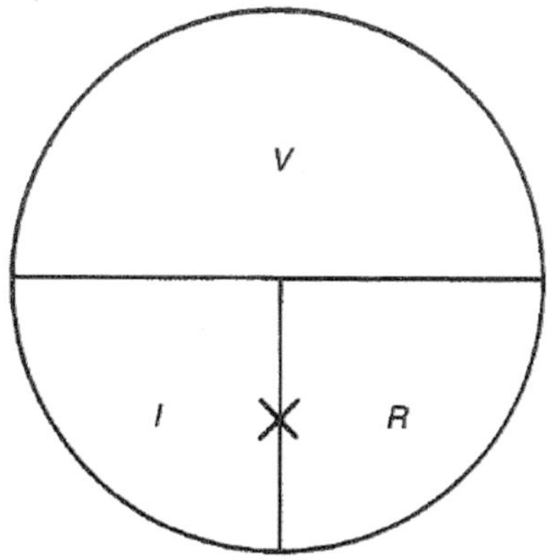

Figure 10.1 Ohm's law circle: *V*, voltage; *I*, current; *R*, resistance. To use the circle, cover the value you want to find and read the other values as they appear in the formula: .

 DOI: 10.1201/9788743810124-11

Part 1: Ohm's Law Problems

1. A doorbell requires 0.2 amperes of current in order to ring. The voltage supplied to the bell is 120 volts. What is its resistance? _____ ohms.

2. A relay used to control a motor is rated at 25 ohms resistance. What voltage is required to operate the relay if it draws a current of 0.25 ampere? _____ volts.

3. An automobile battery supplies a current of 7.5 amperes to a headlamp with a resistance of 0.84 ohms. Find the voltage delivered by the battery. _____ volts.

4. What voltage is needed to light a lamp if the current required is 2 amperes and the resistance of the lamp is 50 ohms? _____ volts.

5. If the resistance of a radio receiver circuit is 240 ohms and it draws a current of 0.6 amperes, what voltage is needed? _________ volts.

6. A television circuit draws 0.15 amperes of current. The operating voltage is 120 volts. What is the resistance of the circuit? _____ ohms.

7. The resistance of the motor windings of an electric vacuum cleaner is 20 ohms. If the voltage is 120 volts, find the current drawn. _____ amperes.

8. The coil of a relay carries 0.05 amperes when operated from a 50-volt source. Find its resistance. _____ ohms.

9. How much current is drawn from a 12-volt battery when operating an automobile horn of 8 ohms of resistance? _____ amperes.

10. Find the resistance of an automobile starting motor if it draws 90 amperes from the 12-volt battery. _____ ohms.

11. What current is drawn by a 5000-ohm electric clock when operated from a 120-volt line? _____ amperes.

12. Find the current drawn by a 50-ohm toaster from a 120-volt line. _____ amperes.

Part 2: DC Electrical Circuits

A fundamental understanding needed in electrical studies is that of basic electrical circuits. All electrical circuits are classified as either series, parallel, or combination (series-parallel) circuits. In this activity, you will review each of these basic electrical circuits and complete a self-test on each type of circuit.

A. Series circuit problems – self test

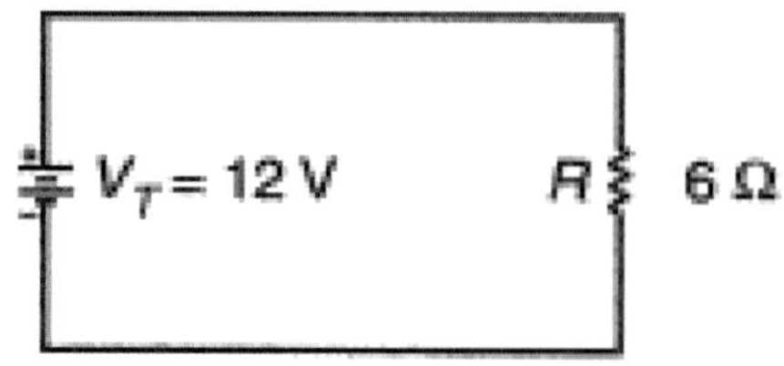

1. Find:

   Total current = _____ amperes.

   Voltage across resistor = _____ volts.

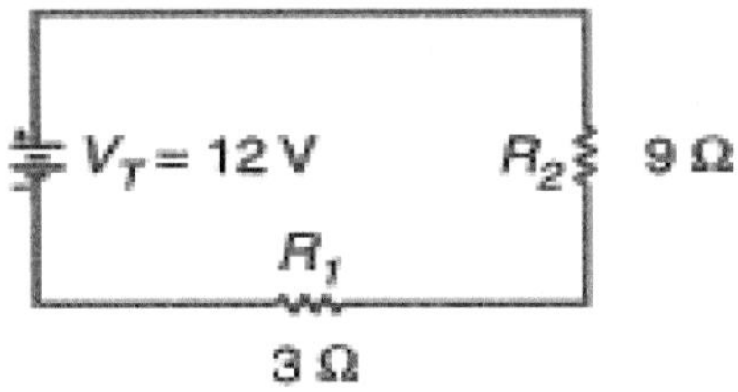

2. Find:

   Total resistance = _____ ohms.

   Current through Resistor 1 = _____ amperes.

   Current through Resistor 2 = _____ amperes.

   Voltage across Resistor 1 = _____ volts.

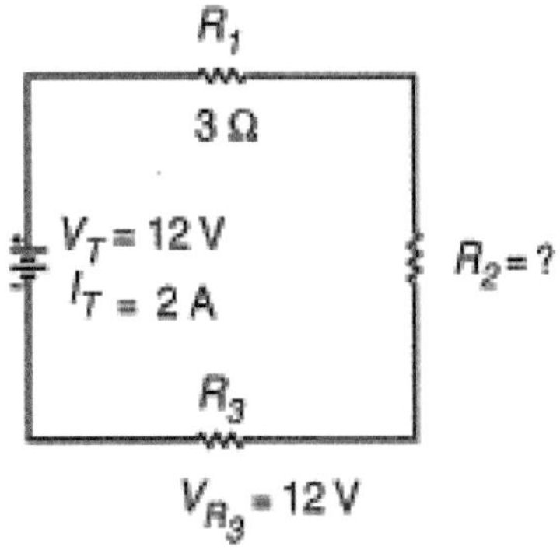

3. Find:

   Voltage across Resistor 1 = _____ volts.

   Voltage across Resistor 2 = _____ volts.

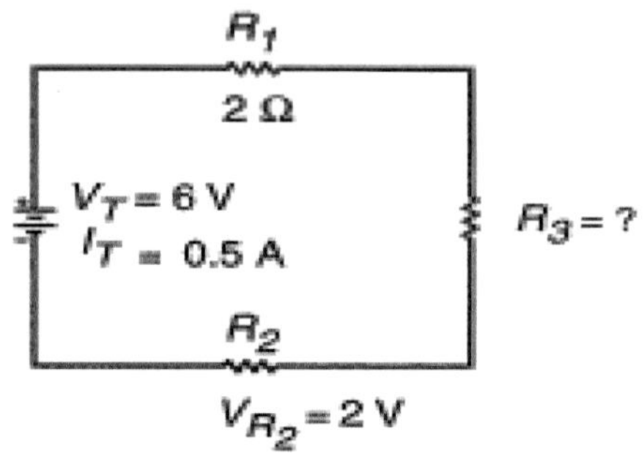

4. Find:

   Voltage across Resistor 1 = _____ volts.

   Resistor 3 value = _____ ohms.

   Voltage across Resistor 3 = _____ volts.

   Resistor 2 value = _____ ohms.

   Current through Resistor 1 = _____ amperes.

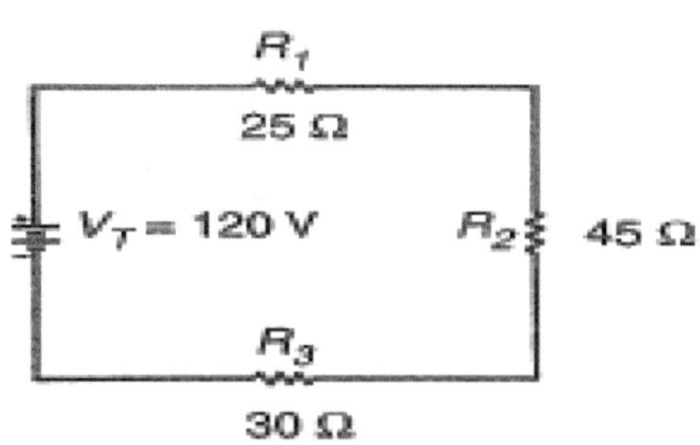

5. Find:

Total current = _____ amperes.

Voltage across Resistor 1 = _____ volts.

Voltage across Resistor 2 = _____ volts.

Voltage across Resistor 3 = _____ volts.

Power converted by Resistor 1 = _____ watts.

Power converted by Resistor 2 = _____ watts.

Power converted by Resistor 3 = _____ watts.

Total power converted by circuit = _____ watts.

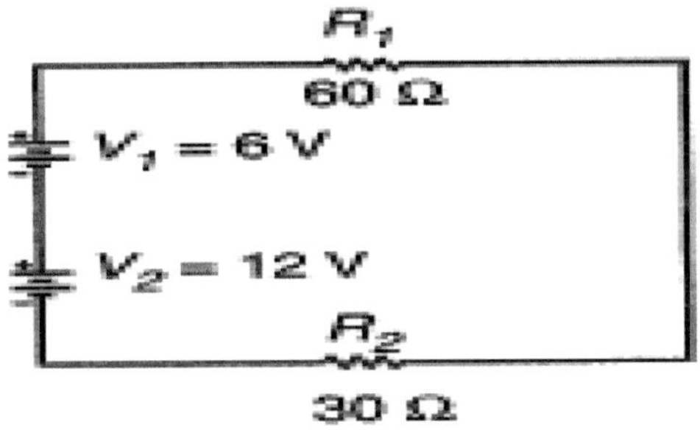

6. Find:

Total voltage = _____ volts.

Total power = _____ watts.

Total resistance = _____ ohms.

Voltage across Resistor 1 = _____ volts.

Total current = _____ amperes.

B. Parallel circuit problems self-test

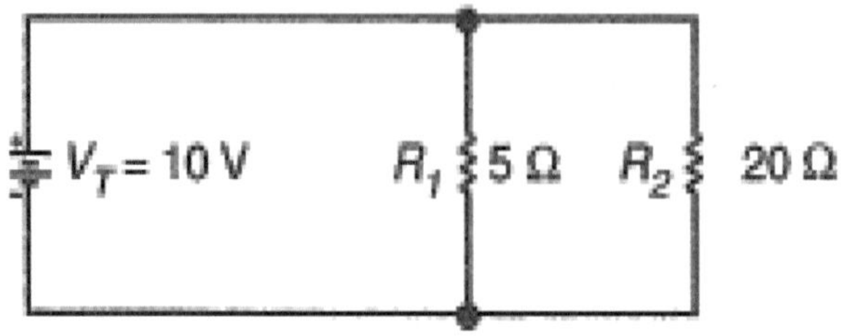

1. Find:

   Total resistance _______ ohms.

   Total current ______ amperes.

   Current through resistor 1 = _____ amperes.

   Current through resistor 2 = _____ amperes.

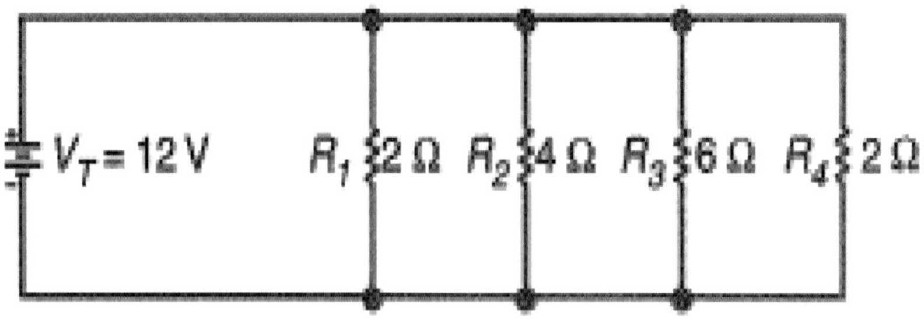

2. Find:

   Total resistance ______ ohms.

   Total current = _____ amperes.

   Current through resistor 1 = _____ amperes.

   Current through resistor 2 = _____ amperes.

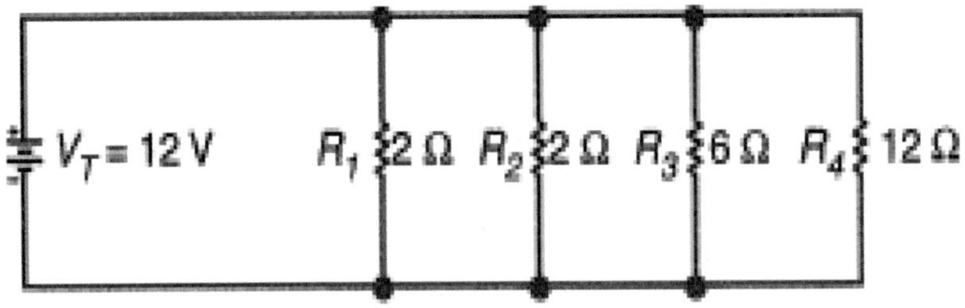

3. Find:

Total resistance ______ ohms.

Total current = _____ amperes.

Current through resistor 1 = _____ amperes.

Current through resistor 2 = _____ amperes.

Current through resistor 3 = _____ amperes.

Current through resistor 4 = _____ amperes.

Total power = _____ watts.

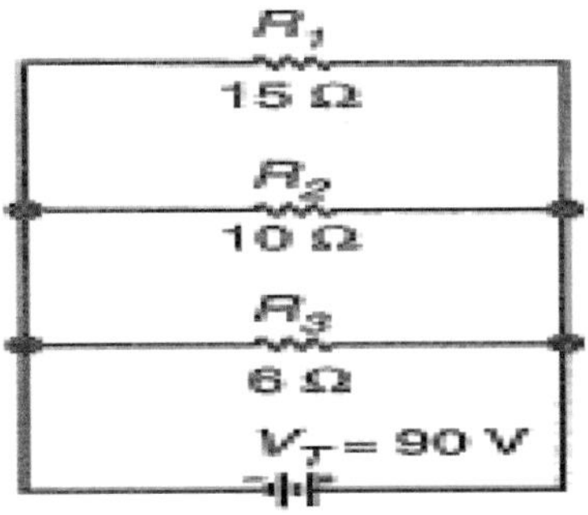

4. Find:

   Total resistance = _____ ohms.

   Total current = _____ amperes.

   Current through resistor 1 = _____ amperes.

   Current through resistor 2 = _____ amperes.

   Current through resistor 3 = _____ amperes.

   Power converted by resistor 1 = _____ watts.

   Power converted by resistor 2 = _____ watts.

   Power converted by resistor 3 = _____ watts.

5. Three resistors of 40, 90, and 75 ohms are connected in parallel across a 120-volt supply.

   Find:

   Total resistance = _____ ohms.

   Total current = _____ amperes.

   Current through each resistor:

C. Combination (series-parallel) circuit problems self-test

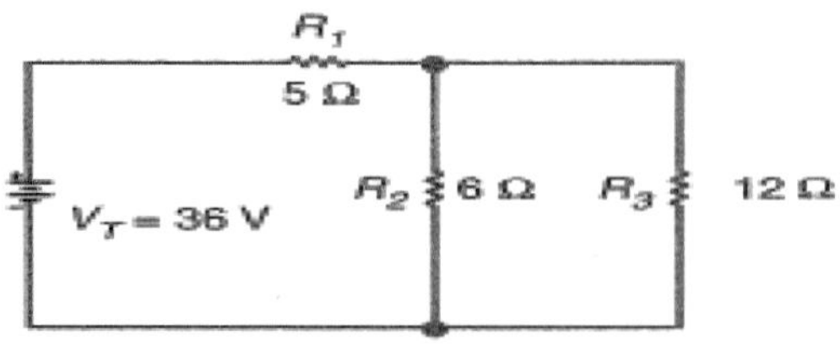

1. Find:

Total resistance = _____ ohms.

Total current = ______ amperes.

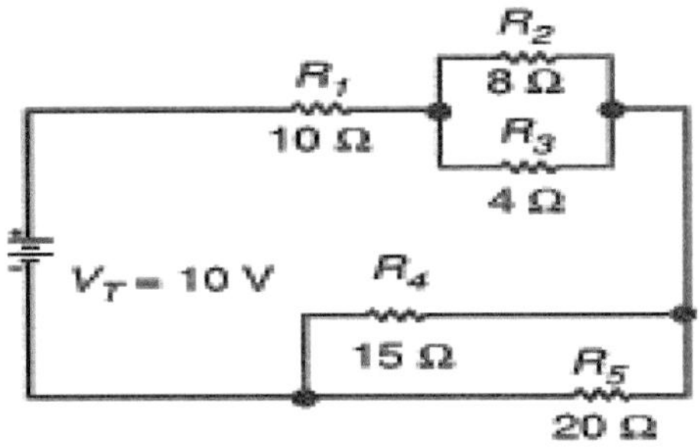

Total power = ______ watts.

2. Find:

Total resistance = ______ ohms.

Total current = _____ amperes.

Total power = _____ watts.

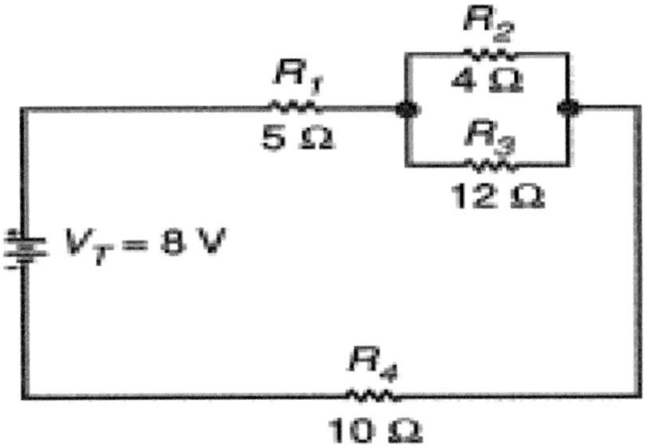

3. Find:

   Total resistance = _____ ohms.

   Total current = _____ amperes.

   Voltage across resistor 1 = _____ volts.

   Voltage across resistor 2 = _____ volts.

   Current through resistor 2 = _____ amperes.

   Current through resistor 3 = _____ amperes.

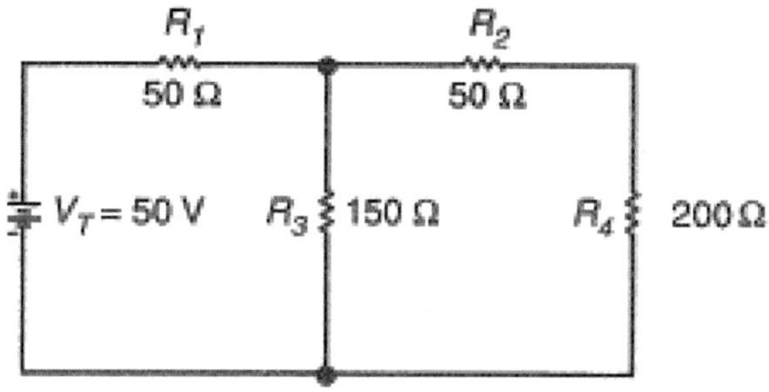

4. Find:

   Total resistance = _____ ohms.

   Total current = _____ amperes.

   Voltage across resistor 3 = _____ volts.

Activity 11

Series DC Circuits

Objectives

A circuit is defined as the complete path or paths through which current flows. All circuits must include a voltage source as well as conductors and/or components through which current flows. There are three broad classifications of all electrical circuits. These are series, parallel, and combination circuits.

The most easily understood circuit is the series circuit. The series circuit exhibits the following electrical characteristics:

- There is only one path for the current.
- Current has the same value everywhere in the circuit.
- The voltage drops, when added, equal the source voltage.
- The total resistance of the circuit is determined by adding the values of all resistors in the circuit.

In this activity, you will:

1. Examine the characteristics of series DC circuits.
2. Supply Ohm's law to series DC circuits.
3. Use a multimeter to make measurements in a series DC circuit.

Equipment and Materials

- Multimeter
- Variable DC power supply

DOI: 10.1201/9788743810124-12

- Resistors
- Connecting wires

Procedure

1. Construct the series circuit illustrated in Figure 11.1.

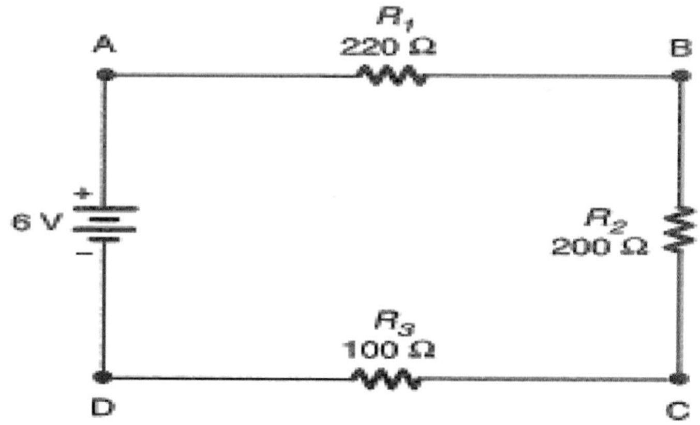

Figure 11.1 Series circuit.

2. In a series circuit, the total resistance is determined by adding the values of all the resistors. When the values of all resistors are added together, the result is the total resistance. Compute the total resistance of the circuit illustrated in Step 1. __________ (computed).

3. Prepare the meter to measure resistance.

4. Disconnect the power supply. Connect the meter in place of the power supply. Measure and record the total resistance of the circuit. _______ (measured).

5. How does the measured value of total resistance compare with the computed value?

6. Disconnect the meter and connect the power supply to its original location in the circuit.

7. Since you now know the total resistance of the circuit as well as the total voltage, use Ohm's law to compute the total current. _____ mA.

8. Prepare the meter to measure direct current (note: the value of current computed in Step 7 will help you select the proper range).

9. Connect the meter at locations, and record the currents at these points.

   Point A: ______ mA. Point C: _______ mA.

   Point B: ______ mA. Point D: ________ mA.

10. How do these currents compare?

11. Since the circuit in Figure 11.1 is a series circuit, the current is the same through each resistor. You should have seen this at Step 9. Since you know the current through each resistor and the value of each resistor, use Ohm's law to compute the voltage drop across each resistor.

12. Prepare the meter to measure DC volts. Measure and record the voltage drop across each resistor.

13. How do the voltage drops computed in Step 11 compare with the voltage drops measured in Step 12?

14. Total the voltage drops in Step 12.

    Total voltage drop =_____ V.

15. How does this total voltage drop compare to the measured source voltage?

Analysis

1. List the characteristics of a series circuit.

2. What is the voltage drop across Resistor 2 in the circuit of Figure 11.2?

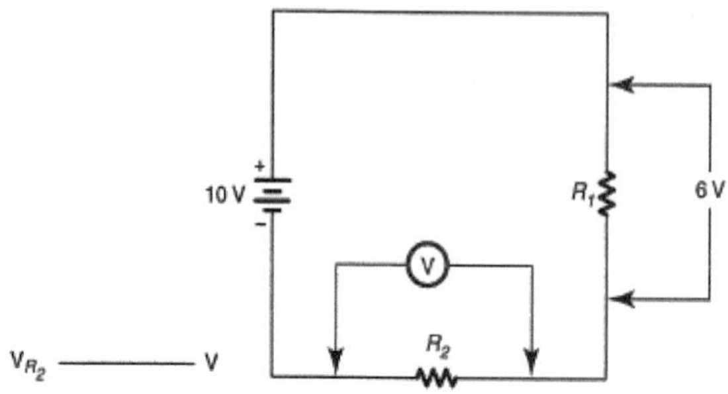

Figure 11.2 Computing the voltage drop across a resistor.

3. If three resistors valued at 6, 8, and 10 Ohms, respectively, were connected in series, what would be their total resistance?

4. In Figure 11.3, if 100 mA of current is flowing through Resistor 1, how much current is flowing through Resistors 2, 3, and 4?

5. How many paths for the current are illustrated in the circuit of Figure 11.3?

6. What is the total voltage drop in the circuit illustrated in Figure 11.3?

7. If three resistors valued at 10 Ohms each were connected in series to a voltage source, what portion of the source voltage would appear across each resistor?

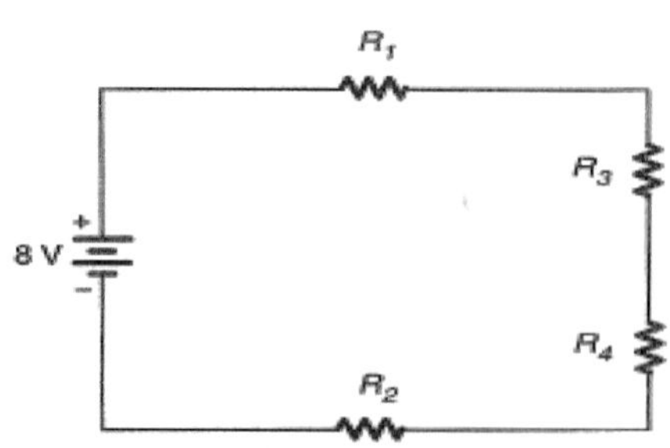

Figure 11.3 Four-resistor series circuit.

Activity 12

Parallel DC Circuits

Objectives

Parallel circuits are circuits that provide two or more paths for current. Most of the light fixtures and wall outlets in your home are connected in parallel circuits. Many of the circuits that cause your radios and television sets to operate are parallel circuits.

The characteristics of parallel circuits are as follows:

- Two or more paths are provided for the current.
- The voltage across each path or branch is the same.
- The sum of the currents in each path or branch will equal the total current.
- The total resistance is computed using the reciprocal formula (see Chapter 4).

In this activity, you will:

1. Examine the characteristics of parallel DC circuits.
2. Apply Ohm's law to parallel DC circuits.
3. Use a multimeter to make measurements in a parallel DC circuit.

Equipment and Materials

- Multimeter
- Variable DC power supply

DOI: 10.1201/9788743810124-13

- Resistors
- Connecting wires

Procedure

1. Construct the circuit illustrated in Figure 12.1. Use either a battery or a power supply.

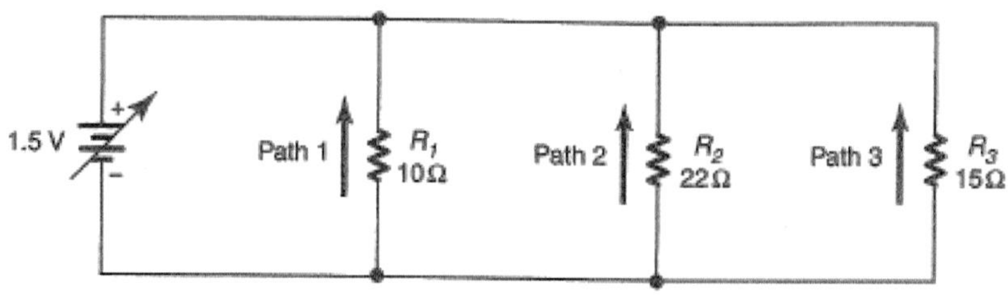

Figure 12.1 Three-branch parallel circuit.

2. How many different paths are provided for the current? Is this a parallel or series circuit?
3. What is the voltage across each path?
4. Compute the current in each path and the total current for the circuit.

   Path 1 = _____ mA.

   Path 2 = _____ mA.

   Path 3 = _____mA.

   Total current = _____ mA.
5. Compute the total resistance for the circuit illustrated in Figure 12.1.
6. Prepare the meter to measure direct current. Measure and record the currents through paths 1, 2, and 3, as well as the total current in the circuit (the meter should be connected as illustrated in Figure 12.2 for each of these measurements. Thus, the current in the path will correctly flow through the meter).

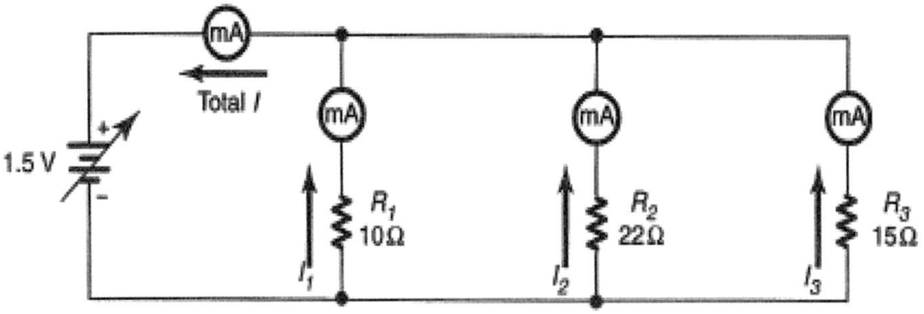

Figure 12.2 Connection of a current meter to measure direct current.

7. How do the values of current in Step 4 compare with the values of current in step?

8. Disconnect the meter from the circuit and restore the circuit to its original condition as shown in Figure 12.1.

9. Prepare the meter to measure DC volts. Measure and record the voltage across each path of the circuit illustrated in Figure 12.1.

   Path 1 = ________ V.

   Path 2 = ________ V.

   Path 3 = ________ V.

10. How do these voltages compare to each other?

11. Disconnect the power supply. Prepare the meter to measure resistance. Connect the multimeter **in place of** the power supply and measure the total resistance of the circuit.

12. How does the measured total resistance in Step 11 compare with the computed total resistance in Step 5?

Analysis

1. List the characteristics of a parallel circuit.

2. What is the current through the circuit illustrated in Figure 12.3?

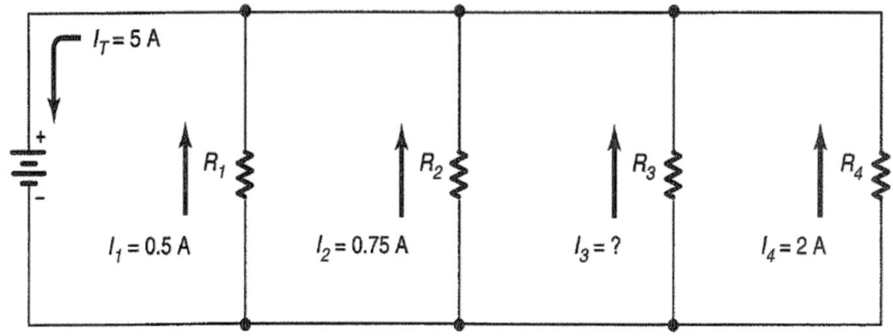

Figure 12.3 Circuit with an unknown branch current.

3. What is the total current in the circuit of Figure 12.4?

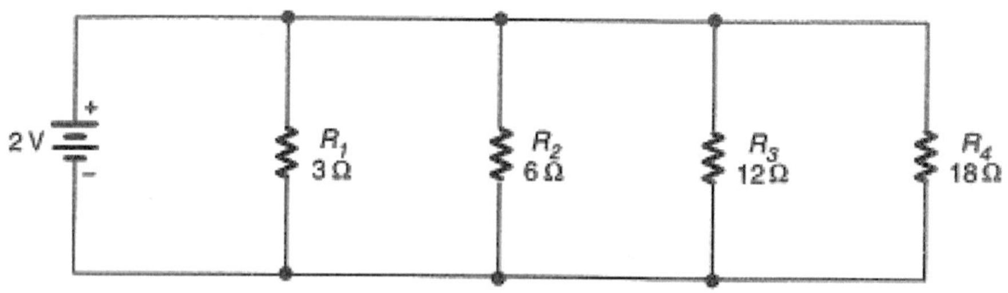

Figure 12.4 Circuit with an unknown total current.

4. What is the voltage across the circuit in Figure 12.4?

5. What is the total resistance of the circuit shown in Figure 12.4?

6. If the resistance of each path of a parallel circuit is equal, how will the current in one path compare with the current in the other paths?

7. In the circuit of Figure 12.5, indicate where the meter should be placed to measure total current. Use an X to indicate placement.

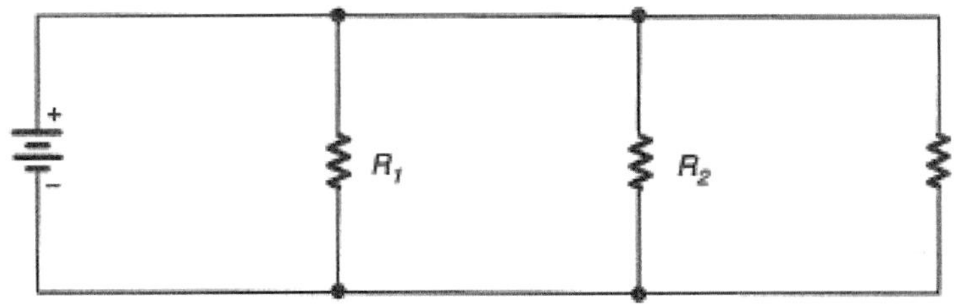

Figure 12.5 Current measurement.

8. In the circuit of Figure 12.5, indicate where the meter should be placed to measure current through Resistor 2. Use a large dot to show the placement of the meter.

Activity 13

Combination DC Circuits

Objectives

The combination circuit is the most used circuit in electronics. Its name is derived from the fact that it is actually a combination of the series and parallel circuits, since it exhibits the characteristics of both. When working with a combination circuit, you must treat its series portion as a series circuit and its parallel portion as a parallel circuit. You must also understand the influence of one portion of a combination circuit upon other portions.

All current, voltage, and power computations and measurements for this circuit are made in the same way as in previous circuits. The total resistance of a combination circuit is determined by first determining the series resistance, then the parallel resistance, and finally their sum.

In this activity, you will:

1. Examine the characteristics of a combination DC circuit.
2. Apply Ohm's law to a combination DC circuit.
3. Use a multimeter to make measurements in a combination DC circuit.

Equipment and Materials

- Multimeter
- Variable DC power supply (or 6 V battery)
- Resistors
- Connecting wires

 DOI: 10.1201/9788743810124-14

Procedure

1. Construct the circuit shown in Figure 13.1.

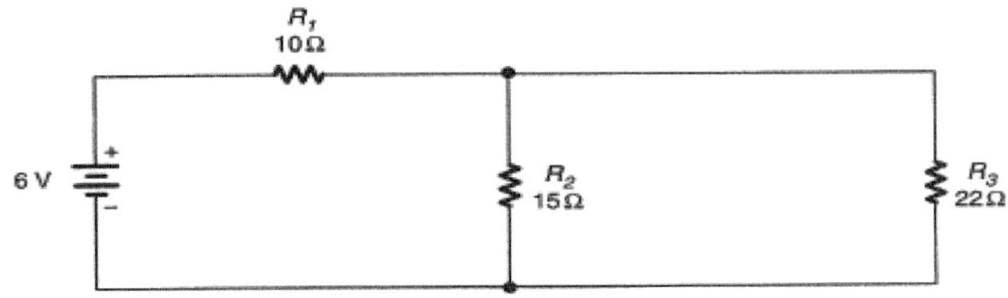

Figure 13.1 Series-parallel combination circuit.

2. Prepare the meter to measure DC voltage. Measure and record the voltages across Resistors 1, 2, and 3.

   V 1 = ________V.

   V 2 = ________V.

   V 3 = ________V.

3. How did the voltage across Resistor 2 compare with the voltage across Resistor 3?

4. How does the sum of the voltage across Resistor 1 and the voltage across Resistor 2 or Resistor 3 compare with the source voltage?

5. Prepare the meter to measure direct current. Measure and record the currents through Resistors 1, 2, and 3 and the total current.

6. How does the total current compare with the current through Resistor 2?

7. How does the sum of the currents through Resistors 2 and 3 compare with the total current?

8. How does the sum compare with the current through Resistor 1?

9. Compute the total resistance for the circuit shown in Figure 13.1 (remember to add the series resistance to the parallel resistance).

   Total resistance = ________ Ohms.

10. Disconnect the power supply from the circuit in Figure 13.1. Prepare the meter to measure resistance. Connect the meter where the power supply **was** connected. Measure and record the total resistance of the circuit.

    Total resistance = _________ Ohms.

11. How did the measured resistance in Step 10 compare with the computed resistance in Step 9?

12. Using Ohm's law, compute the total current for the circuit in Step 1.

    Total current = __________ amperes.

13. How does the current computed in Step 12 compare with the current measured in Step 5?

14. How does the total current in this circuit compare with the current through Resistor 1?

15. Compute the total parallel resistance of Resistors 2 and 3:

    Parallel resistance = __________ Ohms.

16. In Step 7, you found that the sum of the currents through Resistors 2 and 3 equaled the total current. Using the value of total current computed in Step 12 and the parallel resistance computed in Step 15, compute the voltage across the parallel resistors 2 and 3:

    Voltage = _______ volts.

17. How does this voltage compute with the measured values in Step 2?

18. Using the total current computed in Step 12, compute the voltage across Resistor 1:

    Voltage = _________ volts.

19. How does the computed voltage in Step 18 compare with the measured voltage in Step 2?

20. How does the sum of the computed voltages across Resistors 1, 2, and 3 compare with the source voltage?

Analysis

1. How does the total current of a combination circuit compare with the current through its series components?

2. How does the sum of the currents in the parallel paths of a combination circuit compare with the total current of the circuit?

3. How does the sum of the currents in the parallel paths of a combination circuit compare with the current through the series components?

4. How does the sum of the voltages across the series components and the parallel paths of a combination circuit compare with the source voltage?

5. If you subtract the voltage across the parallel paths of a combination circuit from the source voltage, how would the results compare with the voltage across the series components of the circuit?

6. If you subtract the sum of the voltages across the series components of a combination circuit from the source voltage, how would the results compare with the voltage across the parallel paths of the circuit?

7. If you added the sum of the resistance of the series components of a combination circuit to the parallel resistance of the circuit, how would the results compare with the total resistance of the circuit?

Activity 14

Power in DC Circuits

Objectives

Electrical power is probably the most familiar electrical quantity to most people. All our electrical utility bills are computed relative to the amount of power consumed in a given period of time. All electrical appliances have a specified power rating, which indicates how economically they can be operated. A knowledge of electrical power is important to anyone living in this age of energy conservation.

Electrical power is consumed each time a voltage causes current to flow. Power generally appears in the form of heat, light, or motion. The basic unit of measurement for electrical power is the watt. The milliwatt and kilowatt are also commonly used as units of measurement of power. Power is the product of current and voltage in a circuit. That means power (in watts) equals current (in amperes) multiplied by voltage (in volts). Actually, power can be computed when any two of the following electrical quantities are known: voltage (E), current (I), or resistance (R). The three basic formulas used in computing power in watts are discussed in text Chapter 4.

Many electrical components have an electrical power rating. Resistors are an excellent example, having common power ratings of 1/8, 1/4, 1/2, 1 Watt, and so on. This rating indicates the maximum current that can pass through the resistor without damaging the component.

In this activity, you will:

1. Calculate power values in DC circuits by applying the basic power formulas.
2. Use a meter to make measurements that are used to determine power in DC circuits.

DOI: 10.1201/9788743810124-15

Equipment and Materials

- Multimeter
- Variable DC power supply or 6-volt battery
- Lamp and socket, 6 volt
- Resistors
- Connecting wires
- SPST switch

Procedure

1. Construct the circuit shown in Figure 14.1.

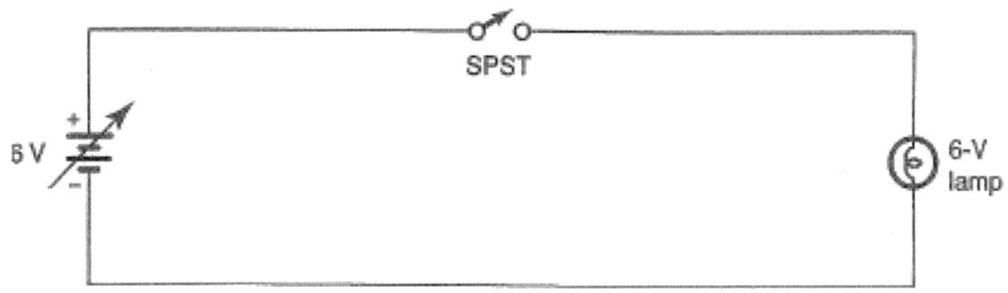

Figure 14.1 Simple six-volt lamp circuit.

2. Adjust the meter to measure current on the highest range.
3. Connect the meter to the circuit illustrated in Figure 14.1. Close the SPST switch and measure and record the current.

   Measured current _____ mA, or _____ A.

4. Open the SPST switch, disconnect the meter, and restore the circuit to its original state, illustrated in Figure 14.1.
5. Prepare the meter to measure DC voltage.
6. Connect the meter to measure the voltage drop across the lamp illustrated in the circuit. Close the switch and record the voltage across the lamp.

   Measured voltage _____ V.

7. You now have measured the current through the lamp, as well as the voltage across the lamp. Compute the power converted by the lamp.

   Computed power = ______ W.

8. An increase in power converted is caused when voltage increases, and resistance remains the same. Notice the relationship between current, voltage, resistance, and power as you solve the following problems:

   a. R = 10 Ohms, I = 2 Amps – P = ____ W and V = _____ volts.

   b. V = 100 Volts, I = 100 mA – P = _____ W and _____ R = _____ Ohms.

   c. I = 4 Amps, R = 5 Ohms – P = _____ W and V = _____ volts.

   d. R = 20 Ohms, V = 10 Volts – P = _____ W and I = _____ amps.

   e. I = 1.5 Amps, V = 90 Volts – P = _____ W and R = _____ Ohms.

9. Construct the circuit shown in Figure 14.2.

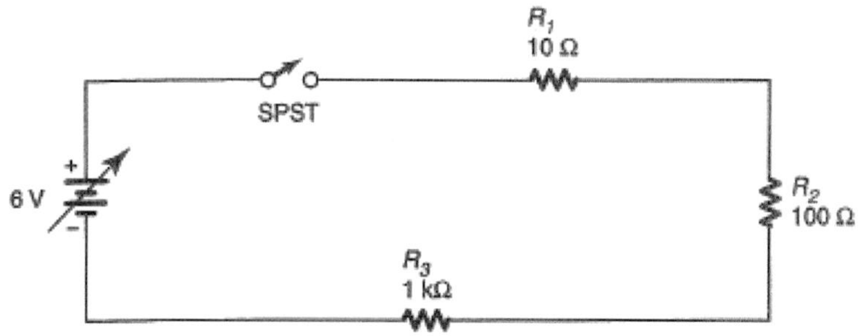

Figure 14.2 Circuit with three resistors.

10. Set the meter to measure direct current in the range or equivalent.

11. Connect the meter to the circuit to measure current, close the SPST switch, and record the current indicated by the meter.

    Measured current = _____ mA.

12. Compute the power converted by the circuit illustrated in Figure 14.2.

    Computed power = _____ W.

13. Open the switch, disconnect the meter, and alter the circuit to the one in Figure 14.3.

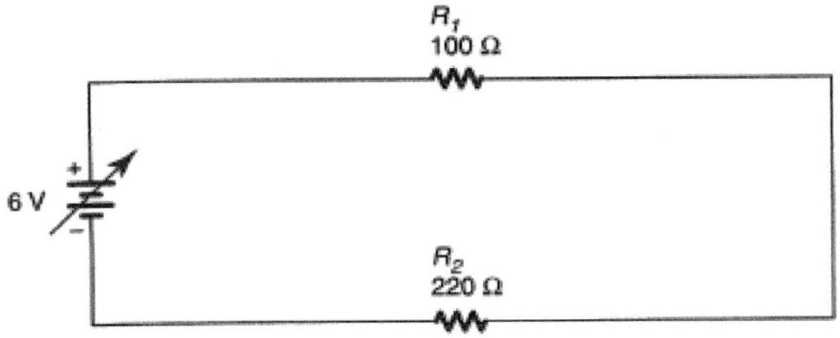

Figure 14.3 Two-resistor circuit.

14. Set the meter to the DC range. Measure and record the voltage across Resistors 1, 2, and 3. Be sure to observe proper meter polarity.

    Voltage across Resistor 1 = _____ V.

    Voltage across Resistor 2 = _____ V.

    Voltage across Resistor 3 = _____ V.

    Voltage across Resistor 4 = _____ V.

15. Compute the power converted by Resistors 1 and 2.

    Computed power for Resistor 1 = _____ V.

    Computed power for Resistor 2 = _____ V.

16. How does the power converted by Resistor 1 compare with the power converted by Resistor 2? Why do they differ?

Analysis

1. How much current flows through a light bulb when 100 V is connected?

2. How much power is converted by a lamp with a filament resistance of 2 Ohms?

3. What is the maximum current that could safely pass through a 1000-Ohm, 1 Watt resistor?

4. What power is converted by a resistor that has 1 mA of current through it?

5. What power is converted by a light bulb that drops 6 V across its filament?

6. What is the maximum voltage drop that could be developed by a 1.2 K Ohm, 1/2 Watt resistor?

Activity 15

Basic AC Electrical Circuit Problems

Objectives

For more practice in solving AC circuit problems, you should complete this activity. It is important to have an understanding of basic AC circuit problems in order to solve more advanced circuit problems. These problems can be referred to when a similar problem is encountered in later activities. The organization of the activity is such that one example of each type of AC circuit calculation is done.

DOI: 10.1201/9788743810124-16

Part 1: Problems – solve for circuits 1 through 8

1. Series inductance:
$L_T = L_1 + L_2 + L_3$.

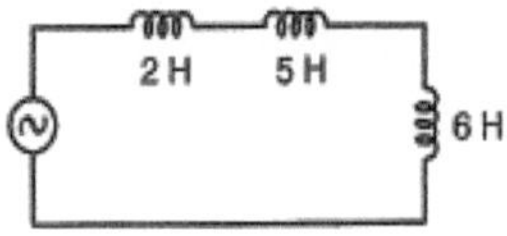

What is the inductance of this circuit?

$L =$ _______ henrys.

2. Parallel inductance:
$\frac{1}{L_T} = \frac{1}{L_1} + \frac{1}{L_2} + \frac{1}{L_3}$.

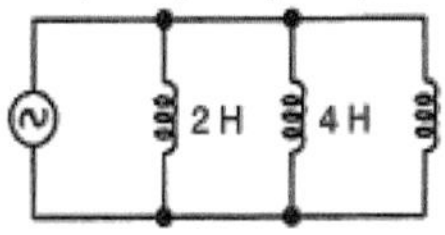

What is the inductance of this circuit?

$L =$ _______ henrys.

3. Inductive reactance: $X_L = 2\pi fL$.

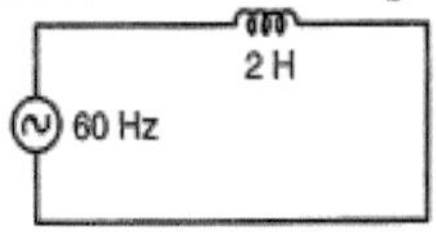

What is the inductive reactance?

$X_L =$ _______ ohms.

4. Series capacitance:
$\frac{1}{C_T} = \frac{1}{C_1} + \frac{1}{C_2} + \frac{1}{C_3}$.

F = farads
µF = microfarads (10^{-6})
pF = picofarads (10^{-12})

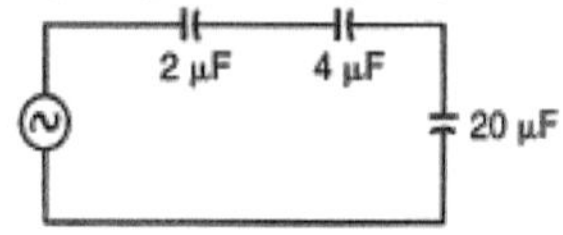

What is the capacitance?

C = _______ farads

5. Parallel capacitance:
$C_T = C_1 + C_2 + C_3$.

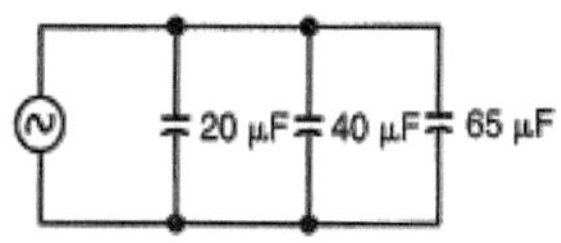

What is the capacitance?

C = _______ µF.

6. Capacitive reactance:
$X_C = \frac{1}{2\pi fC}$.

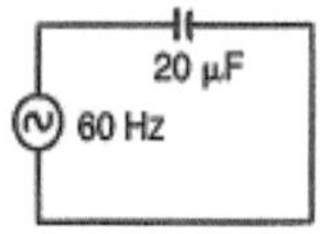

What is the capacitive reactance?

$X_C =$ _______ ohms.

7. Impedance in series RC or RL circuit:
$Z = \sqrt{R^2 + (X_C)^2}$ or
$Z = \sqrt{R^2 + (X_L)^2}$

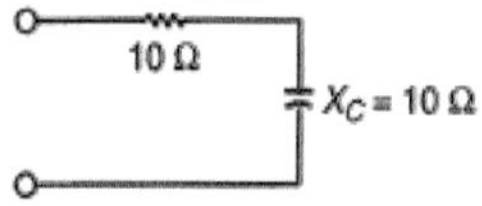

What is the impedance?

Z = _______ ohms.

8. Impedance in series RLC circuits:
$Z = \sqrt{R^2 + (X_L - X_C)^2}$

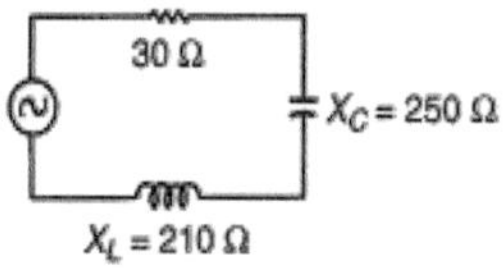

What is the impedance?

Z = _______ ohms.

Part 2: Short answer items

Answer each of the following:

a. The opposition to current caused by inductance is called _____________.

b. The letter symbol for inductive reactance is _____________.

c. Inductive reactance is measured in _____________.

d. The total opposition to current in an AC circuit is called ____________.

e. The letter symbol for impedance is ____________.

f. Impedance is measured in ____________.

g. The relationship in electrical degrees of voltage and current in an AC circuit is called ____________.

h. In a pure inductive circuit, current ____________ voltage by -____________.

i. In an AC circuit containing both resistance and inductance, current ____________voltage by _______ through ________.

j. Maximum effective voltage multiplied by maximum effective current equals ______ power.

k. Apparent power is measured in ____________.

l. The power consumed by the resistance in an AC circuit is called the ______ power.

m. True power is measured in _________.

n. The ratio of true power to the apparent power is called ____________.

o. Power factor is expressed as a ________ or a ____________.

p. Apparent power multiplied by power factor equals ____________

Part 3: Basic AC measurement values

There are several basic AC values that are commonly used in electrical measurement and problem-solving. These units are derived from AC sine wave relationships. The following AC sine wave values are often converted from one value to the other: Effective value, peak value, average value, and peak-to-peak value. Refer to text Chapter 4 for the equations that define their relationships.

In this activity, you should review the preceding AC sine wave relationships and complete the self-test.

Basic AC values – self test

Solve the following problems by placing the correct answer in the blank space.

1. Compute the following effective values:

   Peak voltage = each of the following:

   4 Vac = _____ V rms

   12 Vac = _____ V rms

   6 Vac = ______ V rms

   10 Vac = ______ V rms

   1.5 Vac = ______ V rms

   2 Vac = ______ V rms

   5 Vac = ______ V rms

   9 Vac = ______ V rms

2. Compute the following peak and peak-to-peak values:

   Rms voltage = each of the following:

   3 Vac = ______ V peak and _______ V p-p

   8 Vac = ______ V peak and _______ V p-p

   7 Vac = ______ V peak and _______ V p-p

   9 Vac = ______ V peak and _______ V p-p

   10 Vac = ______ V peak and _______ V p-p

   15 Vac = ______ V peak and _______ V p-p

   11 Vac = ______ V peak and _______ V p-p

   18 Vac = ______ V peak and _______ V p-p

Part 4: Trigonometry problems for AC circuits

In this activity, you will use a calculator to find sine, cosine, and tangent values. You will also find the values of angles when a sine, cosine, or tangent value is given. A knowledge of trigonometry is particularly important in dealing with AC circuits. Refer to **Appendix B** (Figure B.18) and **Appendix C** to complete this activity.

Trigonometry problems
Find the values of the following functions

1. sin 138 degrees = ______

2. sin 62 degrees = ______

3. sin 212 degrees = ______

4. cos 152 degrees = ______

5. tan 141 degrees = ______

6. sin 222 degrees = ______

7. cos 230 degrees = ______

8. tan 201 degrees = ______

9. sin 318 degrees = ______

10. cos 290 degrees = ______

11. tan 310 degrees = ______

12. inv sin 0.4226 = ______

13. inv cos 0.4848 = ______

14. inv tan 9.5114 = ______

15. inv tan 0.2309 = ______

16. inv sin 0.9613 = ______

17. inv cos 0.2924 = ______

18. inv sin – 0.8660 = ______

19. inv cos – 0.6820 = ______

20. inv tan – 3.7321 = ______

Activity 16

Measuring AC Voltage

Objectives

Alternating current is the most common form of current presently used in the United States. It is called alternating because it changes its direction periodically. The most frequently used unit of time associated with AC is the second.

The number of AC cycles per second is known as frequency. Frequency is the number of times in 1 second that the AC signal moves from zero, reaches a peak in one direction, changes its direction, peaks in the opposite direction, and goes back to zero. These AC cycles per second (cps) are called hertz (Hz). The standard frequency of alternating current and voltage used in the United States is 60 Hz. The period of this standard frequency is 0.0166 seconds.

Since alternating current, voltage, and power are constantly changing, two types of electrical values are used in AC measurement. These are the instantaneous and the effective values. The instantaneous values are those used to describe the value of AC current, voltage, or power at any specified instant. The most common instantaneous value is the peak value: the maximum value of voltage, current, or power during any cycle. The effective values are more well-known, since these values are used to describe the amount of voltage, current, and power that can be counted upon to produce light, heat, motion, or work of an electrical nature. The effective values of AC produce the same amount of work as DC values. Generally, the effective values are also called rms (root-mean-square) values.

Ohm's laws are used to compute AC values when the opposition to current is resistance only. Instantaneous values of voltage and current must be used to determine instantaneous power. Effective values must be used to determine effective power. The same procedure must be followed when determining voltage or current with a known resistance.

DOI: 10.1201/9788743810124-17

Instantaneous values of voltage, current, and power are converted to effective values, or vice versa, by using the mathematical constants 0.707 or 1.41. Refer to text **Chapter 4** for equations.

In this activity, you will:

1. Study the characteristics of alternating current (AC).
2. Use a multimeter to measure AC voltage.

Equipment and Materials

- Multimeter
- DC power supply or battery
- AC source – 0 to 30 Vac, variable adjust or similar
- SPST switch
- DPST switch
- Lamp with socket – 6 volt
- Resistors
- Connecting wires

Procedure

1. Measure and record the resistance of the filament of the lamp. R = _____ Ohm.
2. Construct the circuit.

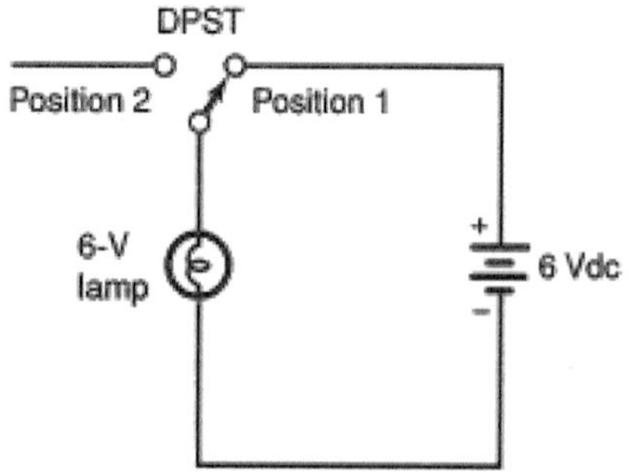

3. Place the switch in position 1 and measure the DC voltage across the lamp. Record this voltage. ____________Vdc.

4. Place the switch in position 2 and disconnect the multimeter.

5. Connect a variable AC power supply to the circuit shown (note: be sure to adjust the AC power supply to **zero**).

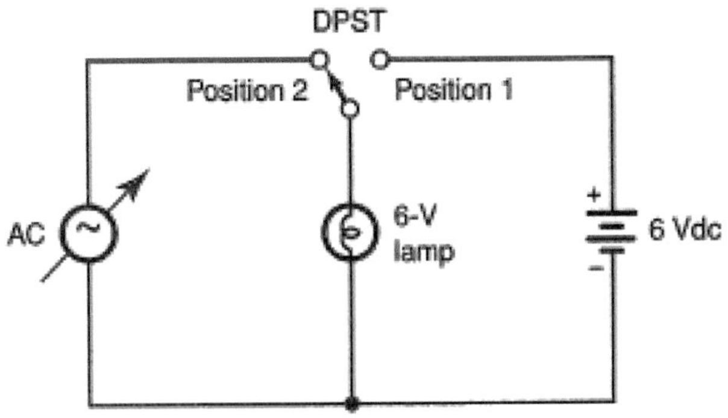

A variable AC voltage is obtained as shown from the circuit.

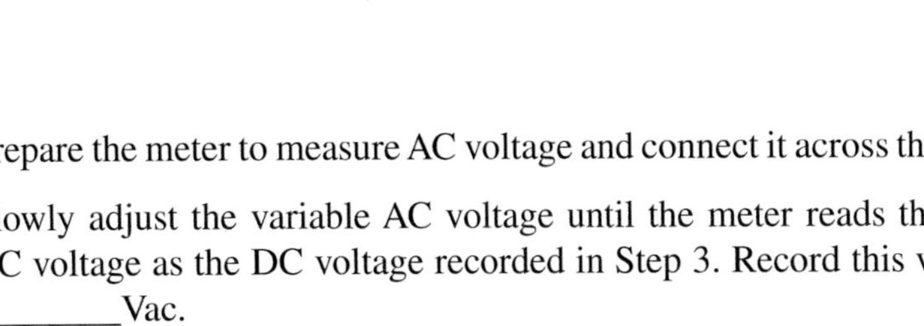

6. Prepare the meter to measure AC voltage and connect it across the lamp.

7. Slowly adjust the variable AC voltage until the meter reads the same AC voltage as the DC voltage recorded in Step 3. Record this voltage. ________Vac.

8. Slowly change the switch from position 2 to position 1 and back several times. How does this action affect the brightness of the bulb? ________ ________

9. Using the data from Steps 1 and 3, compute the following DC values when the switch is in position 1.

   Current through the lamp = ______ A.

   Power dissipated by the lamp ________ W.

10. Using the data from Steps 1 and 7, compute the following AC values when the ___________ DPST switch is in position 2.

    Current through the lamp = ______ A.

    Power dissipated by the lamp = ______ W.

11. How did the DC data in Step 9 compare with the AC data in Step 10?

12. What conclusions can you reach when you compare AC effective values with identical DC values?

13. Disconnect the circuit shown in Step 5 and construct the circuit.

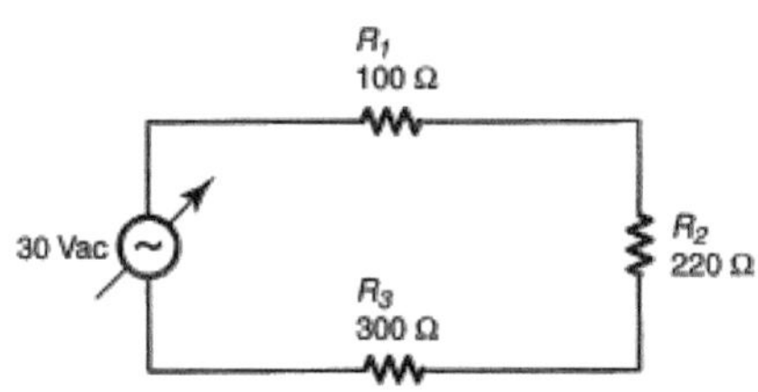

14. Using a multimeter to measure AC voltage, complete the table in Figure 16.1. Use the proper formulas for computations.

Resistors	Measured RMS voltage (*V*) across resistor	Computed peak voltage across resistor = *V* × 1.41	Computed p-p voltage across resistor = *V* × 2.82	Computed RMS power (P) across resistor $= \frac{V^2}{R}$	Computed peak power across resistor = *P* × 1.41
R_1					
R_2					
R_3					

Figure 16.1 Complete the table.

Analysis

1. Define the following terms:

   a. Hertz: ____________________

   b. Frequency: ____________________

   c. Period: ____________________

   d. Cycle: ____________________

   e. Effective AC values: ____________________

   f. Instantaneous AC values: ____________________

2. Draw a sine wave in the space below. Indicate its four quadrants, positive peak, negative peak, and time base (in degrees).

3. Convert the following effective values to peak and peak-to-peak values.

   6 V rms = ______ V peak; ______ V p-p

   10 V rms = ______ V peak; ______ V p-p

4. Convert the following peak values to effective values.

   12 W peak = ______ W rms

   2 mW peak = ______ W rms

5. How does the power produced by 10 Vdc across a 10 Ohm resistance compare with the power produced by 10 Vac rms across the same resistance?

6. Why is a cycle said to consist of 360 degrees?

7. Compute the periods for the following AC frequencies:

   10 K Hz = _______ seconds

   1.2 K Hz = _______ seconds

   60 K Hz = _______ seconds

8. What is the standard frequency of the alternating current used in the United States?

9. What is alternating current?

Activity 17

Inductance and Inductive Reactance

Objectives

Inductance is an opposition to alternating current. It does not oppose direct current. Because AC encounters many forms of opposition not encountered by DC, all quantities that oppose AC are called **impedance**. The letter in mathematical formulas represents these opposing quantities. One of the impedance quantities that opposes AC is inductance.

Inductance is the characteristic of an AC circuit to oppose any increase or decrease in current. Inductance is represented by L in mathematical formulas and is measured in Henry (H). Physically, inductance is embodied in an inductor, which is no more than a coil. The impedance caused by opposition other than resistance in AC circuits is known as **reactance**. In mathematical formulas, the letter represents reactance. Reactance brought about by inductance is represented by XL and is measured in ohms.

In this activity, you will:

1. Examine the effects of AC on inductive reactance.
2. Make computations concerning total inductance in series and parallel.

Equipment and Materials

- Multimeter
- Audio signal generator
- Variable AC power source
- Inductor – 4.5 H, iron-core (or any value from 2 to 15 H)
- Resistor

DOI: 10.1201/9788743810124-18

- Switch (SPST)
- Lamp and socket – 6 volt
- Connecting wires

Procedure

1. Measure and record the resistance of the inductor and the 6 V lamp filament.

   Inductor DC resistance = _______ Ohms.

   Lamp filament resistance = ______ Ohms.

2. Construct the circuit.

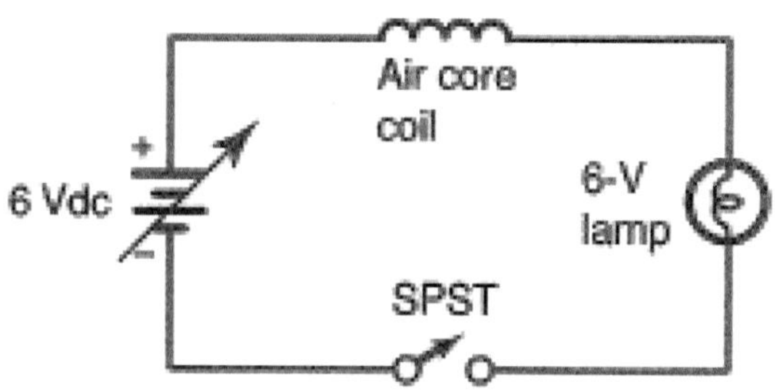

3. Prepare a multimeter to measure DC voltage. Close the switch, measure, and record the voltage across the lamp and the coil.

   coil ______ Vdc.

   (lamp) ______ Vdc.

4. How does the sum of the voltages across the coil and the lamp compare with the source voltage?

   Why? ____

5. Using the data from Steps 1 and 3, compute the current through the coil and lamp.

   ________ A.

6. Replace the DC power supply with the AC power source as shown.

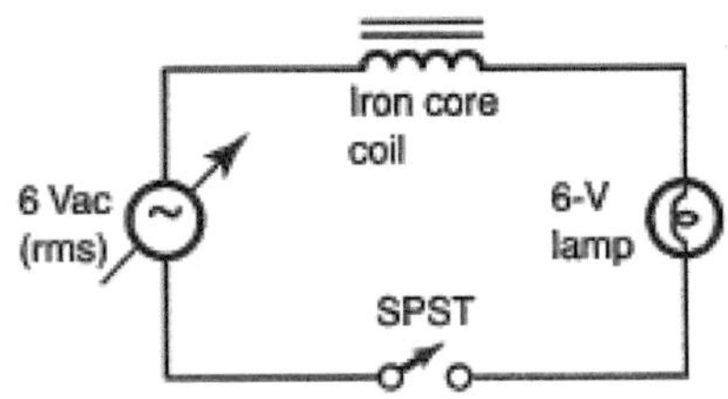

7. Prepare the meter to measure AC voltage. Close the SPST switch. Measure and record the voltage across the lamp and coil. V coil = ______ Vac.

   V (lamp) = ______ Vac.

8. How does the sum of these voltages compare with the source voltage? Why?

9. How did the sum of the voltages in Step 4 compare with the sum of the voltages in Step 8? Why was this relationship different?

10. Construct the circuit shown.

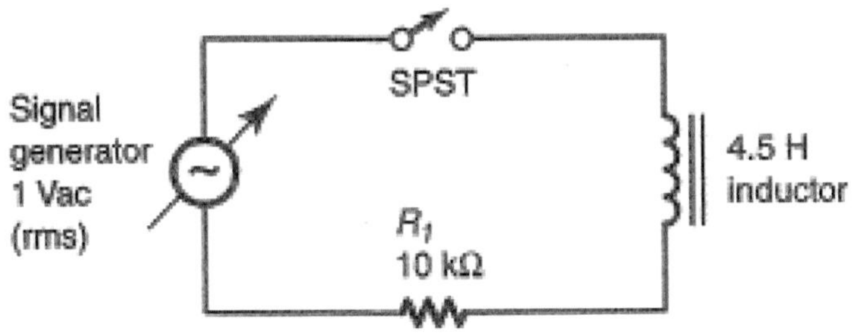

11. Prepare the meter to measure AC voltage and connect it across R1.

12. Close the switch and complete Figure 17.1. Adjust the signal generator to produce the AC frequencies indicated (note: the input voltage from the signal generator must be maintained at a constant level for each different frequency).

AC frequency input	Voltage across R_1	Circuit current (I) $= \frac{V_{R_1}}{R_1}$	Inductive reactance (X_L) computed = $2\pi \cdot f \cdot L$
100 Hz			
200 Hz			
300 Hz			
400 Hz			
500 Hz			
1 kHz			
10 kHz			

Figure 17.1 Complete the table.

13. What is the relationship between AC frequency and circuit current?

____

____

14. What is the relationship between AC frequency and inductive reactance?

____

____

15. What is the relationship between inductive reactance and circuit current?

____

____

____

Analysis

1. What factors determine the inductance of an inductor?

2. What factors determine the value of inductive reactance?

3. Why does inductance cause AC current to lag voltage?

4. What is the total inductance when inductors of 4 and 3 H are connected in series with no mutual inductance factor?

5. What is the total inductance when inductors of 4 and 3 H are connected in parallel with no mutual inductance factor?

6. Assume the inductors in Question 4 are connected to aid with a mutual inductance of 0.6 H. What is the total inductance?

7. Assume the inductors in Question 4 were connected to oppose with a mutual inductance of 0.86 H. What is the total inductance?

8. Assume the inductors in Question 5 were connected to aid with a mutual inductance of 0.2 H. What is the total inductance?

9. Assume the inductors in Question 5 were connected to oppose with a mutual inductance of 0.9 H. What is the total inductance?

10. What is mutual inductance?

11. If two inductors valued at 8 and 10 H are connected in parallel, which would allow the most AC current if the AC frequency were 1000 Hz?

12. Why will an inductor oppose AC more than DC for any given voltage?

Activity 18

Capacitance and Capacitive Reactance

Objectives

Capacitance is the property of an AC circuit to oppose any increase or decrease in voltage. It is present any time two conductors or plates are separated by a dielectric material. Capacitance is represented by the letter C in mathematical formulas and is measured in farads (F). Components designed to add capacitance to a circuit are known as **capacitors** and are rated according to their capacitance in farads and their working voltage (maximum DC voltage that can be placed across a capacitor without doing damage to its dielectric). Capacitors, known for their ability to store an electrical charge, can be connected in series or parallel.

Capacitance in an AC circuit causes current to **lead** the voltage. The impedance to alternating current due to capacitance is known as capacitive reactance (Xc) and is measured in ohms. Capacitive reactance is computed by using the formula discussed in Chapter 4 of the text.

Capacitors can be further classified as electrolytic or non-electrolytic. If electrolytic capacitors are used in a circuit, the polarity printed on the body of the capacitor must be observed.

In this activity, you will:

1. Study capacitance and capacitive reactance.
2. Examine how AC affects capacitive reactance.
3. Make simple computations concerning total capacitance in series and parallel.

DOI: 10.1201/9788743810124-19

Equipment and Materials

- Multimeter
- Signal generator
- Battery – 6 volt
- Resistors
- Capacitor – 0.01 uF
- Switch (SPST)
- Connecting wires

Procedure

1. Prepare the multimeter to measure resistance in the range or its equivalent.
2. Measure and record the resistance of the capacitor and the resistor.

   Resistor resistance = _______.

   Capacitor resistance = _______.

3. How do these resistances compare?
4. From the data shown in Step 1, which component would allow a direct current to flow and which would not?
5. Construct the circuit.

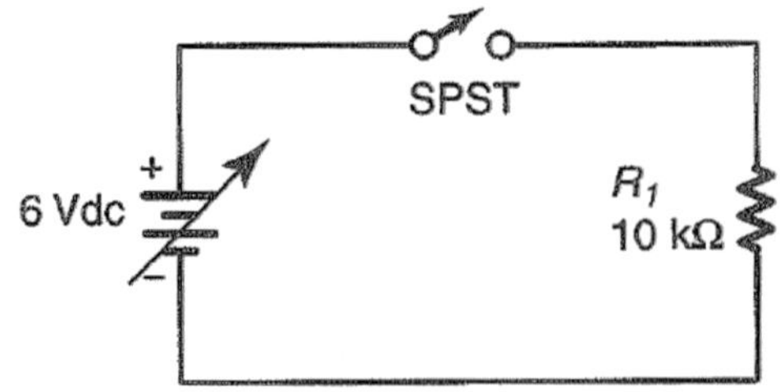

6. Prepare the meter to measure direct current in the range. Connect the meter into the circuit to measure current. Close the SPST switch and record the current.

7. Current _______ mA.

8. Alter the circuit in Step 4 to the circuit shown.

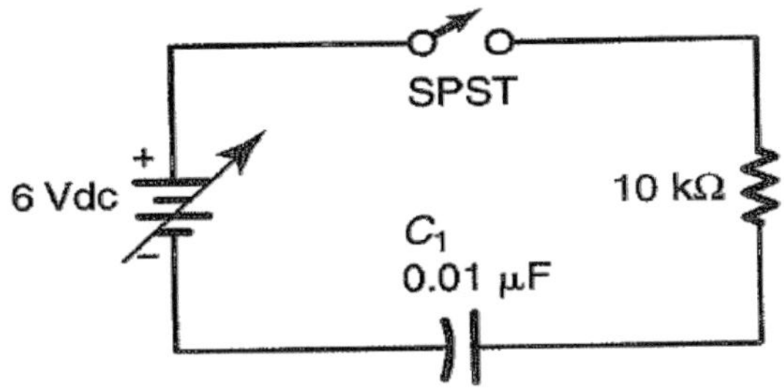

9. Connect the meter to measure DC (start in the highest range). Close the switch and record the current.

   Current = _______ mA.

10. How did the current recorded in Step 5 compare with the current recorded in Step 7?

11. How do you account for the current difference in Steps 5 and 7?

12. Alter the circuit above illustrated so it is like the one below.

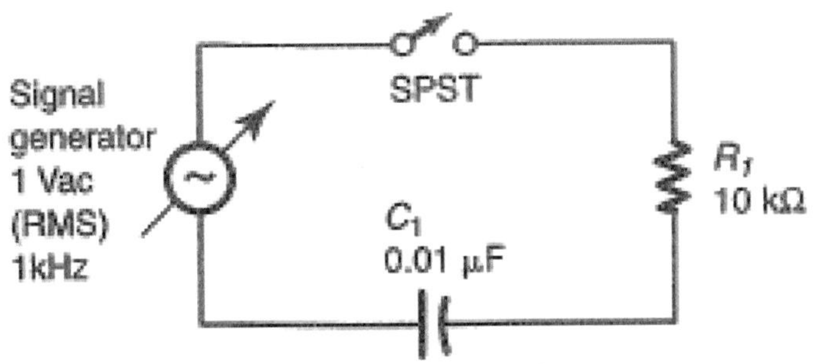

13. Prepare the meter to measure AC volts, starting in the highest range.
14. Connect the meter across, close the SPST switch, and record the voltage.

    Volta ge across/R1 = _____ Vac.

15. Using the resistance as measured in Step 1 and the voltage measured in Step 12, compute the current in this circuit.

    Current = _____ mA.

16. How did the current computed in Step 13 compare with the current measured in Step 7?
17. Since the two circuits are similar, how do you account for the difference in the currents in Steps 7 and 13?
18. With the meter connected as in Step 12, adjust the signal generator to the frequencies shown in Figure 18.1. Complete the figure for each frequency.

Frequency (Hz)	Measured voltage across R_1	Computed current $I = \frac{V_{R_1}}{R_1}$	Computed X_C
300			
500			
700			
900			
1100			
1300			
1500			
1700			
1900			
2500			
3000			

Figure 18.1 Complete the table.

19. From the data recorded in the table of Figure 18.4, what is the relation between frequency, current, and capacitive reactance?

    As frequency increases, the voltage across R1 ________, current ________, and XL ________ (enter either increases or decreases).

Analysis

1. What is capacitive reactance?

2. What is meant by a capacitor's working voltage?

3. What variables determine the capacitance of a capacitor?

4. If you connected in series two capacitors, valued at 4 uF each, what would be their total capacitance?

5. If you connected the two capacitors of Question 4 in parallel, what would their total capacitance be?

6. What is the relation between AC frequency and Xc?

7. How does capacitance affect the phase relation between alternating current and voltage?

8. How does the relation between AC frequency and Xc compare with the relation between AC frequency and XL?

Activity 19

Series RL Circuits

Objectives

Series RL circuits are encountered in many power systems. When an AC voltage is applied to this type of circuit, the current will be the same through each component. However, the voltage drops across each component will be distributed according to the relative value of resistance and inductive reactance in the circuit. The vector sum of the voltage drops will equal the source voltage. Therefore, in calculating voltage distribution in the circuit, study the relationships in Chapter 4 in the text.

In this laboratory activity, you will observe the characteristics of a series AC circuit that has resistance and inductance. The inductance characteristic is associated with many power systems that have windings, such as motors, generators, and transformers.

Equipment and Materials

- Multimeter
- AC voltage source
- Resistor
- Inductor – 8 H

Procedure

1. Construct the series RL circuit shown in Figure 19.1.

DOI: 10.1201/9788743810124-20

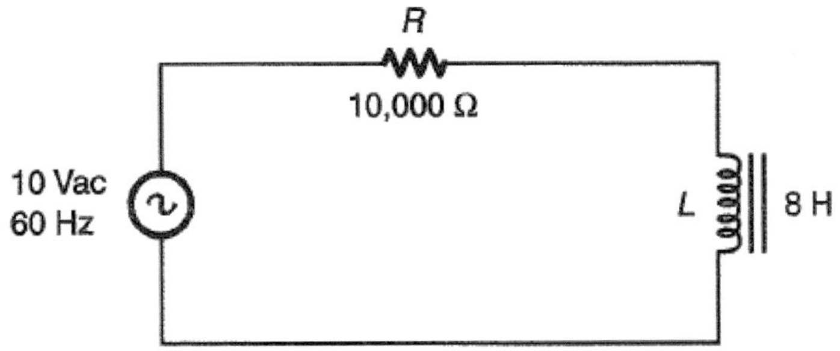

Figure 19.1 Series RL circuit.

2. Calculate the following:

   a. Inductive reactance = ______.

   b. Impedance = ______.

3. With a meter, measure the following values:

   a. Voltage across the resistor = ______ volts AC.

   b. Voltage across the inductor = ______ volts AC.

   c. Current through the circuit = ______ amperes AC.

4. Determine the impedance of the circuit:

5. Complete the impedance triangle of Figure 19.2 for the circuit you constructed by using the calculated values of XL and Z :

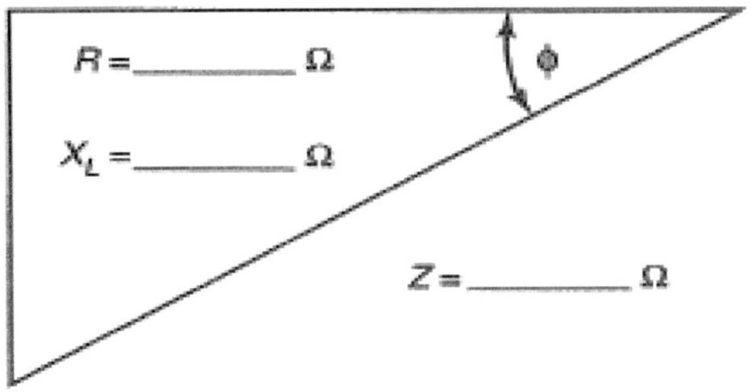

Figure 19.2 Impedance triangle.

6. Calculate the voltage drops in the circuit:

7. Show that the voltages must be added vectorially by using the values you calculated: ______volts AC.

8. Complete the following voltage triangles of Figure 19.3 for this circuit using the calculated, then the measured values:

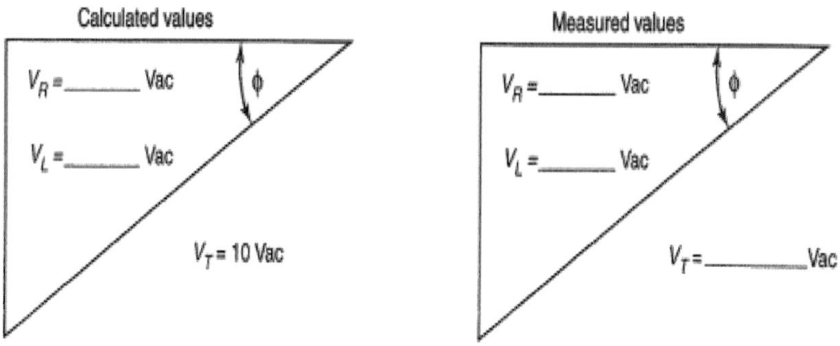

Figure 19.3 Voltage triangles.

9. Calculate the power converted in the circuit:

   a. True power (watts) = _____ watts.

   b. Apparent power (volt-amperes) = _____volt amperes.

   c. Reactive power _____volt-amperes-reactive.

10. Complete the power triangle of Figure 19.4 for the circuit using your calculated values of true power, apparent power, and reactive power.

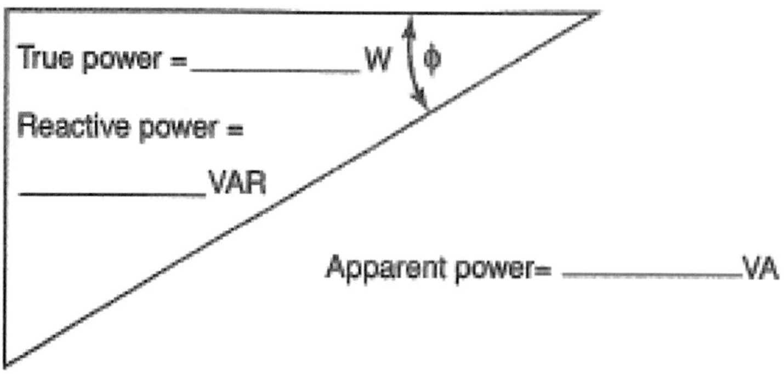

Figure 19.4 Power triangle.

11. Why must the voltage in a series RL circuit be added vectorially?
12. Compare the calculated and measured values of the circuit using Figure 19.5.

Value	Calculated	Measured
I_T		
Z		
V_R		
V_L		

Figure 19.5 Comparison of calculated and measured values.

13. What are some factors that could cause a difference in your measured values and the calculated values?
14. Determine the phase angle of the circuit by using trigonometric methods as discussed in Chapter 4 of the text:
15. Determine the power factor of the circuit as:

    Power factor = true power/apparent power

16. Determine the cosine of the phase angle. This value should be equal to the current power factor = ________.

Activity 20

Series RC Circuits

Objectives

Series RC circuits are used in many electrical systems. When an AC voltage is applied to this type of circuit, the characteristics are similar to those of a series RL circuit. You should recall that in a capacitive circuit, current leads voltage (leading phase angle) and in an inductive circuit, voltage leads current (lagging phase angle). Therefore, the vectorial relationships show the reactive values in the opposite direction.

In this laboratory activity, you will observe the characteristics of a series AC circuit that has resistance and capacitance.

Equipment and Materials

- AC voltage source
- Resistor
- Capacitor
- Multimeter

Procedure

1. Construct the series RC circuit shown in Figure 20.1.

 DOI: 10.1201/9788743810124-21

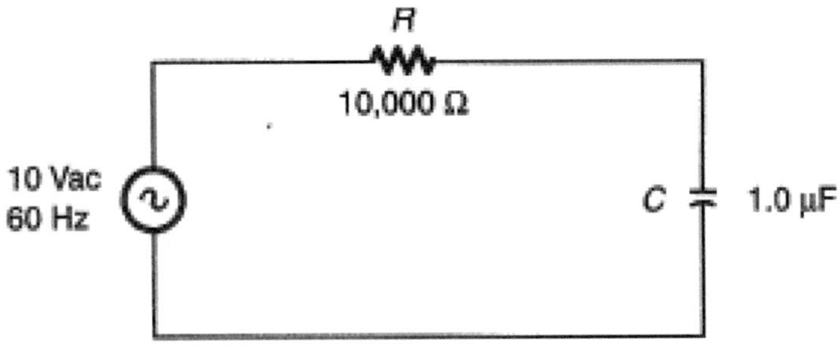

Figure 20.1 Series RC circuit.

2. Calculate the following:

   a. Capacitive reactance = _____ ohms.

   b. Impedance = ______ ohms.

3. Measure the following values:

   a. Voltage across the resistor = ______ volts AC.

   b. Voltage across the capacitor = _____ volts AC.

   c. Current through the circuit = ______ amperes AC.

4. Determine the impedance of the circuit:

5. Complete the impedance triangle of Figure 20.2 for the circuit you constructed by using the calculated values of Xc and Z.

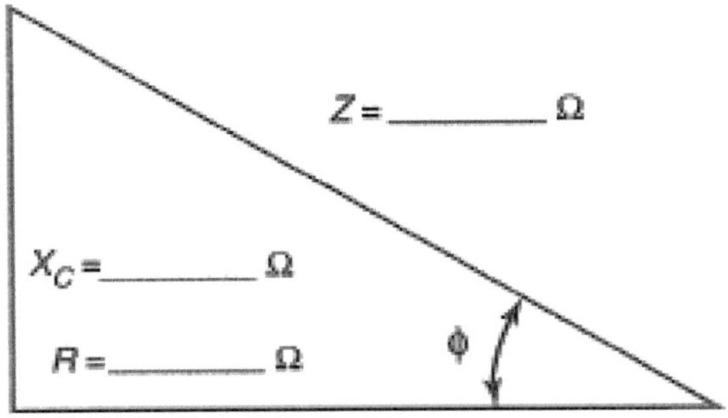

Figure 20.2 Impedance triangle.

6. Calculate the voltage drops in the circuit:

7. Show that the voltages must be added vectorially by using the values you calculated:

8. Complete the voltage triangles of Figure 20.3 for this circuit using the calculated, then the measured values of VT, VR, and Vc.

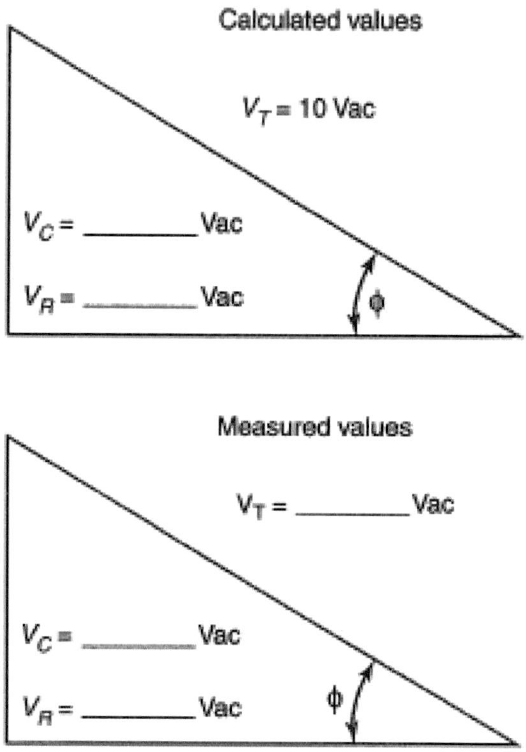

Figure 20.3 Voltage triangles.

9. Calculate the power converted in the circuit.

   a. True power (watts) = ____ watts.

   b. Apparent power (volt amperes) = _____ volt amperes.

   c. Reactive power = _____ volt amperes-reactive.

10. Complete the power triangle of Figure 20.4 for the circuit by using your calculated values of true power, apparent power, and reactive power.

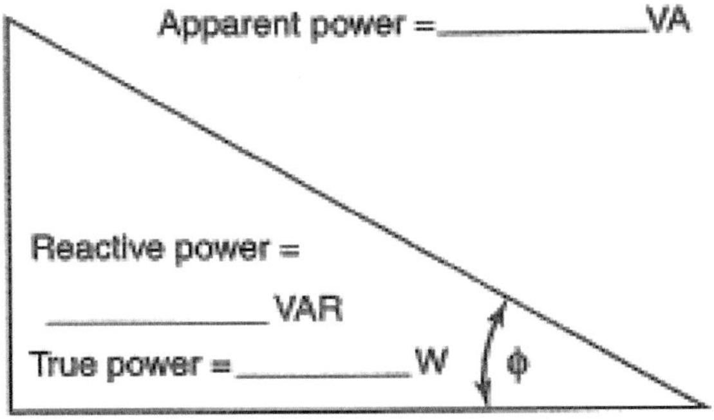

Figure 20.4 Power triangle.

Analysis

1. Compare the calculated and measured values of the circuit in Figure 20.5.

Value	Calculated	Measured
I_T		
Z		
V_R		
V_C		

Figure 20.5 Comparison of calculated and measured values.

2. Determine the phase angle of the circuit by using trigonometric methods:

3. Determine the power factor of the circuit as:

   Power factor = true power/apparent power = ______.

4. Determine the cosine of the phase angle. This value should equal the power factor of the circuit.

   Cosine = _______.

Activity 21

Series RCL Circuits

Objectives

In this activity, you will observe the characteristics of an AC circuit that has resistive, capacitive, and inductive components. Since the effects of the capacitance and inductance are out of phase with each other, we must analyze these circuits by using their vector relationships.

Equipment and Materials

- AC voltage source
- Resistor – 10,000 Ohm
- Capacitor – 1.0 uF
- Inductor – 8 H
- Multimeter

Procedure

1. Construct the series RCL circuit shown in Figure 21.1.

DOI: 10.1201/9788743810124-22

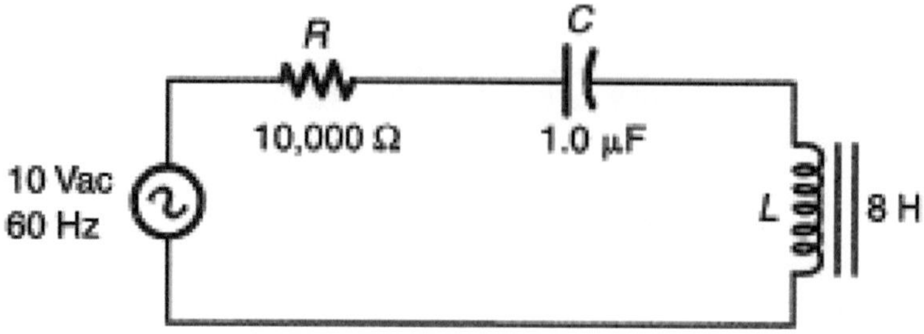

Figure 21.1 Series RCL circuit.

2. Calculate the following:
   a. Capacitive reactance
   b. Inductive reactance
   c. Total reactance
   d. Impedance

3. With a meter, measure the following values:
   a. Voltage across the capacitor = ______ volts AC.
   b. Voltage across the inductor = ______ volts AC.
   c. Voltage across the resistor = ______ volts AC.
   d. Voltage across the circuit = ______ amperes AC.

4. Determine the impedance of the circuit:

5. Complete the impedance triangle of Figure 21.2 for the circuit you constructed by using the calculated values.

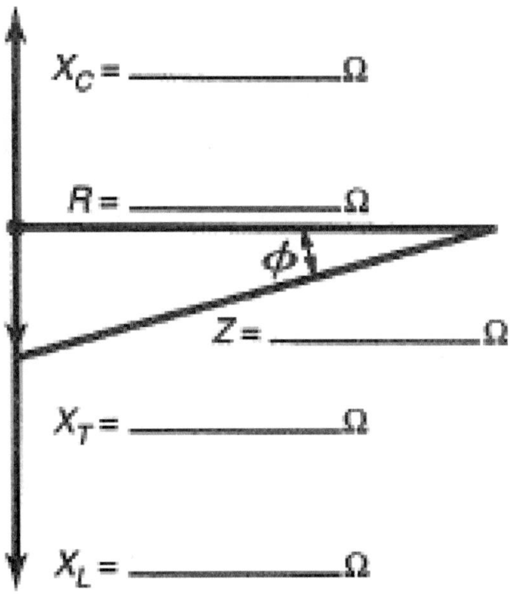

Figure 21.2 Impedance triangle.

6. Calculate the voltage drops in the circuit:

7. Show that the voltages must be added vectorially by using the values you calculated.

8. Complete the voltage triangles of Figure 21.3 circuit using the calculated, then the measured values of V_T, V_R, V_L and V_C.

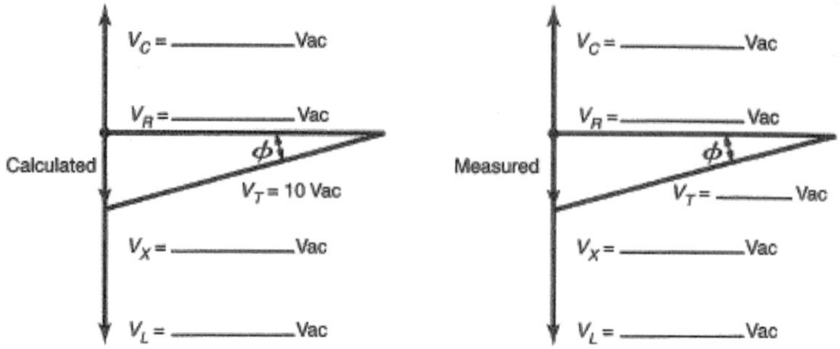

Figure 21.3 Voltage triangles.

9. Calculate the power converted in the circuit:
   a. True power = ______ W.
   b. Apparent power = ______ VA.
   c. Capacitive reactive power = ______ VAR.
   d. Inductive reactive power VAR.
   e. Total reactive power VAR.

10. Complete the power triangle of Figure 21.4 for the circuit by using your calculated values of true power, apparent power, and reactive power:

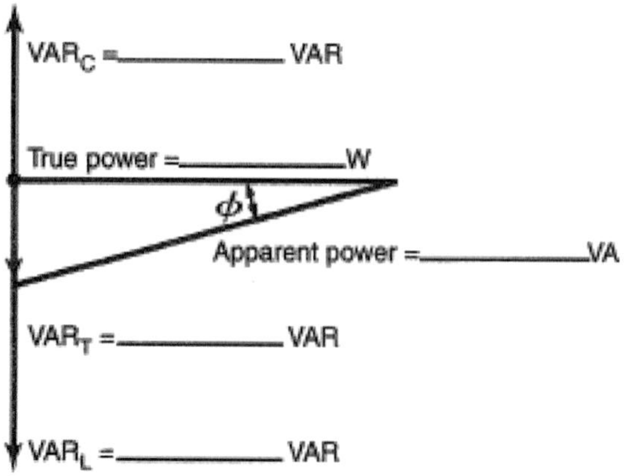

Figure 21.4 Power triangle.

Analysis

1. Compare the calculated and measured values of the circuit in Figure 21.5.

Value	Calculated	Measured
I_T		
Z		
V_R		
V_C		
V_L		

Figure 21.5 Comparison of calculated and measured values.

2. Determine the phase angle of the circuit.

   Phase angle = ______ degrees.

3. Determine the power factor of the circuit as:

   Power factor = true power/apparent power = ______.

4. Determine the cosine of the phase angle = ______.

   This value should equal the power factor of the circuit.

5. How does the circuit of this experiment show power factor correction?

6. As total reactive power increases, the phase angle (increases/decreases).

Activity 22

Parallel RL Circuits

Objectives

Another category of basic AC circuits is the parallel type. Parallel circuits are employed for many industrial control applications. The basic calculations necessary to study parallel AC circuits are somewhat different from those of series circuits. Since the impedance of a parallel circuit is less than the individual values of resistance, inductive reactance, or capacitive reactance, the impedance relationships must be modified. Also, a vector relationship exists between the currents that flow in the branches of the circuit.

In this activity, you will observe the impedance and current relationships of a parallel resistive-inductive (RL) circuit with AC applied to it.

Equipment and Materials

- AC voltage source
- Resistor – 1 K Ohm
- Inductor – 8 H
- Multimeter

Procedure

1. Construct the parallel RL circuit shown in Figure 22.1.

DOI: 10.1201/9788743810124-23

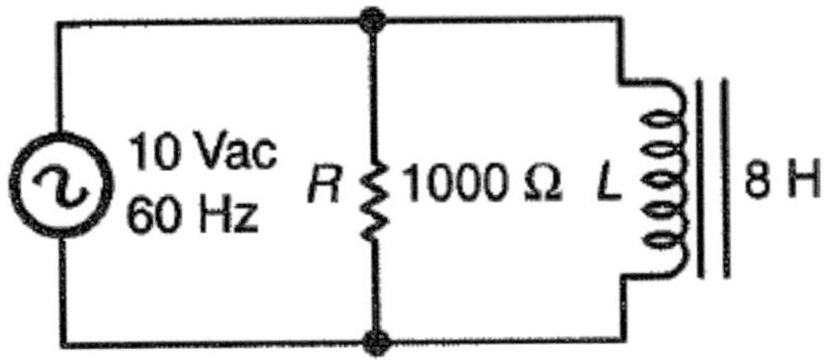

Figure 22.1 Parallel RL circuit.

2. Calculate the following:

   a. Inductive reactance = ______.

   b. Current through the resistor = ______ A.

   c. Current through the inductor = ______ A.

   d. Total current = ______ A.

   e. Impedance = ______ Ohms.

3. Complete the current triangle of Figure 22.2 for this circuit using the values from Step 2.

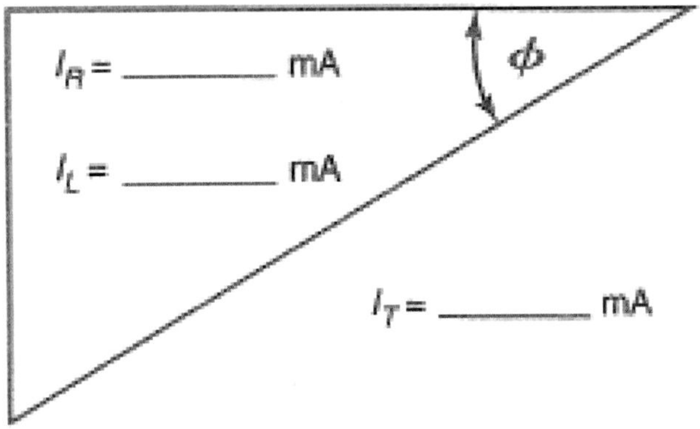

Figure 22.2 Current triangle.

4. With a meter, measure the following values:

   a. Current through the resistor = ______ A.

   b. Current through the inductor = ______ A.

   c. Total current A.

5. Since the impedance is less than the resistance or inductive reactance of the circuit, an **admittance** diagram must be used. In this diagram, admittance, conductance, and inductive susceptance are shown. The unit of measurement is the mho or Siemens (S). Note that "mho" is "ohm" spelled backward.

6. Calculate the values to complete the admittance triangle of Figure 22.3.

   a. Admittance = ______ S.

   b. Conductance = ______ S.

   c. Inductive susceptance = ______ S.

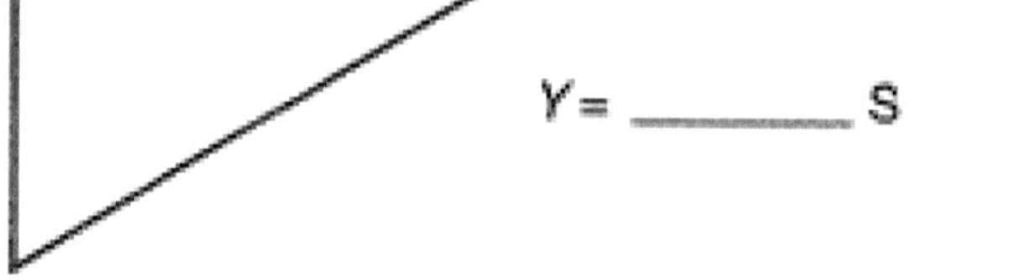

Figure 22.3 Admittance triangle.

7. Calculate the power converted in the circuit:

   a. True power (watts) = ______ W.

   b. Apparent power (volt amperes) = ______ VA.

   c. Reactive power = ______ VAR.

8. Complete the power triangle of Figure 22.4 for the circuit.

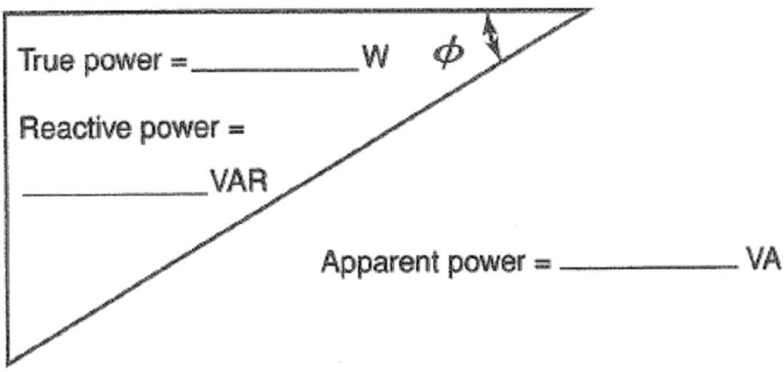

Figure 22.4 Power triangle.

Analysis

1. Compare the calculated and measured values of the circuit in Figure 22.5.

Value	Calculated	Measured
I_R		
I_C		
I_T		

Figure 22.5 Comparison of calculated and measured values.

2. Determine the phase angle of the circuit:

   Phase angle = ______ degrees.

3. Determine the power factor of this circuit.

   Power factor = true power/apparent power = ______ .

4. Compare the power factor of this circuit with the value of the cosine function of the triangle.

5. Define the following terms associated with parallel AC circuits:

   a. Admittance:

   b. Conductance:

   c. Inductive susceptance:

   d. Capacitive susceptance:

6. What are some sources of error that account for some difference between calculated and measured values in this activity?

Activity 23

Parallel RC Circuits

Objectives

In this laboratory activity, you will observe the characteristics of a parallel RC circuit with AC applied. This type of circuit is similar to the RL circuit. However, the effect of capacitance is opposite to that of inductance. Similar vector relationships exist in this circuit.

Equipment

- Multimeter
- AC voltage source
- Resistor – 1 K Ohm
- Capacitor – 1.0 uF

Procedure

1. Construct the parallel RC circuit shown in Figure 23.1.

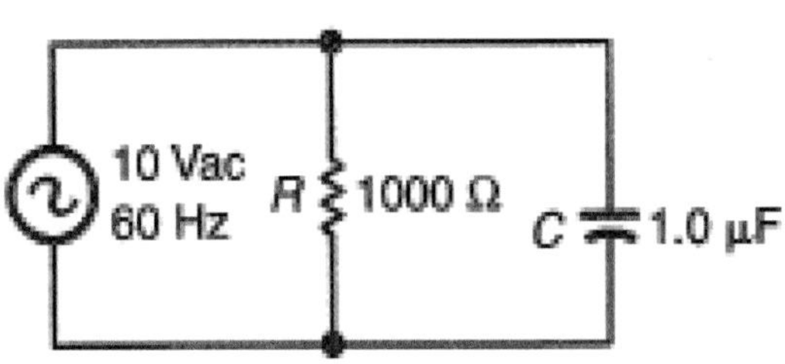

Figure 23.1 Parallel RC circuit.

DOI: 10.1201/9788743810124-24

2. Calculate the following:

   a. Capacitive reactance = ______ Ohms.

   b. Current through the resistor = ____ A.

   c. Current through the capacitor ____A.

   d. Total current = ____ A.

   e. Impedance = ______ Ohms.

3. Complete the following current triangle of Figure 23.2 for this circuit using the values from Step 2.

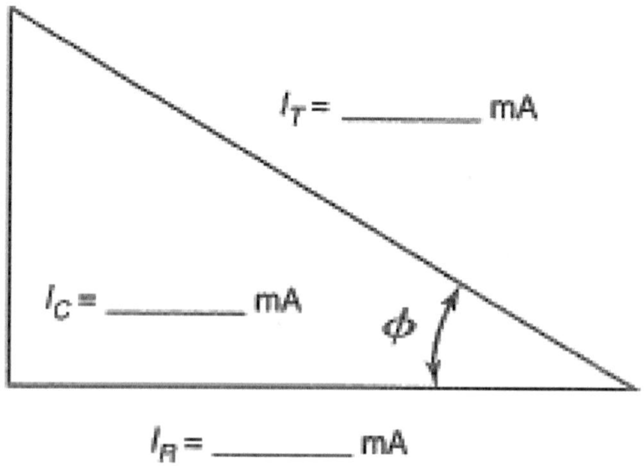

Figure 23.2 Current triangle.

4. Measure the following AC current values:

   a. Current through the resistor = ____ A.

   b. Current through the capacitor = ____A.

   c. Total current = ____ A.

5. Calculate the following values:

   a. Admittance = ____ S.

   b. Conductance = ____ S.

   c. Capacitive susceptance S.

6. Complete the admittance triangle of Figure 23.3 for the circuit:

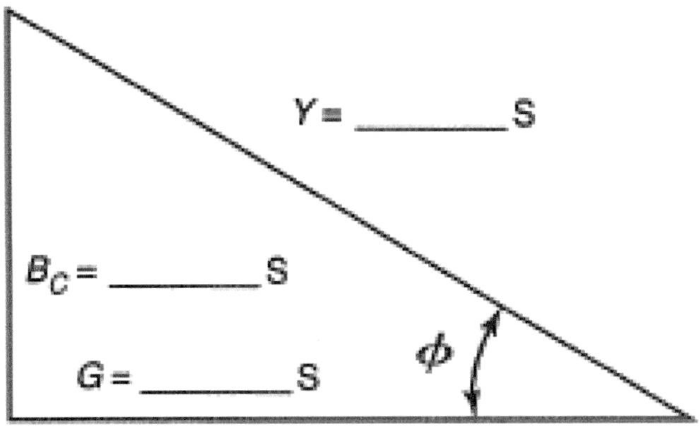

Figure 23.3 Admittance triangle.

7. Calculate the power converted in the circuit:

   a. True power = ______ W.

   b. Apparent power = ______ VA.

   c. Reactive power = ______ VAR.

8. Complete the power triangle of Figure 23.4 for the circuit:

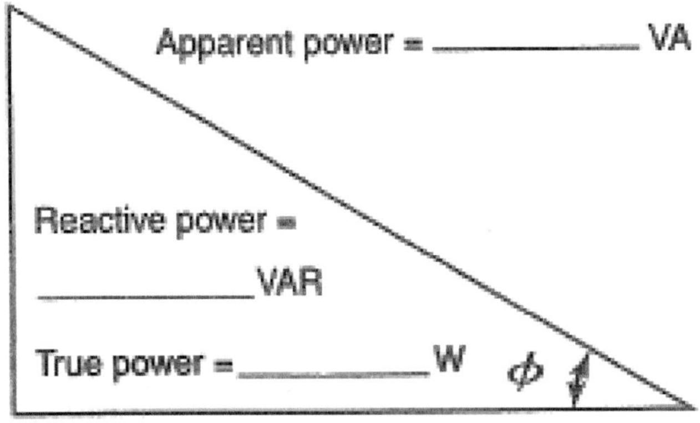

Figure 23.4 Power triangle.

9. This concludes the experiment.

Analysis

1. Insert the calculated and measured values for the circuit of this experiment into the following table. Compare the values.

Value	Calculated	Measured
I_R		
I_C		
I_T		

2. Determine the phase angle of the circuit:

   Phase angle = ______ degrees.

3. Determine the power factor of the circuit:

   Power factor = true power/apparent power = ____ .

4. If an 8-Henry inductor was connected in parallel with the circuit of this activity, compute the following:

   a. X_L = ______ Ohms.

   b. V/X_L = ____ A.

   c. $Ix = I_L/I_C$ = _____ A.

   d. I_T = ____ A.

   e. V/I_T = ____ Ohms.

   f. Phase angle = ______ degrees.

   g. Power factor = _______.

5. How do the values of power factor for the original circuit and those of the preceding differ? Why

6. How do the values of total current differ in the two circuits? Why?

Activity 24

Working with Permanent Magnets

Objectives

In this activity, you will learn the properties of magnetic fields. It is important to understand the effects of magnetism since the operation of many types of machines depends on magnetic fields.

Equipment and Materials

- Permanent magnets (2)
- Iron filings
- Paper

Procedure

1. Obtain two permanent magnets, a sheet of paper, and some iron filings.
2. Place one magnet on a flat surface with a sheet of paper placed over it.
3. Carefully pour iron filings onto the paper above the magnet.
4. Carefully lift the paper and place the second magnet under the paper with the two north poles about one inch apart. Make a sketch of the magnetic field pattern.

DOI: 10.1201/9788743810124-25

5. Place the north and south poles about one inch apart. Make a sketch of the magnetic field patterns.

Analysis

1. Discuss the laws of magnetism.

Activity 25

The Nature of Magnetism

Objectives

Magnetism is one of the longest-known natural forces. It was first discovered and used by ancient cultures as a curiosity. Many believed that this force was magic and, therefore, to be feared. The first magnets used were natural magnets called lodestones. They were first put to practical use in navigation. Someone discovered that when these devices were suspended by a string and allowed to move freely, they would always align themselves to a point to the north. Thus, natural magnetism was first used for compasses.

Much later, it was discovered that magnetism could be used to create an electric current, and an electric current could be used to create a magnetic field. This relationship makes a knowledge of magnetism extremely important.

In this activity, you will examine the characteristics of natural magnetism.

Equipment and Materials

- Permanent magnets (2)
- Magnetic compass

Procedure

1. Place a permanent magnet on the surface in front of you.

 DOI: 10.1201/9788743810124-26

2. Examine the needle of the compass. You will find that the north-seeking pole (south pole) of the compass needle is painted or otherwise colored to distinguish it from the north pole.
3. Use the compass to identify the north pole and south pole of the permanent magnet (remember, like poles repel; unlike poles attract). Notice the north and south poles of your permanent magnet.
4. If the north pole of one magnet is brought near the north pole of another magnet, what happens? Why?
5. If the south pole of one magnet is brought near the south pole of another magnet, what happens? Why?
6. If the north pole of one magnet is brought near the south pole of another magnet, what happens? Why

Analysis

1. Describe the reaction of like poles and unlike poles of magnetic fields.

2. Why can a compass be used to detect the presence of a magnetic field?

3. Define the following terms:
   a. Flux:
   b. Magnetic field:
   c. Magnetic lines of force:

4. In what direction does magnetic flux travel externally around a magnet?

5. Sketch the magnetic field of a permanent magnet.

Activity 26

Making an Electromagnet

Objectives

In this activity, you will learn the properties of an electromagnet by making one that will actually work. It is important to understand the effects of electromagnetism since the operation of many types of machines depends on electromagnetic fields.

Equipment and Materials

- Bolt – 3/8 to 3/4 inch in diameter and about 4 inches long
- Solid copper wire – 10 to 20 feet of No. 18 to No. 24 diameter
- Lantern battery – 6 volt
- Compass

Procedure

1. Use the bolt as an iron core. Start at one end and wind the wire around the core in the same direction. Leave 6 to 10 inches of wire on each end of the wire for connecting the battery. Wrap tape around the wire to secure it in place. The finished coil should have between 100 and 200 turns of wire.
2. Momentarily connect the ends of the coil to a 6 V lantern battery. It may be necessary to remove about one-half inch of insulation from the ends of the wire. This will allow good electrical contact to be made. Do not keep the battery connected for over a few seconds at a time.

 DOI: 10.1201/9788743810124-27

3. Test the electromagnet with a compass to see if it has a north polarity at one end and a south polarity at the other end.

4. Try picking up metal objects with the electromagnet.

Analysis

1. Discuss the construction and operation of an electromagnet. Be sure to discuss ways to make the electromagnet stronger.

UNIT II

Electrical Distribution Basics

Activity 27

Problems for Electrical Distribution Systems

Problems

1. A 1500-turn primary winding has 120 volts applied. The secondary voltage is 20 volts. How many turns of wire does the secondary winding have?

2. When a 200-ohm resistance is placed across a 240-volt secondary winding, the primary current is 30 amperes. What is the value of the primary voltage of the transformer?

3. A transformer has a 20:1 turns ratio. The primary voltage is 4800 volts. What is the secondary voltage?

4. A transformer has a 2400-volt primary winding and a 120-volt secondary. The primary winding has 1000 turns. How many turns of wire does the secondary winding have?

5. A 240-volt to 4800-volt transformer has a primary current of 95 amperes and a secondary current of 4 amperes. What is its efficiency?

DOI: 10.1201/9788743810124-29

6. A transformer has a 20 KVA power rating. Its primary voltage is 120 volts, and its secondary voltage is 480 volts. What are the values of its maximum primary current and secondary current?

7. Convert the following diameters of electrical conductors into mils.

   a. 1/8 inch = ______ mils.

   b. ¼ inch = ______ mils.

   c. 3/8 inch = ______ mils.

   d. ½ inch = ______ mils.

   e. 0.2 inch = ______ mils.

   f. 0.28 inch = ______ mils.

   g. 0.15 inch = ______ mils.

   h. 0.66 inch = ______ mils.

8. Calculate the cross-sectional area of the following electrical conductors (in circular mils).

   a. 3/16 inch diameter = _______.

   b. 0.5 inch square conductor = ______.

   c. 0.38 inch diameter = ______.

   d. ½ inch × 3/8 inch rectangular conductor = ______.

9. Calculate the resistance of 480 feet of copper conductor that is 0.28 inches in diameter.

10. Use the proper formula to calculate the resistances of the following electrical conductors.

    a. 380 feet of No. 10 AWG copper.

    b. 2500 feet of No. 000 AWG copper.

    c. 230 feet of 500 MCM copper.

   d. 1900 feet of No. 8 AWG aluminum.
   e. 220 feet of No. 10 AWG aluminum.

11. Use the proper table to calculate the resistance of the following electrical conductors.
   a. 600 feet of No. 0000 AWG copper.
   b. 500 feet of No. 1 AWG aluminum.
   c. 300 feet of No. 12 AWG copper.
   d. 200 feet of 250 MCM aluminum.
   e. 1800 feet of No. 10 AWG copper.

12. In order to carry the following currents, what are the minimum sizes of conductors needed? No more than three conductors will be used in the same raceway.
   a. 380 amperes using THW copper.
   b. 95 amperes using THW aluminum.
   c. 180 amperes using RH copper.
   d. 230 amperes using TW copper.
   e. 230 amperes using copper.
   f. 650 amperes using RHW copper.
   g. 580 amperes using T copper.
   h. 82 amperes using TH aluminum.

13. Find the equivalent size of THW aluminum conductors for the following copper conductors.
   a. 500 MCM, TW.
   b. 750 MCM, TH
   c. No. 00 AWG, R.
   d. No. 4 AWG, T.
   e. No. 8 AWG, RHW.

14. Three 20,000-watt loads are connected to a three-phase, wye-connected system. The power factor of the system is 0.9. Find the minimum size of RH copper feeder conductors needed to supply this load.

15. Determine the branch-circuit rating and size of conductors needed for the following systems:

    a. 2500-watt, 240-volt heater.

    b. 1800-watt, 120-volt washer.

    c. 12,000-watt, 240-volt range.

    d. 800-watt, 120-volt toaster.

16. Calculate the maximum distance a 120-volt, 30-ampere branch circuit can extend from the power source when a 3-kW appliance is connected to the circuit the voltage drop should be limited to 2%.

17. A 120-volt 20-ampere branch circuit (using No. 12 copper conductors) extends 150 feet. Loads connected at 30-foot intervals each draw 3.5 amperes. With all loads connected, calculate the voltage at the last outlet.

18. Calculate the longest distance a single-phase 20-ampere, 120-volt branch circuit can extend from the power source the voltage drop should be limited to 2%.

19. A single-phase 120-volt load is rated at 40 kilowatts. The feeder circuit will be 400 feet of RH copper conductors the voltage drop should be limited to 2%.

20. A 240-volt three-phase delta system converts 30 kW of power per phase (balanced). The system power factor is unity (1.0). The feeders will be 400 feet of RHW copper conductor the voltage drop should be limited to 2%.

Activity 28

Transformer (Self Test)

Problems: Complete the following

Identify the parts of the transformer shown.

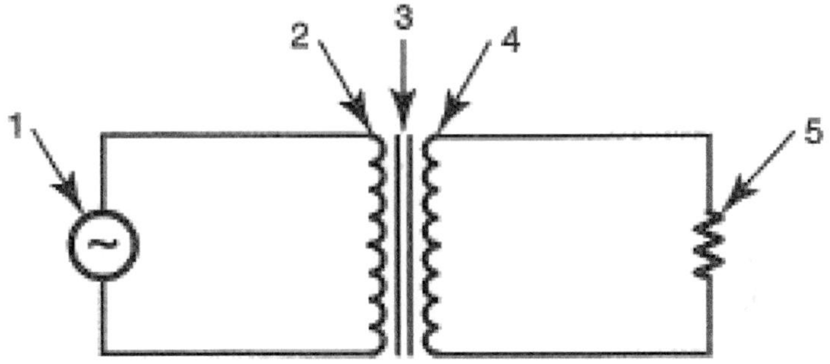

1. __________

2. __________

3. __________

4. __________

5. __________

DOI: 10.1201/9788743810124-30

6. What is the secondary voltage of the transformer shown?

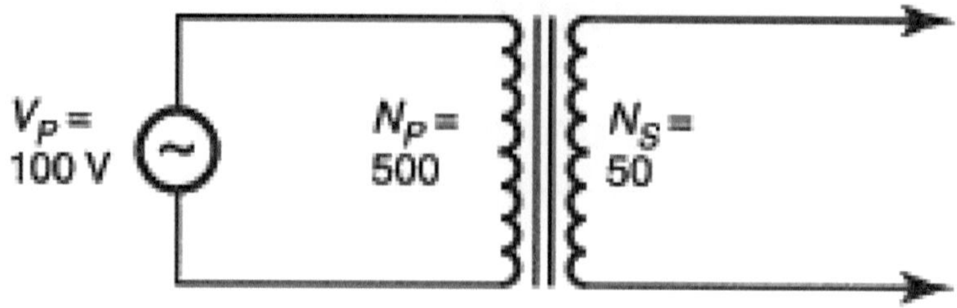

7. If the primary current of the transformer shown in Figure 28.2 is 5 A, what would the secondary current equal?

8. A transformer that only has one winding is called: ________________.

9. The major reason for using transformers to increase voltage for power transmission is to reduce ________________.

10. An electrical device consisting of two coils very close together yet electrically insulated from each other is ________________

11. A transformer coil connected to a source of AC voltage is the ________________winding.

12. A transformer coil connected to the load is the ________________ winding.

13. If there are fewer turns in the secondary winding than in the primary winding, the device is a(n) ____________________ transformer.

14. If there are more turns in the secondary winding than in the primary winding, the device is a(n) ____________________ transformer.

15. To reduce the opposition to magnetic lines of force between the primary and secondary, the windings may be wound on a(n) ____________________.

16. If 120 V are applied to a primary of a step-down transformer with a 10:1 turns ratio, the voltage induced in the secondary is ________ V.

17. If 120 V is applied to the primary of a step-up transformer with a turns ratio of 1:2, the voltage induced in the secondary is ______________V.

18. If the voltage across the complete secondary winding of a center-tap transformer is 240 V, the voltage from one outside conductor to the center tap is _____V.

19. A transformer with more than one secondary winding is called ___________transformer.

20. The ratio of power output to power input of a transformer is called ____________________.

Activity 29

Transformer Operation

Objectives

In this activity, you will learn the properties of a transformer by checking the operation of a working transformer. It is important to understand the operation of transformers since they are used in power distribution systems.

Equipment and Materials

- Multimeter
- Transformer

Procedure

1. Obtain a transformer and a multimeter. Record all available data on your transformer on a sheet of paper.
2. Identify the primary windings of the transformer. These primary windings will be connected to an AC voltage source. Do **not** plug it in until you are ready to make the necessary measurements. Remove the power cord when the transformer is not in use. Be very aware of safety. It is best to make measurements with a low AC voltage applied to the primary (from 6 to 15 V). **If a 120-Vac source is used, be very careful.**

Safety

Safety is always a primary factor for laboratory activities. This lab should only be completed with supervision. You may complete

DOI: 10.1201/9788743810124-31

certain steps of the "Procedure" and the "Analysis" for better understanding, without applying voltage.

3. Some transformers have one secondary winding. Many transformers have three or more secondary windings. Count the number of secondary windings on the transformers.

   Number of secondary windings _________.

4. Make sure that none of the wires from the secondary windings touch.
5. Prepare the meter to measure AC voltage. Begin with the highest AC voltage range for each secondary voltage measurement.
6. Carefully connect the primary winding to an AC source. Measure the voltage across each secondary winding. Record these voltages.
7. Disconnect the AC voltage from the transformer primary. With no voltage applied, make the following resistance measurements with a multimeter.

   a. Resistance of primary winding = ______ Ohms.

   b. Resistance of each secondary winding = ______ Ohms.

Analysis

1. Discuss your readings and the basic operation of a transformer.

Activity 30

Transformer Turns Ratio

Procedure

1. Review the diagrams shown in Figure 30.1. Observe that there are 100-turn windings and 500-turn windings shown in various combinations with 10 volts AC applied to the primary winding of each illustration.

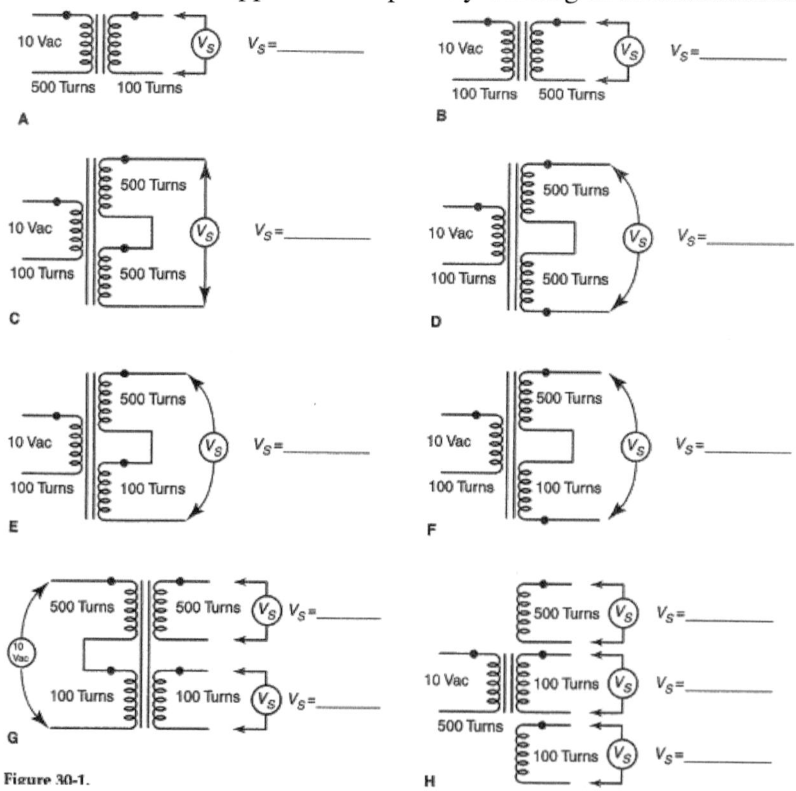

Figure 30.1 Calculate secondary voltages.

DOI: 10.1201/9788743810124-32

2. Using the transformer connections shown in Figure 30.1, calculate the indicated secondary voltages of each combination. Be sure to observe the polarity dot, which indicates the beginning of the winding.
3. If multiple-winding transformers such as those shown in Figure 30.1 are available in the laboratory, perform secondary voltage measurements with different winding combinations and a constant voltage input.

Analysis

1. Briefly discuss the turns ratio, voltage ratio, and current ratio of single-phase transformers.

Activity 31

Transformer Problems

Procedure

Solve the following problems dealing with transformer operation. Record all your answers on a separate sheet of paper. Show all of your work.

1. A 1500-turn primary winding has 120 V applied. The secondary voltage is 20 V. How many turns of wire does the secondary winding have?

2. When a 200 Ohm resistance is placed across a secondary winding, the primary current is 30 A. What is the value of the primary voltage of the transformer?

3. A transformer has a 20:1 turns ratio. The primary voltage is 4800 V. What is the secondary voltage?

4. A transformer has a 2400 Volt primary winding and a 120 Volt secondary. The primary winding has 1000 turns. How many turns of wire does the secondary winding have?

DOI: 10.1201/9788743810124-33

5. A 240 Volt to 4800 Volt transformer has a primary current of 95 A and a secondary current of 4 A. What is its efficiency?

6. A single-phase transformer has a power rating of 20 kVA. Its primary voltage is 120 V, and its secondary voltage is 480 V. What are the values of its maximum primary current and secondary current?

Activity 32

Auto Transformer Principles

Objectives

The auto transformer has only one winding. Therefore, it requires less copper wire to construct than a conventional single-phase transformer of similar size. The major disadvantage of an auto transformer is the necessity of having a direct connection to one of the primary input lines.

In this laboratory activity, you will observe the operational principles of an auto transformer. You can use either an auto transformer or a secondary winding of a center-tapped transformer. You should notice that this device can be constructed as a step-up or step-down transformer.

Equipment and Materials

- Multimeter
- Variable AC power source
- Center-tapped transformer
- Resistor – 1000 Ohm, 10 K Ohm, 10 Watt
- AC current meter

Procedure

1. Construct the auto transformer circuit shown in Figure 32.1.

DOI: 10.1201/9788743810124-34

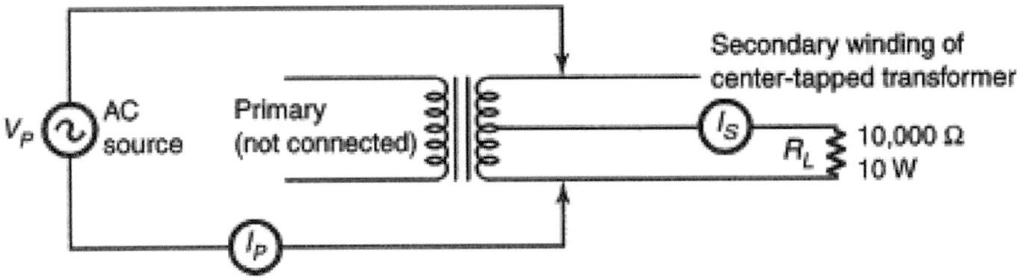

Figure 32.1 Auto transformer step-down connections.

2. Apply rated voltage to the outer winding terminals of the center-tapped transformer. Measure the voltage across the load, primary current, and secondary current.

3. Change the value of load resistance to 1000 ohms and repeat the voltage and current measurements.

4. Modify the circuit as shown in Figure 32.2. The AC input voltage should remain the same as before.

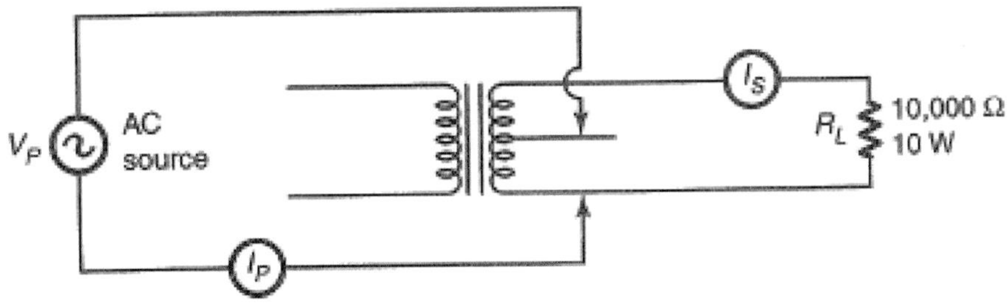

Figure 32.2 Auto transformer step-up connections.

5. Again, measure and record the voltage and current values.

6. Change the value of load resistance to 1000 ohms and repeat the measurements.

Analysis

1. What type of auto transformer was constructed in Step 1?

2. What type of auto transformer was constructed in Step 4?

3. From Step 2, verify that Vp × Ip = Vs × Is ---

4. From Step 3, verify that Vp × Ip = Vs × Is ---

5. From Step 5, verify that Vp × Ip = Vs × Is ---

6. From Step 6, verify that Vp × Ip = Vs × Is ---

7. What are some applications of auto transformers?

8. How does decreased load resistance affect the primary current?

9. Are the primary and secondary power values (Vp × Ip and Vs × approximately equal? Why?

Activity 33

Electrical Conductors and Insulators

Objectives

Two very important parts of industrial power distribution systems are conductors and insulators. Conductors of various types are designed to carry electrical current. Conductors can be in the form of solid wires, stranded cables, cords, bus bars, or many other types. The amount of current that a conductor can safely handle is mainly determined by its diameter. Examples of several conductor tables are included in Chapter 6 in your textbook.

The diameter of a conductor is measured with an American Wire Gage (AWG). The sizes from No. 18 through 1000 MCM (1000 circular mils) are shown in Figure 6.17 of your textbook.

The insulation used on electrical conductors is mainly determined by the type of location where it will be used and the type of enclosure or wireway where it will be placed. The symbols R, H, W, T and N indicate rubber, heat-resistance, moisture-resistance, thermoplastic, and nylon, respectively. A table of some insulation types is shown in Figure 6.24 of your textbook.

In this activity, you will observe several types of conductors and insulation. You should use the samples of conductors provided by your instructor.

Equipment and Materials

- American Wire Gage (AWG)
- Conductor samples

DOI: 10.1201/9788743810124-35

Procedure

1. Obtain an American Wire Gage and several conductors from your instructor.

2. Measure the diameter of each conductor with the AWG and record the results in the following spaces. Also, record whether each conductor is copper or aluminum.
   Sizes of conductors:

   a. ____

   b. ____

   c. ____

   d. ____

   e. ____

   f. ____

3. Refer to Figures 6.20 and 6.21 in your textbook to determine the amount of current each conductor can carry. Record your answers in the following spaces.
   Ampacity of conductors:

   a. ____ amperes (raceway) ____ amperes (free air)

   b. ____ amperes (raceway) ____ amperes (free air)

   c. ____ amperes (raceway) ____ amperes (free air)

   d. ____ amperes (raceway) ____ amperes (free air)

   e. ____ amperes (raceway) ____ amperes (free air)

   f. ____ amperes (raceway) ____ amperes (free air)

4. List the type of insulation on each conductor and record it in the following spaces. Type of insulation:

   a. _____

   b. _____

   c. _____

   d. _____

   e. _____

   f. _____

Analysis

1. What is the area in circular mils of the following conductors? (Use Figure 6.17 of your textbook).

   a. No. 16 = _____ cmil.

   b. No. 0 = _____cmil.

   c. No. _____cmil.

   d. _____cmil.

   e. _____cmil.

   f. No. _____cmil.

2. How many wires are used to construct the conductors listed in the preceding question?

   a. _____

   b. _____

   c. _____

   d. _____

   e. _____

   f. _____

3. What are the types of the following insulations?

   a. RH

   b. RHW

   c. THW

4. What are the ampacities of the following conductors in a raceway?

   a. No. 12 (RH) copper = _______ A.

   b. No. 4 (TW) aluminum ______A.

   c. 250 MCM (THW) aluminum ______A.

   d. No. 6 (T) copper ______A.

   e. No. 000 (R) aluminum ______A.

   f. Copper ______A.

5. What is the DC resistance of the following lengths of conductors?

   a. 500 feet of No. 8 copper = ______ Ohms.

   b. 800 feet of No. 12 aluminum = _______ Ohms.

   c. 5000 feet of 250 MCM aluminum = _______ Ohms.

   d. 1500 feet of 500 MCM copper = ______ Ohms.

   e. 400 feet of No. 6 copper = _______ Ohms.

   f. 8000 feet of 1000 MCM aluminum = _______ Ohms.

Activity 34

National Electrical Code (Self Test)

Objectives

The National Electrical Code (NEC) is a very important document to understand. All industrial equipment and wiring must conform to the NEC standards. The NEC is not difficult to use. The user should become familiar with the comprehensive index contained in the NEC and the organization of the various sections. For instance, if you wish to review the standards related to system grounding, you should look in the index and locate this term. The index will refer you to the appropriate sections in the NEC that discuss system grounding.

In this activity, you will need to review the latest edition of the NEC. Once you have learned how to use it, you should complete the self-test on the NEC.

Self Test

Using a recent copy of the NEC, find the correct responses to complete each of the statements below. Place the article number/page number where the answer was obtained in parentheses at the end.

1. A new edition of the NEC is published every _________ years. (___)

2. A No. 8 copper conductor in free air with THW insulation is capable of conducting _____ amperes of current. (___)

DOI: 10.1201/9788743810124-36

3. When selecting nonmetallic-sheathed cable for general use, you should specify type ____. (___)

4. If you were wiring for an air conditioner, you should refer to Article ____ in the NEC. (___)

5. The diameter of electrical conductors is measured with a(n) ____ gauge. (___)

6. Line-to-ground shock hazards can be reduced by utilizing __________. (___)

7. The definition of a "ground" can be located in Article ______ of the NEC. (___)

8. Article 402 of the NEC specifies that type CF fixture wire has _______ insulation. (___)

9. If you were installing a three-phase motor with associated control circuits, you should refer to Article ___ of the NEC. (___)

10. Article 810 of the NEC provides electrical wiring specifications for ______. (___)

11. Underground feeder cables come in sizes ____ to ____inclusive. (___)

12. The full-load current of a one-horsepower, 120-volt DC motor as listed in the NEC is ______ amps. (___)

13. The full-load current of a 30-horsepower, three-phase AC induction motor operating on 208 volts is ___ amps (see information at the bottom of the table for conversion). (___)

14. According to Article 210 of the NEC, grounded conductors in a branch circuit should be color coded _________or ____________ (__)

15. The general lighting load of an office building is specified to be ___watts per square foot of floor space. (___)

Activity 35

Power Distribution Substation Visit

Objectives

After completing this laboratory activity, you will be able to identify and explain the function of equipment used in electrical power substations.

Procedure

Visit any distribution substation. You should be able to observe the construction and operation of the following equipment. Take notes describing what you see.

1. High voltage conductors.

2. Wood poles and cross-arms.

3. High voltage insulators.

4. High voltage transformers.

5. Voltage regulators.

DOI: 10.1201/9788743810124-37

6. Oil-filled circuit breakers.

7. Bus systems.

8. Lightning arresters.

9. Air break switches.

10. Metal-clad switchboards.

Analysis

1. Discuss each of the 10 items of equipment listed above.

Activity 36

Distribution for Electrical Lighting

Objectives

In Chapter 7 of your textbook, you learned how to wire a power distribution panel. This activity will illustrate the proper methods for wiring electrical lighting circuits for different areas of a building. Some lighting fixtures are controlled from one point by one switch. Other lighting fixtures can be controlled from two or more points by a switch at each point.

Electrical lighting circuits are typical branch circuits. The most common type of lighting circuit is a 120-volt branch, which extends from the power distribution panel to the light fixture or fixtures of an area of a building. The path for electrical distribution is controlled by one or more switches, which are usually placed in small metal enclosures inside a wall and covered by rectangular plastic covers.

Switches are always placed in a wiring circuit so that they open or close a hot wire that distributes electrical power to the lighting fixtures that they control. A switch that accomplishes control from one location is a simple single-pole, single-throw (SPST) switch. Control from two locations, such as an outside door and an inside door of a home, is accomplished by two three-way switches. When control of a lighting fixture from more than two points is desired, two three-way switches and one or more four-way switches are used. For instance, control of one light from five points could be accomplished by using a combination of two three-way switches and three four-way switches. The three-way switches are always connected to the power panel and to the light fixture, with the four-way switches between them. Using three-way and four-way switch combinations, it is possible to achieve control of a light from any number of points.

In this activity, you will study circuits that achieve control of one light from one, two, and three points, respectively. Each involves a different type

DOI: 10.1201/9788743810124-38

of switching combination to accomplish control of a lighting fixture. The same type of circuits are commonly wired in the home to provide the types of lighting desired for each room.

In this laboratory activity, you will also study a light dimmer circuit for controlling an incandescent light. Light dimmers are relatively simple to install, and their installation is commonly done in the home to control dining room lighting.

A light dimmer is ordinarily placed in a lighting circuit as a substitute for a switch that turns a light on and off. The light dimmer provides a wide range of light intensity in a room by changing the adjustment of a potentiometer that is mounted on the wall. This light dimmer takes the place of a switch that does not provide the convenience of light intensity variation.

Equipment and Materials

- Multimeter
- Single-phase, three-wire AC power source
- Switch boxes (3)
- Single-pole, single-throw (SPST) switch
- Three-way switches (2)
- Four-way switch
- Power distribution panel
- Octagonal box
- Porcelain lamp receptacle
- Light bulb – 60-watt
- Box connectors
- Solderless connectors
- No. 12-2 WG wire (4 pieces cut to desired lengths)
- No. 12-3 WG wire (2 pieces)
- Wire strippers
- Cable-ripping tool

- Screwdriver
- Needle-nose pliers
- Light dimmer

Safety

Safety is always a primary factor for laboratory activities. This lab should only be completed with supervision.

You may complete certain steps of the "Procedure" and the "Analysis" for better understanding without applying voltage.

Procedure

Section A: Light controlled from one location

1. Use the power distribution panel studied in **Chapter 7** of your text, wired to deliver power to a branch circuit. Also, obtain three metal switch boxes and a hexagonal metal holder for a porcelain or plastic lamp receptacle. These should be mounted on a board.
2. Prepare the distribution panel for operation as outlined in the activity from Chapter 7.
3. Run a 120-volt branch circuit using a No. 12-2 WG cable from one of the plug fuse terminals and the neutral/ground terminal strip to a metal switch box. Both ends of the cable should be secured by box connectors. About 1/2 inch of insulation should be stripped from the ends of each conductor before connecting these to the terminal strips.
4. Leave sufficient lengths of wire extending from the box connectors to allow easy connection to the terminal screws. Remember that the hot (black) wire should be connected to the fuse terminal, and the neutral (white) wire and the safety ground (bare or green) should be connected to the neutral/ground terminal strip. Refer to Figure 36.1 to wire a branch circuit that controls a light from one location.

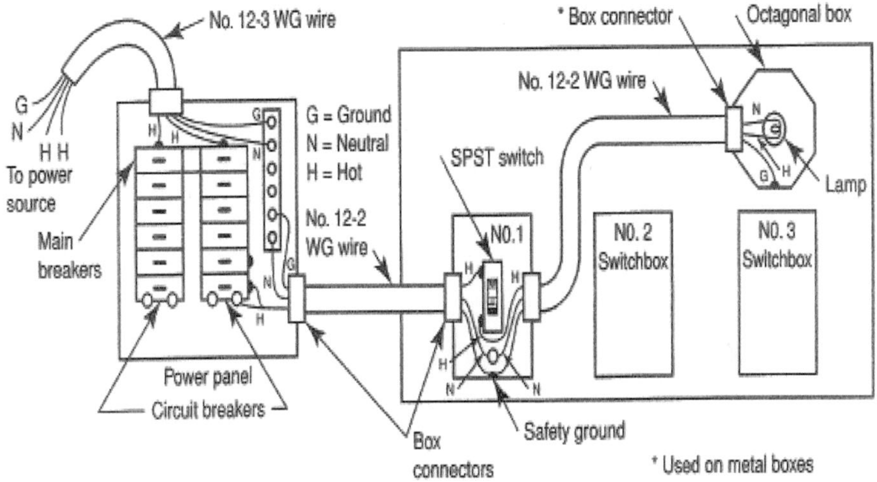

Figure 36.1 Wiring of a branch circuit for one light.

5. Obtain a single-pole, single-throw (SPST) switch for mounting in a plastic or metal switch box. Use an ohmmeter (set on lowest resistance range) to check across the switch terminals. What occurs when the switch is turned on and off?

6. Connect a No. 12-2 WG cable from the switch box to the lamp receptacle box. Use box connectors at both ends. Prepare the wire as outlined in Step 3.

7. Place the SPST switch above the switch box. Connect the hot wire to one terminal of the switch.

8. Place the other hot wire in the switch box onto the SPST switch terminal. The neutral wires inside the switch box should be secured together by using a solderless connector. The safety ground conductors should be connected directly to the metal switch box or with wire connectors in plastic. Tighten the screws that fasten the switch to the switch box.

9. Using a porcelain or plastic light receptacle with a 60-watt light bulb, prepare the light fixture for operation. The hot wire should be connected to the darker colored screw terminal, and the neutral wire to the lighter colored screw terminal. The safety ground conductor should be

connected directly to the octagonal box or connected to the other safety ground in plastic. Do not tighten the screws that fasten the receptacle onto the hexagonal box yet.

10. Before connecting the input of the power distribution panel to the power source, connect an ohmmeter across the light terminals. Turn the switch on and off. What occurs?
11. Now secure the lamp receptacle with screws and connect the power source to the distribution panel.
12. Turn on the power source and check the circuit for operation.
13. Have your instructor or supervisor check your circuit for correct operation and wiring methods.

Approval _______________

14. Turn off the power source.

Section B: Light controlled from two locations

1. Disconnect the SPST switch from the switch box.
2. Move the No. 12-2 WG cable from the right side of the switch box at location 1 on the previous diagram to the right side of the switch box at location 2.
3. Run a No. 12-3 WG cable from the right side of switch box 1 to the left side of switch box 2. The wires between the switches are called **travelers.**
4. Obtain two three-way switches. Use an ohmmeter to locate the **hinge points** of these switches. The hinge point is illustrated in Figure 36.2. Be sure to check properly to determine the hinge points of each three-way switch.

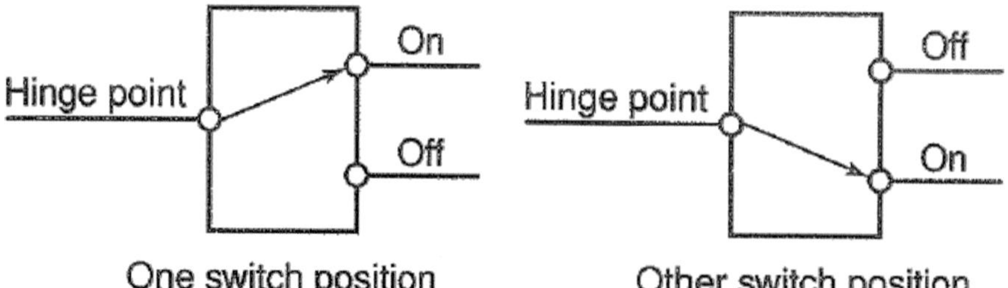

Figure 36.2 Switch the positions of the three-way switch.

5. Place the three-way switches above the switch boxes and connect the terminal screws as shown pictorially in Figure 36.3. Observe the position of the hinge points.

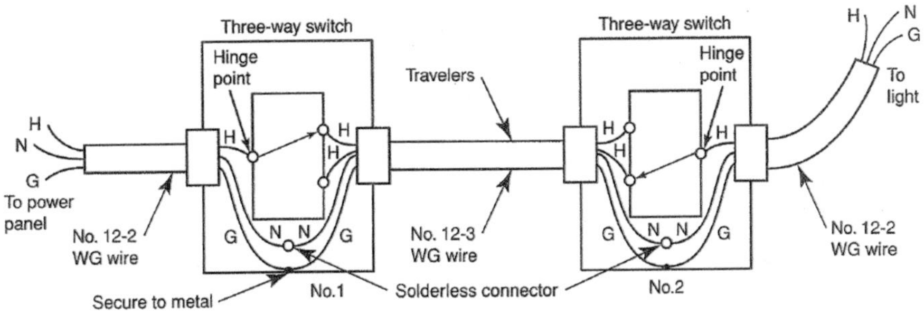

Figure 36.3 Connecting three-way switches.

6. Secure the switches onto the switch boxes by using screws.

7. Turn on the power source and check the circuit for operation.

8. Have your instructor or supervisor check your circuit for correct operation and wiring methods.

Approval ________________________

9. Turn off the power source.

Section C: Light controlled from three locations

1. Disconnect the three-way switch at position 2.

2. Move the cable from the light to the right of switch box 2.

3. Obtain a four-way switch. Use an ohmmeter to check the internal continuity of the switch. The four terminals are interconnected as shown in Figure 36.4. Verify these connections using the ohmmeter.

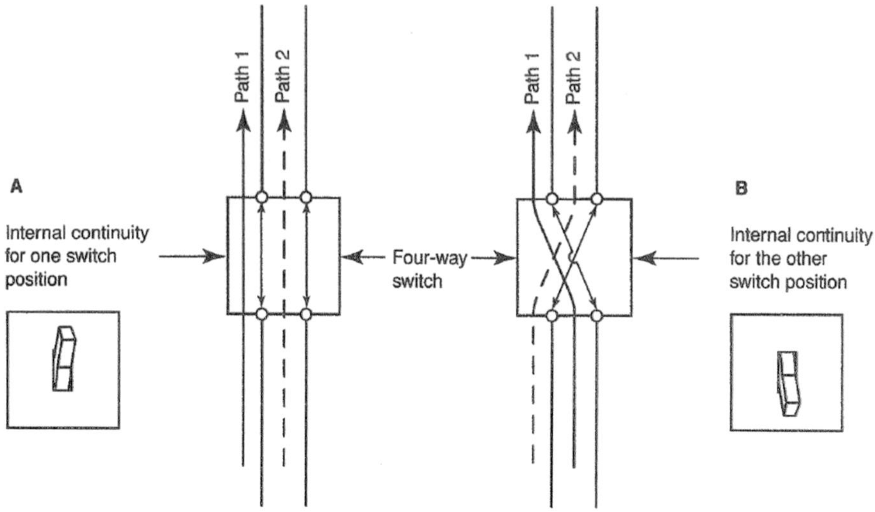

Figure 36.4 Wiring of three- and four-way switches.

4. Wire the switch combination shown pictorially in Figure 36.5.

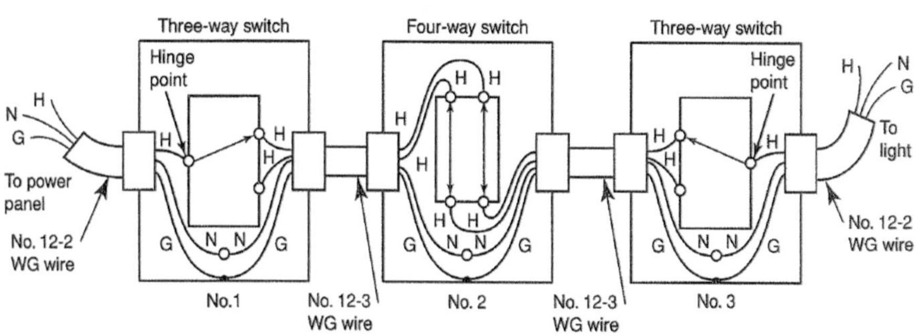

Figure 36.5 Wiring a three-way switch combination.

5. Secure the switches onto the switch boxes with screws.
6. Turn on the power source and check the circuit for operation.
7. Have your instructor or supervisor check your circuit for correct operation and wiring methods.

Approval ______________________________

8. Turn off the power source and disconnect the switches and the light. Leave the wires connected to the power source.

Section D: Light dimmer circuit

1. Wire the power distribution panel to supply a 120-volt branch circuit as shown in Figure 36.6. Connect a No. 12-2 WG branch circuit into one of the switch boxes on the lighting circuit board.

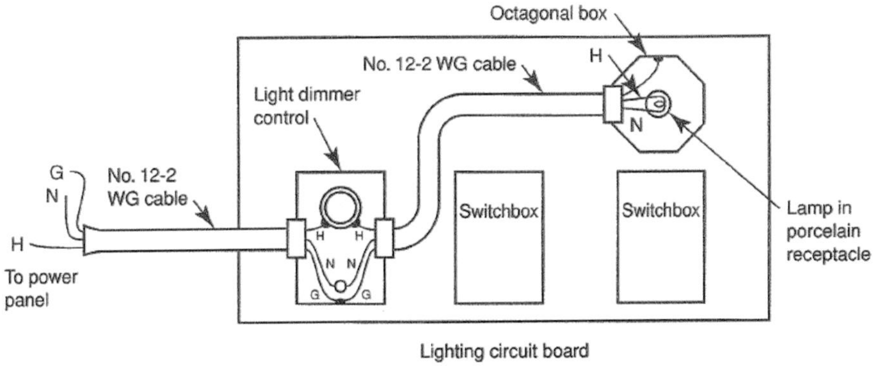

Figure 36.6 Wiring of the light dimmer circuit.

2. Run a No. 12-2 WG wire from the switch box to the lamp box.
3. Place the light dimmer near the switch box and connect it in series with the hot lines (black wires) of the branch circuit.
4. Connect the neutrals together using a solderless connector.
5. Connect the ground wires together and secure them to the switch box. Fasten the dimmer onto the switch box.
6. Connect the No. 12-2 WG cable from the switch box to the porcelain light receptacle and attach the hot and neutral wires of the cable to the terminals of the receptacle.
7. Attach the receptacle to the octagonal box.
8. Connect the No. 12-3 WG wire from the power source to the distribution panel.
9. Turn on the power source and check the light dimmer for operation.

10. Have your instructor or supervisor check your circuit for correct operation and wiring methods.

    Approval ____________________________

11. Disassemble the light dimmer circuit and return all equipment to the proper storage areas.

Analysis

1. What size conductor should be used for lighting circuits in the home?

2. Make a simplified sketch showing how a light could be controlled from four points.

3. Why must the hot and neutral wires connected to the lamp receptacle be placed on the proper terminals?

4. What is the purpose of the safety ground in lighting circuits?

5. How many and what type of switches would be required to control a light from 10 points?

6. Explain how a light dimmer is connected to a branch circuit.

7. What determines the power rating requirements of a light dimmer?

8. What are some reasons for using a light dimmer to control light intensity?

9. Could the light dimmer used in this activity be used with a light controlled from more than one point? Why?

Activity 37

Distribution for Duplex Outlets and Door Chimes

Objectives

In this activity, you will learn the proper method of wiring two common types of branch circuits. One of the most used branch circuits is for duplex power outlets. The other type of branch circuit you will study is a door chime circuit.

Duplex power outlets are connected in parallel throughout the various rooms of a building. When a small appliance or room lamp is plugged into the duplex outlet, 120 volts is applied to the appliance or lamp. The branch circuit for the duplex outlets begins at the power distribution panel and extends to the first duplex outlet. The number of duplex outlets that can be connected in parallel with the first outlet is determined by the power requirements of the room or area of the building. The number of outlets is also limited by the current rating of the circuit breaker used to protect the branch circuit.

Door chimes are used in many homes and commercial buildings. The branch circuit used to deliver electrical power to door chimes is a special type of circuit. This branch circuit uses a step-down transformer to reduce the 120 volts from the power distribution panel to a lower voltage, such as 12 volts. The 12 volts are used to energize the solenoids or electromagnetic coils of the door chimes. Two momentary-on push-button switches are used in this activity to represent the push buttons at the front door and back door of a building. One push button energizes one of the solenoids, which represents the front door sound, and the other energizes the second solenoid, which represents the back door sound.

 DOI: 10.1201/9788743810124-39

Equipment and Materials

- Multimeter
- Single-phase, three-wire power source
- Neon circuit tester
- Duplex power outlets (2)
- Outlet boxes (2)
- Door chimes
- Push-button switches-momentary-on (2)
- Power distribution panel
- Transformer – 120-volt primary and 12-volt secondary
- Octagonal box
- No. 12-2 WG nmc cable (3 pieces)
- No. 26 single-strand wire (for door chimes)
- Screwdriver
- Wire strippers
- Knife
- Cable-ripping tool
- Box connectors
- Solderless connectors
- Needle-nose pliers

Safety

Safety is always a primary factor for laboratory activities. This lab should only be completed with supervision.

You may complete certain steps of the "Procedure" and the "Analysis" for better understanding without applying voltage.

Procedure

1. Refer to Figure 37.1 to wire the branch circuits for duplex power outlets and door chimes.

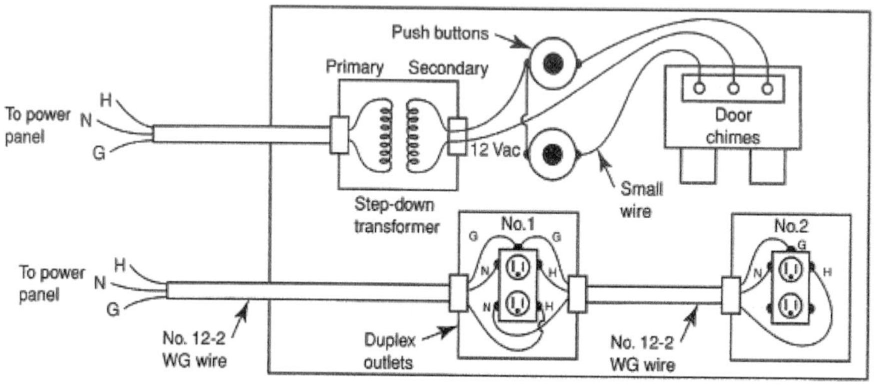

Figure 37.1 Wiring of branch circuits.

2. Run a No. 12-2 WG cable from the power distribution panel through a box connector into the metal outlet box of the No. 1 duplex outlet. Ensure that the hot wire connects through a breaker and the neutral wire is grounded in the power panel.

3. Run a piece of No. 12-2 WG cable from outlet box 1 to outlet box 2. Use box connectors for each end of the cable.

4. Place a duplex outlet above the outlet box and connect the No. 12-2 WG cable to the screw terminals of the outlet. The hot wires should always be connected to the darker colored terminal and the neutral wires to the lighter colored terminal. The safety ground wire connects to the green colored terminal. These precautions are very important for safety reasons.

5. Connect the wires of the other No. 12-2 cable to the terminals of the duplex power outlet 1. Use the same procedure to secure the outlet to the box by using screws.

6. Now connect the wires of outlet box 2 to a duplex outlet using the same procedure. Secure this outlet to the outlet box. This completes the duplex power outlet circuit wiring.

7. Run a No. 12-2 WG cable from the power distribution panel to the primary winding of the step-down transformer. Use solderless terminals to connect the cable to the transformer wires. This branch circuit will energize the door chimes.

8. Use the small wire to connect the door chimes and push-button switches to the low-voltage secondary winding terminals of the transformer. This will apply 12 volts to the door chimes when the circuit is energized.

9. Connect the input of the power distribution panel to the single-phase, three-wire source.

10. Turn on the power source. Be very careful.

11. Test the duplex power outlets with a neon circuit tester.

12. Test the door chimes by pressing the push-button switches.

13. Have your instructor or supervisor check your circuits for operation and wiring methods.

    Approval: ____________________

14. Turn off the power source and disassemble the branch circuits.

Analysis

1. Why are door chimes controlled by a low voltage?

2. Why must the hot wires of duplex outlet circuits be connected to the darker colored terminals of the outlet?

3. Explain the method used to identify electrical wiring cable, for example, No. 10-2 WG nmc.

4. Are duplex outlets connected in series or parallel? Why?

Activity 38

Three-phase Transformer Schematics

Objectives

Upon completion of this activity, you will be familiar with the connections of five common types of transformer connections.

Procedure

Complete each of the following diagrams.

1. **Delta-delta connection**

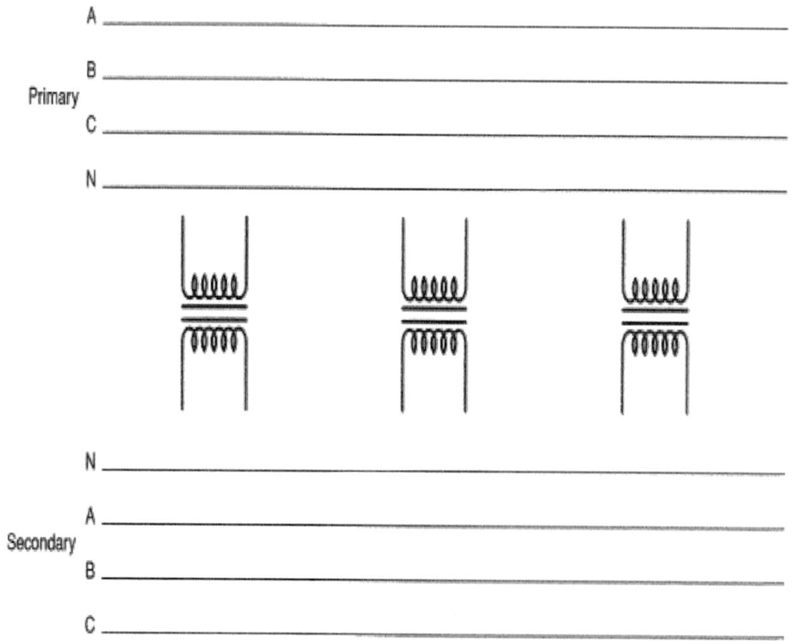

DOI: 10.1201/9788743810124-40

2. Delta-wye connection

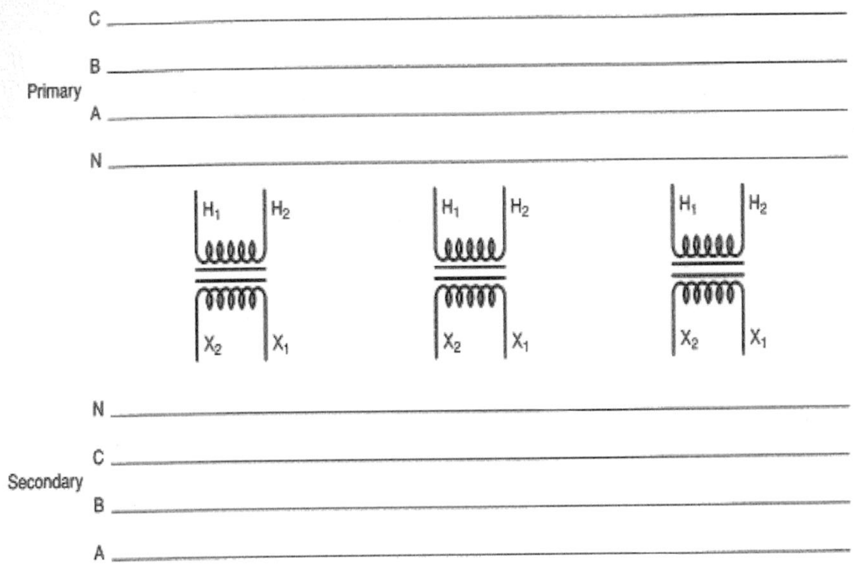

3. Wye-wye connection

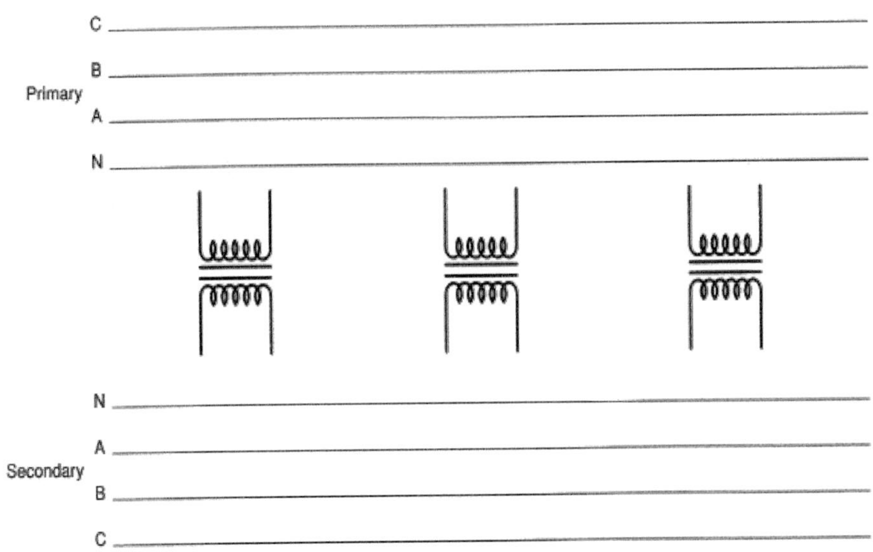

4. Wye-delta connection

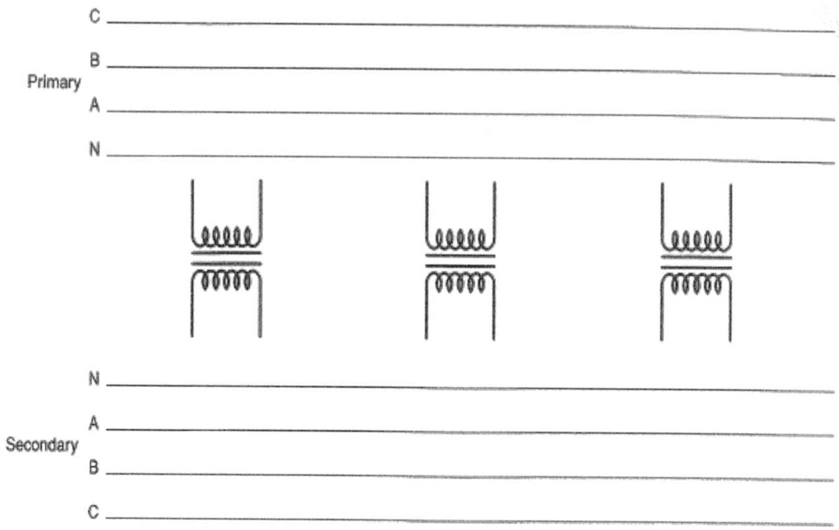

5. Open delta connection

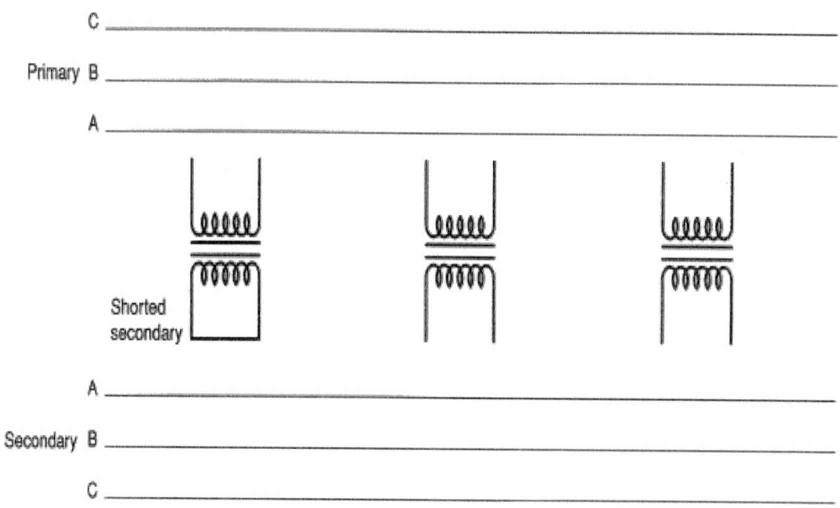

Activity 39

Three-phase Transformer Calculations

Problems

Given:

- A three-phase transformer connected in a delta-wye configuration.
- A 2:1 turns ratio.
- Primary line voltage = 600 V.
- Primary phase voltage = 600 V.
- Primary line current = 200 A.

Calculate:

1. I P (Pri) = ______A.

2. VL (Sec) = ______V.

3. VP (Sec) = _______V.

4. IL (Sec) = ______ A.

DOI: 10.1201/9788743810124-41

5. IP (Sec) = ______ A.

6. PP (Sec) = ______ kW.

7. PT (Sec) = _______ kW.

UNIT III

Electric Motor Basics

Activity 40

Electric Motor Principles

Objectives

In this activity, you will study the operation of an electric motor. Motors convert electrical energy into mechanical energy in the form of rotary motion. The type of motor you will study is called a universal motor. It will operate with either alternating current or direct current applied. The universal motor is one of the most used motors. Applications of the universal motor include electric drills, electric saws, and household items such as mixers, blenders, and sewing machines. The universal motor is one of the few types of variable-speed motors that operate on alternating current. Most types of AC motors operate at a constant speed. Thus, the primary advantage of a universal motor is its speed control capability. By varying the applied voltage, the speed of the motor will change.

The universal motor has several basic parts. One part is the rotor, or armature, which is constructed of many windings of copper wire and which is embedded in a laminated metal core assembly. A second part is the stator, or field, which is the metal enclosure and at least one set of field coils that are mounted on laminated metal pole pieces. The armature windings are connected to the power source and the field windings through a brush/commutator assembly. The brushes make contact with the rotating commutator segments to apply electrical power to the armature. The commutator segments, which are made of copper, provide a conductive path through the armature windings. The field coils are also constructed of copper windings that develop a concentrated magnetic field since they are actually electromagnets. The interaction of the electromagnetic fields of the field windings and the armature windings causes a force to be developed that will produce rotation.

DOI: 10.1201/9788743810124-43

In this activity, you will use DC voltage to energize the universal motor. Keep in mind that this type of motor will also operate with alternating current applied to it. You will examine several of the important characteristics of the universal motor in the following procedure.

Equipment and Materials

- Multimeter
- Universal motor
- Variable DC power supply

Safety

Safety is always a primary factor for laboratory activities. This lab should only be completed with supervision.

You may complete certain steps of the "Procedure" and the "Analysis" for better understanding without applying voltage.

Be sure to wear eye protection while electric motors are in operation. Also, be very careful when working with high voltages.

Procedure

1. The connection diagram for a universal motor is shown in Figure 40.1. Note that the armature and field coils are connected in series with the power.

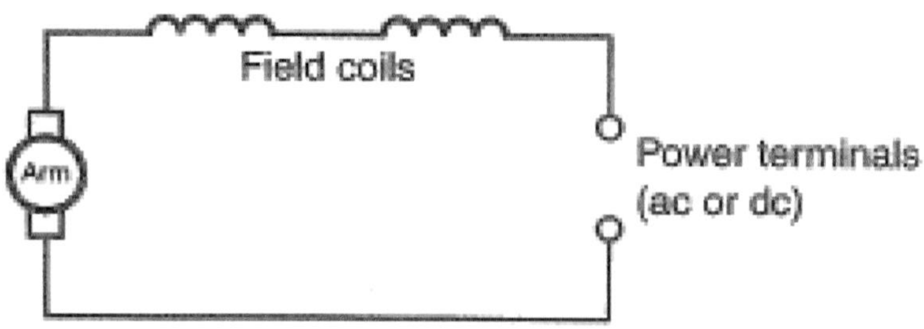

Figure 40.1 Universal motor schematic diagram.

2. Connect the power terminals of the universal motor to the variable DC power supply. The voltage should be set at zero.

3. Turn on the power supply and make the necessary measurements to complete Figure 40.2. The current should be measured with the meter in series with either one of the power terminals.

Motor Voltage, Current, and Speed Characteristics

Applied voltage	*Current*	*Relative speed*
10		
20		
30		
40		

Figure 40.2 Complete the chart.

4. Set the power supply to 30 volts DC. Then turn the motor off. Keep the ammeter in series with the motor terminals.

5. Turn the motor on and record the initial current (starting current) and the running current.

   Starting current = ___A; Running current = ___ A.

6. While the motor is running on 30 Vdc, carefully apply a load to its shaft. What is the effect on the current as the load is increased?

7. Look toward the external shaft of the motor and observe the direction of rotation. The direction of rotation is: clockwise or counterclockwise (circle one).

8. Reverse the motor terminal connections to the power supply. Does the direction of rotation change?

Analysis

1. Complete the following statements, which refer to the operation of a universal motor, by using the words increases or decreases.

   a. As applied voltage increases, current _______

   b. As applied voltage increases, speed __________

   c. As load applied to the shaft of the motor increases, speed ______

2. In Step 5, why were the starting and running currents different?

3. How can the direction of rotation of a universal motor be reversed?

4. What are the basic parts of a universal motor and the function of each part?

Activity 41

Series-wound DC Motors

Objectives

The series-wound DC motor is constructed of low-resistance field coils connected in series with the armature coils. The field coils have a few turns of large-diameter wire. This arrangement permits a current flow that is the same throughout the circuit. Like all motors, the series-wound machine slows down with increased loads and speeds up with decreased loads. Its speed, like other DC motors, is inversely proportional to its magnetic field strength. The counter-electromotive force, or cemf, which is generated when the armature turns through the magnetic field, is directly related to speed. As a result, these three factors, speed, cemf, and field strength, are extremely important characteristics in the operation of a series-wound DC motor.

In this activity, you will analyze the operation of a series-wound DC motor. This type of motor is commonly designed so that it will operate from either a DC or an AC voltage source. Such machines are called universal motors. This activity is set up in such a way that you can analyze: (1) a commercially manufactured DC series motor, such as a small universal motor used in an electric drill or an automobile starter, or (2) a DC series motor that has been constructed using a rotating machinery unit.

Equipment and Materials

- Multimeter
- Series-wound DC motor
- DC power supply
- Tachometer

DOI: 10.1201/9788743810124-44

Safety

Safety is always a primary factor for laboratory activities. This lab should only be completed with supervision.

You may complete certain steps of the "Procedure" and the "Analysis" for better understanding without applying voltage.

Be sure to wear eye protection while electric motors are in operation. Also, be very careful when working with high voltages.

Procedure

1. Examine the connection diagram for a series-wound DC motor, shown in Figure 41.1.

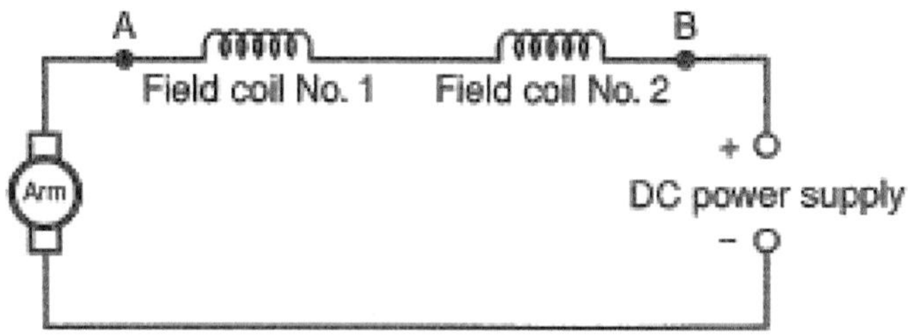

Figure 41.1 Connection diagram for a series-wound DC motor.

2. Obtain a DC series motor or construct one using the rotating machinery unit.

3. Measure the resistance of the armature. This will be a low resistance, so use the lowest resistance range of the meter.

   R_A = ________ Ohms.

4. Measure the resistance of the field coils (point A to point B on the diagram). Make sure the coils are isolated from other parts of the circuit. It may be necessary to remove a wire at point A.

   R_F = ________ Ohms.

5. Prepare the multimeter to measure direct current and connect it in series with the motor circuit.

6. Gradually increase the DC voltage applied to the machine until you reach its rated voltage. What is the effect on current and speed?

7. With the rated voltage applied, turn the power supply off and then back on momentarily. What is the effect on the current as speed increases toward its maximum?

8. Carefully apply a load on the motor shaft. What is the effect on current and speed?

9. With a tachometer, measure the motor speed at its rated voltage.

Speed = _______ RPM.

10. Record the current indicated at the rated voltage of the machine.

Current = ________ A.

11. Turn the power supply off and use the necessary procedure to reverse the direction of rotation of the motor.

12. Turn on the power supply and measure the speed and current of this machine with its direction of rotation reversed.

Speed = ________ RPM. Current = ________ A.

13. Try to apply a heavy enough load to stall the motor. Very quickly record the maximum current the motor draws.

Maximum current = ________ A.

14. Allow the motor to run for a few seconds to cool off. Then, turn off the power.

Analysis

1. What effect would having more than two field coils have on the operation of this motor?

2. What effect does increased voltage have on motor speed? Why? (see Step 6).

3. In Step 7, what happened to motor speed and current as the load was increased? Why?

4. How is the direction of rotation of the DC series motor reversed?

5. Was there any change in motor speed or current when the direction of rotation was reversed? Should there have been?

6. Show in the graph of Figure 41.2, how a speed-versus-load curve should look for a series DC motor.

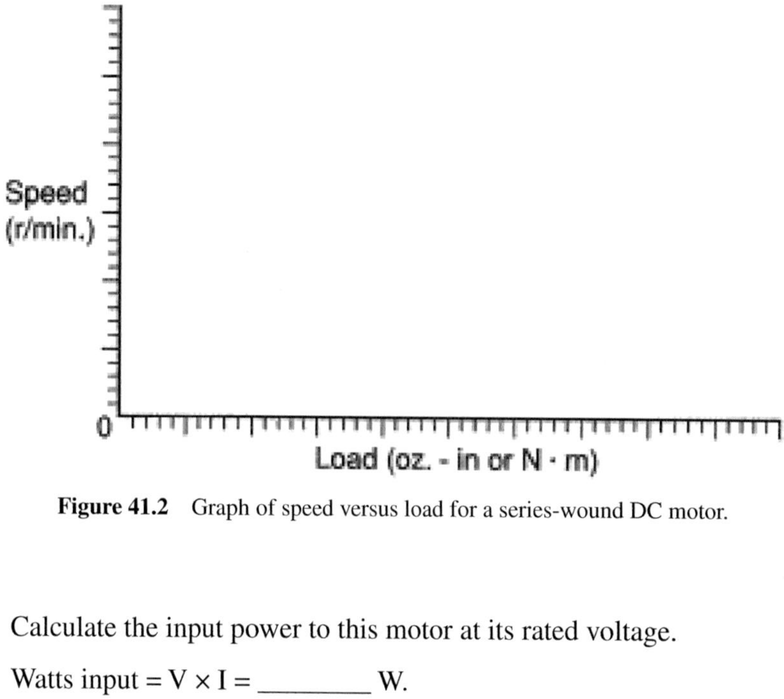

Figure 41.2 Graph of speed versus load for a series-wound DC motor.

7. Calculate the input power to this motor at its rated voltage.

   Watts input = V × I = ________ W.

8. Convert the watts (W) input to horsepower (hp).

   HP = W/746 = ________ HP.

Note that this calculated value is not the output horsepower rating of the machine. The horsepower output would be somewhat lower due to the energy losses of the machine.

Activity 42

Shunt-wound DC Motors

Objectives

The construction of DC motors is based on the method of connecting the field windings and armature coils together. In the shunt-wound DC motor, for example, the field coils are connected in parallel with the armature. This is achieved by making the shunt field a fairly high resistance. Accordingly, it is wound of many turns of fine wire. The armature, however, offers less resistance and permits a larger current to flow. Its construction is of a larger wire with fewer turns of wire.

In the laboratory activity, you will analyze either (1) a commercially manufactured shunt-wound DC motor or (2) a shunt-wound DC motor that has been constructed in the laboratory using a rotating machinery unit. You should observe the general construction and operational characteristics of the shunt-wound DC motor.

Equipment and Materials

- Multimeter
- Shunt-wound DC motor
- Variable DC power supply
- Field rheostat (0–500 Ohm, high wattage)
- Armature rheostat (0–25 Ohm, high wattage)
- Tachometer

DOI: 10.1201/9788743810124-45

Safety

Safety is always a primary factor for laboratory activities. This lab should only be completed with supervision.

You may complete certain steps of the "Procedure" and the "Analysis" for better understanding without applying voltage.

Be sure to wear eye protection while electric motors are in operation. Also, be very careful when working with high voltages.

Procedure

Section A: Shunt-wound DC motor characteristics

In this section of the laboratory activity, you will observe the operational characteristics of a shunt-wound DC motor. You will observe the effect of load on the speed, counter-electromotive force (cemf), current, and torque of this machine.

1. The connection diagram for a shunt-wound DC motor is shown in Figure 42.1. Set up this circuit.

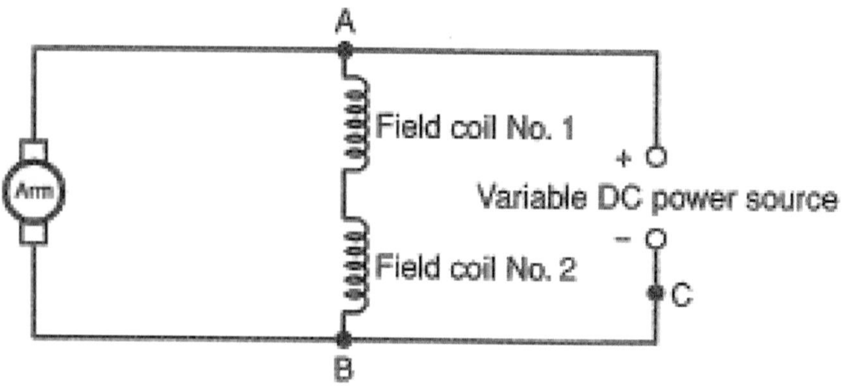

Figure 42.1 Connection diagram for a shunt-wound DC motor.

2. Measure the resistance of the field coils (point A to point B on the diagram). You will have to remove a wire at point A. It will be necessary to use the lowest resistance range on your meter.

   R_F = ________ Ohms.

3. Measure the resistance of the armature circuit (across the brushes). Make sure the power supply is not connected to the armature, and the wire at point A is still removed. This will isolate the armature resistance from other circuits.

   R_A = ________ Ohms.

4. Prepare a meter to measure a high value of DC current. Put the current meter in series with the power supply (at point C on the diagram). This meter will measure the total current drawn by the motor.

   R_T = ________ Ohms.

5. Record the rated voltage value of this motor.

   Rated voltage = ________ volts DC.

6. Turn on the power supply and slowly increase the voltage to the rated voltage of the motor. What is the effect on the current and speed of the motor?

7. Record the total current drawn by the motor at its rated voltage.

   IT = ________ amperes DC.

8. With a tachometer, measure and record the speed of the motor.

   Speed = ________ RPM.

9. Carefully apply a load to the motor shaft to slow its speed. What is the effect of the increased load on speed and total current?

10. Increase the load as much as possible, or until rotation stops or slows down appreciably. Quickly record the maximum current drawn by the motor.

    Maximum current = ________ amperes DC.

11. Turn off the power supply and perform the necessary procedure to reverse the direction of rotation of the motor.

12. With the direction of rotation reversed, increase the applied DC voltage to the rated voltage value of the motor.

13. Measure and record the current and speed.

    Current = ________ amperes DC.

    Speed = ________ RPM.

14. Turn off the power supply.

Section B: Speed control of shunt-wound DC motors

In this section, you will observe two methods of controlling the speed of a DC shunt-wound motor. These methods are very simple. The first method utilizes a rheostat in series with the armature circuit. The second method employs a rheostat in series with the field circuit. By measuring current through each rheostat and calculating power dissipation, you should be able to decide which speed control method is most desirable.

1. Modify the DC shunt motor as shown in Figure 42.2. Be sure the wattage ratings of the rheostat are high enough.

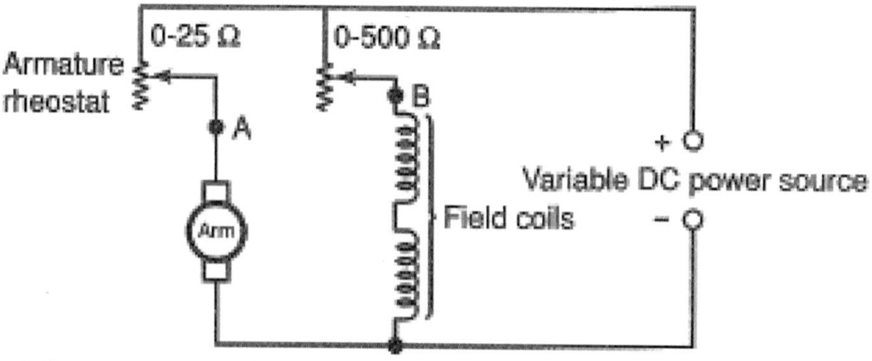

Figure 42.2 Shunt-wound DC motor speed-control test circuit.

2. The rheostats should be connected in series with the armature and field circuit as shown in the diagram. Adjust each rheostat to the zero resistance position.

3. Adjust the variable DC power supply from zero volts to the rated voltage of the motor to check it for operation.

4. Turn off the power supply and connect a DC current meter in series with the armature circuit at point A on the diagram.

5. With rated voltage applied, turn on the motor and observe the maximum starting current indicated on the meter.

   Maximum starting current = ________ amperes DC.

6. Measure and record the running current of the motor with no load applied.

   Running current = ________ amperes DC.

7. Adjust the armature rheostat from zero toward its maximum resistance while observing the speed of the motor. As armature resistance increases, speed ________. With a tachometer, measure the range of speed.

   Speed range = ________ to ________ RPM.

8. Adjust the armature rheostat back to zero resistance. Carefully apply a load to the motor shaft. Measure and record the maximum armature current as the load is increased.

   Maximum armature current = ________ amperes DC.

9. Turn off the power supply. Now, remove the armature current meter and connect the meter in series with the field circuit at point B on the diagram.

10. With rated voltage applied, turn on the motor and observe the maximum field current during starting.

    Maximum field current = ________ amperes DC.

11. Record the field current indicated while the motor is running with no load applied.

    Field current = ________ amperes DC.

12. Adjust the field rheostat from zero toward its maximum resistance while observing the speed of the motor. As field resistance increases, speed ________. With a tachometer, measure the range of speed.

    Speed range = ________ to ________ RPM.

13. Adjust the field rheostat back to zero resistance. Carefully apply a load to the shaft of the motor. Measure and record the maximum field current as the load is increased.

    Maximum field current = ________ amperes DC.

14. Turn off the power supply. Return all materials to the storage area.

Analysis

1. How does the field resistance of a shunt-wound DC motor compare to that of a DC series motor?

2. How does increased load on a shunt-wound DC motor affect:

   a. Speed of the motor?

   b. Cemf developed by the armature?

   c. Total current drawn by the motor?

   d. Torque developed by the motor?

3. In Section A, Step 11, how was the direction of rotation of the motor reversed?

4. Did reversing the direction of rotation have any effect on total current and speed (see Section A, Steps 7, 8, and 13)? Why?

5. How do the starting current and running current drawn by the armature compare (see Section B, Steps 5 and 6)? Why?

6. Using the maximum armature current value from Section B, Step 5, and the measured armature resistance, determine the minimum power rating of an armature rheostat.

   Maximum power rating = ________ W.

7. In Section B, Step 7, as armature resistance changes, why does speed change as it does?

8. How do the starting current and running current drawn by the field compare (see Section B, Steps 10 and 11)? Why?

9. In Section B, Step 12, as field resistance changes, why does speed change as it does?

10. Using the maximum field current obtained in Section B, Step 13, and the measured field resistance, determine the minimum power rating of a field rheostat for speed control.

    Minimum power rating = ________ W.

11. Which type of speed control (armature or field) is most desirable? Why?

Activity 43

Single-phase AC Capacitor-start Motors

Objectives

Single-phase induction motors are not self-starting without some auxiliary method used to provide initial movement of the rotor. One method is the use of a starting winding. Single-phase induction motors that employ auxiliary windings for starting can have a capacitance connected in series. A capacitor-start induction motor is made to operate as though a two-phase AC source supplied the power to produce the rotating magnetic field. A capacitor is employed to provide initial starting torque and to develop a strong, uniform magnetic field.

In this activity, you will be able to analyze the operational characteristics of either a commercially manufactured capacitor motor or one that has been constructed in the laboratory. You will see the effect the capacitor has on the starting torque of the motor.

Equipment and Materials

- Multimeter
- Single-phase AC power source
- AC current meter
- Capacitor-start motor
- Tachometer

DOI: 10.1201/9788743810124-46

Safety

Safety is always a primary factor for laboratory activities. This lab should only be completed with supervision.

You may complete certain steps of the "Procedure" and the "Analysis" for better understanding without applying voltage.

Be sure to wear eye protection while electric motors are in operation. Also, be very careful when working with high voltages.

Procedure

1. A connection diagram for a single-phase AC, capacitor-start induction motor is shown in Figure 43.1. Set up this circuit.

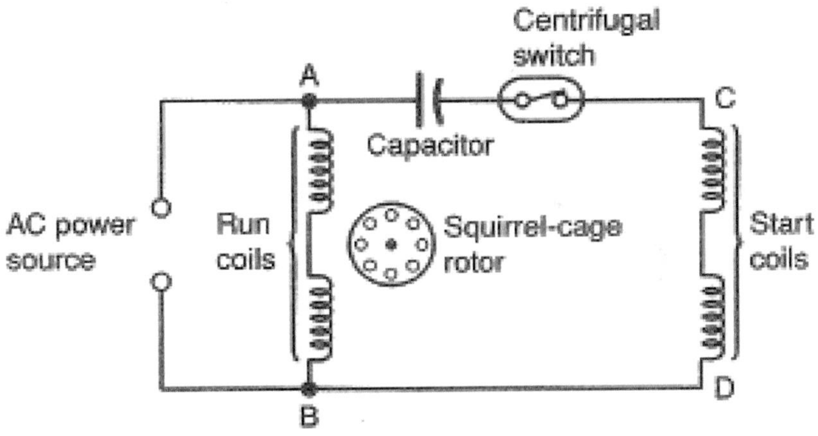

Figure 43.1 Connection diagram for a single-phase AC capacitor-start motor.

2. Count the number of stator coil sets (poles) that the motor you are analyzing has in the run and start circuits.

   Number of running-coil sets = ______.

   Number of starting-coil sets = ______.

3. With a meter, measure the resistance of the run windings (point A to point B on the diagram). Temporarily disconnect the run windings to make this measurement.

   R of run winding = ______ Ohms.

4. Measure the resistance of the start windings (point C to point D on the diagram).

   R of start winding = ______ Ohms.

5. Connect an AC current meter in series with one of the power lines. Apply rated voltage to the motor and observe the initial current drawn by the motor.

   Starting current = ______ A.

6. Record the running current of the motor.

   Running current = ______ A.

7. Carefully apply a load to the shaft of the motor. What is the effect on speed and current?

8. With a tachometer, measure the speed of rotation.

   Speed = ______ RPM.

9. With the motor in its original circuit configuration, perform the necessary procedure to reverse the direction of rotation. Measure and record the starting current, running current, and speed.

   Starting current = ______ A.

   Running current = ______ A.

   Speed = ______ RPM.

10. Turn off the power supply. Return all materials to the storage area.

Analysis

1. Why does the addition of a capacitor in the start windings of an induction motor improve its starting torque?

2. Describe the following types of capacitor motors:
   a. Capacitor-start induction motor.
   b. Capacitor-start, capacitor-run motor.
   c. Permanent-capacitor motor.

3. From the data of Step 2, compare the number of starting and running coil sets of the motor.

4. From the data of Steps 3 and 4, compare the resistances of the start and run windings of the motor.

5. From the data of Steps 5 and 6, compare the starting current and running current values.

6. How does increased load affect the speed and current of the motor? (see Step 7)

7. How is the direction of rotation of the capacitor-start motor reversed?

8. What are some applications of capacitor motors?

Activity 44

Shaded-pole AC Motors

Objectives

A unique method of making a single-phase AC motor self-starting is by a process called pole shading. The shading poles are contained within the field poles of the machine. Generally, about one-third of the field pole is enclosed by the shading coil. When the magnetic flux around the pole starts to build up, it cuts across the shading coil. This induces a voltage into the shading coil that produces a current. The area inside the shading coil then has a reverse polarity compared with the remaining pole face. As a result, there is a shifting effect created across the pole face from the unshaded to the shaded side. The shifting effect is significant enough to cause the rotor to turn. This turning action then makes the motor self-starting. The starting torque of the shaded pole motor is very small, with some of the largest ratings of this type of motor around horsepower.

The shaded-pole motor is mainly used for very small torque applications, such as fan or blower motors. In this laboratory activity, you will be able to analyze either a commercially manufactured, AC shaded-pole motor or one that has been constructed in the lab.

Equipment and Materials

- Multimeter
- Shaded-pole AC motor
- AC power source
- Tachometer
- AC current meter

DOI: 10.1201/9788743810124-47

Safety

Safety is always a primary factor for laboratory activities. This lab should only be completed with supervision.

You may complete certain steps of the "Procedure" and the "Analysis" for better understanding without applying voltage.

Be sure to wear eye protection while electric motors are in operation. Also, be very careful when working with high voltages.

Procedure

1. The connection diagram for a shaded-pole AC motor is shown in Figure 44.1. Set up the circuit.

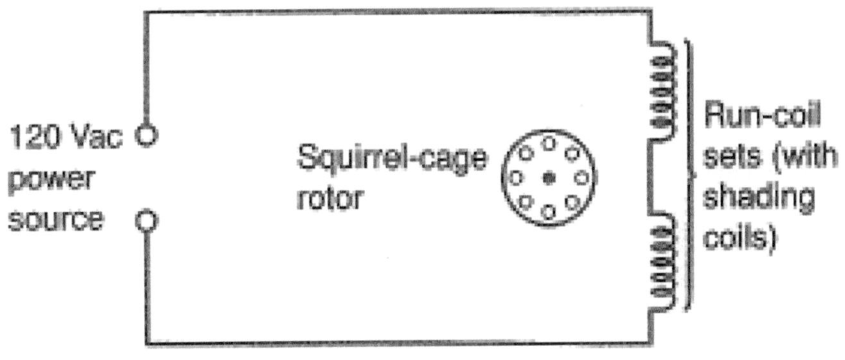

Figure 44.1 Connection diagram for a shaded-pole AC motor.

2. Measure the resistance of the stator windings.

   Stator winding resistance = _____ Ohms.

3. Connect an AC current meter in series with the power line. Apply rated voltage to the motor and observe the initial current it draws when turned on.

   Starting current = _____ A.

4. Record the current drawn by the motor with no load applied.

   Running current = _____ A.

5. Carefully apply a load to the shaft of the motor. What is the effect on speed and current?

6. With a tachometer, measure the speed of rotation.

   Speed = _____ RPM.

7. Record the number of poles or coil sets your motor has.

   _____ poles.

8. Turn off the power supply. Return all materials to the storage area.

Analysis

1. Explain how the shaded-pole induction motor produces starting rotation.

2. What are some applications of shaded-pole motors?

3. How is the synchronous speed of all AC induction motors determined?

4. What is the synchronous speed of the motor you analyzed?

5. What is meant by the term slip as applied to AC induction motors?

6. From the data of Step 6, determine the slip percentage of the motor you analyzed.

7. How do the starting and running currents of the shaded-pole motor compare? Why?

8. What is the effect of increased load on motor speed and current?

9. How could the direction of rotation of a shaded-pole motor be reversed?

10. What would happen if one of the shading coils became open-circuited?

Activity 45

Three-phase AC Synchronous Motors

Objectives

Three-phase AC synchronous motors operate at a constant speed regardless of the mechanical load applied. In addition to the constant speed characteristic, this type of motor also has the ability to operate at leading or lagging power factors. By changing the DC excitation voltage applied to the rotor, the power factor at which the motor operates will vary. When a synchronous motor operates at a leading power factor, it improves the overall power factor of the system on which it is operating. This operating mode produces a capacitive effect on the system. Therefore, the machine is referred to as a **synchronous capacitor**.

In this laboratory activity, you will have an opportunity to analyze the operational characteristics of a commercially manufactured, three-phase AC synchronous motor, or one that has been constructed in the lab using a rotating machinery unit. The construction of this motor is the same as that of a three-phase AC alternator. Therefore, a three-phase alternator, such as one used in an automobile, could be substituted for an operational three-phase AC synchronous motor for a laboratory analysis.

Equipment and Materials

- Multimeter
- Three-phase AC synchronous motor
- Three-phase AC power source
- Variable DC power source
- AC current meter

 DOI: 10.1201/9788743810124-48

- DC current meter
- Tachometer

Safety

Safety is always a primary factor for laboratory activities. This lab should only be completed with supervision.

You may complete certain steps of the "Procedure" and the "Analysis" for better understanding without applying voltage.

Be sure to wear eye protection while electric motors are in operation. Also, be very careful when working with high voltages.

Procedure

1. A simplified connection diagram for a three-phase AC synchronous motor is shown in Figure 45.1. Normally, the stator coils are connected in a wye configuration. Set up this circuit.

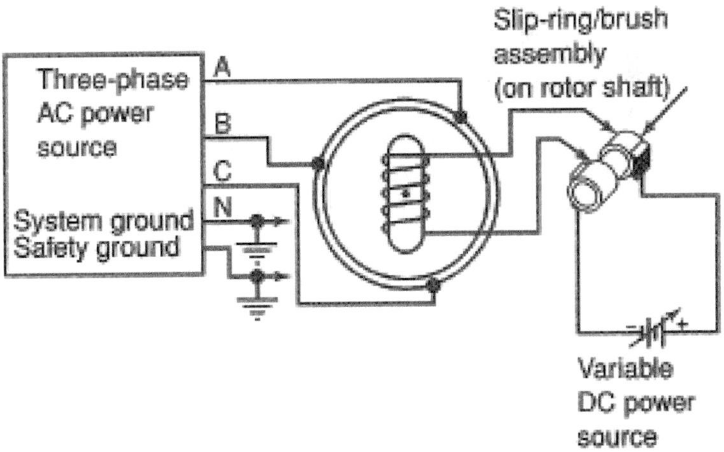

Figure 45.1 Connection diagram for a three-phase AC synchronous motor.

2. Measure the following resistances across the power lines **before** connected:

A – B = _________ Ohms.

B – C = _________ Ohms.

A – C = _________ Ohms.

3. Connect the synchronous motor to a three-phase AC power source in order to supply the rated voltage to the stator coils of the motor.

4. Place an AC current meter in series with one of the power lines.

5. Connect the terminals from the slip-ring/brush assembly of the rotor to a variable DC power source. Adjust this DC voltage to zero.

6. Determine the number of stator coil sets (poles) that your motor has.

   Stator coil sets = ________.

7. Apply three-phase AC power to the machine. Does it start? Why?

8. If the machine does not start, and if it does not have damper windings, it may be necessary to bring its speed up to near its synchronous speed by connecting another motor to its shaft to supply the necessary RPM to reach synchronous speed.

9. With the synchronous motor operational, record the AC current indicated on the meter.

   Current = ________ amperes AC.

10. Gradually increase the DC voltage applied to the rotor. What is the effect on speed and AC current?

11. Turn the DC voltage down to zero once more. Connect a DC current meter in series with the power source.

12. Now measure the speed of rotation with a tachometer.

    Speed = ________ RPM.

13. If available, record the rated DC voltage of the machine you are analyzing.

    ________ volts DC at ________ power factor.

14. Vary the DC voltage applied to the rotor in increments of approximately 2 volts until about 50 % above the rated DC voltage is reached. At each DC voltage value, record the DC current through the rotor and the AC current drawn by the stator. Record your data in Figure 45.2.

DC voltage	Rotor current (DC)	Stator current (AC)
0		

Figure 45.2 Rotor current versus stator current.

15. With the rated voltage applied to the rotor, measure the speed of rotation with a tachometer.

    Speed at rated DC voltage = _________ RPM.

16. Carefully apply a load to the shaft of the machine. What is the effect of increased load on AC stator current, DC rotor current, and speed?
17. Turn the DC voltage back to zero volts DC. Now, gradually increase the DC voltage until the rotor is synchronized in step with the revolving stator field.
18. Turn off the DC, then the AC power source. Return all materials to the storage area.

Analysis

1. Describe the basic construction of a three-phase synchronous motor.

2. What is the speed of the following three-phase synchronous motors?

   *6 pole = ________ RPM.

   *12 pole = ________ RPM.

   *24 pole = ________ RPM.

   *Refers to the total number of stator poles.

3. How does a three-phase synchronous motor differ from a three-phase induction motor?

4. What are some methods of starting three-phase synchronous motors?

5. What is pull-in torque? Pull-out torque?

6. When is a three-phase synchronous motor considered to be under-excited? Overexcited?

7. Verify your observation of the number of stator poles in Step 6 with the speed of rotation measured in Step 15.

   Synchronous speed = ___________ RPM.

8. What effect does increased rotor current have on the stator current of the synchronous motor?

9. What effect does increased load have on: (see Step 16).

   a. speed?

   b. DC rotor current?

   c. AC stator current?

10. What effect does increasing the DC rotor voltage (from zero) have on the speed of the motor?

11. With the data recorded in Figure 45.2, make a graph of DC rotor current versus AC stator current. Use Figure 45.3. Also, mark the point on the curve that corresponds to the highest power factor operation of the motor.

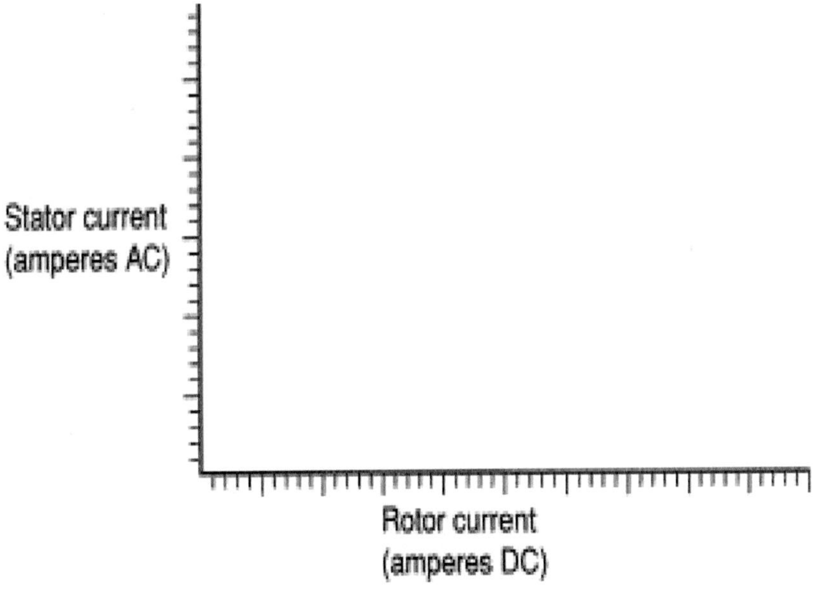

Figure 45.3 Graph of DC rotor current versus AC stator current.

Activity 46

AC Synchronous Motors

Objectives

The AC synchronous motor used in automated industrial systems is commonly classified as a constant-speed device. It has extremely rapid starting, stopping, and reversing characteristics. Motors of this type only necessitate a simple clockwise/stop/counterclockwise rotary switch for control. The starting and running currents of a synchronous motor are identical, which is unique for AC motors. This characteristic means that a motor of this type can withstand a high inrush of current when direction changes occur. As a general rule, the synchronous motor can even be stalled without damaging the motor.

Motors of this type are commonly used as drive mechanisms for industrial systems in machinery operations. Starting begins within one and one-half cycles of the line frequency, and stopping occurs in five mechanical degrees of rotation. This motor represents a unique part of all servomechanisms used in precision rotary control applications today.

In this activity, a simple test circuit will be constructed so that you can observe an AC synchronous motor's starting, stopping, and current flow characteristics.

Equipment and Materials

- AC ammeter – 0–5 A
- AC synchronous motor – Superior Electric Type SS150 or equivalent motor rated at no more than 200 in - oz.
- Capacitor – 3.75 uF, 400 Vdc
- Resistor – 250 Ohm, 25 Watt

DOI: 10.1201/9788743810124-49

- SPST switches (3)
- Isolated AC power source – 120 V
- Piece of wood

Safety

Safety is always a primary factor for laboratory activities. This lab should only be completed with supervision.

You may complete certain steps of the "Procedure" and the "Analysis" for better understanding without applying voltage.

Be sure to wear eye protection while electric motors are in operation. Also, be very careful when working with high voltages.

Procedure

1. Construct the AC synchronous motor test circuit of Figure 46.1. Ensure that the motor is mounted securely.

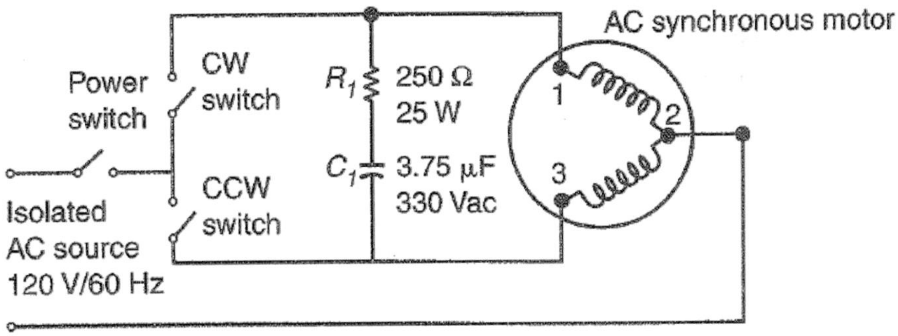

Figure 46.1 AC synchronous motor test circuit.

2. Turn on the CW switch for clockwise rotation.
3. Momentarily turn on the power switch. What is the direction of rotation of the motor?
4. With a tachometer, measure the speed of rotation.

Speed = ___________ RPM.

5. Momentarily turn off the power switch and observe the stopping action of the synchronous motor. You may want to run several trials to see the quickness of the stopping action. How does this compare with other AC motors?

____

6. Note each time that the motor is turned on, how quickly it comes to speed.

7. With an AC ammeter, measure the running current and starting current of the synchronous motor.

   Running current ______ A.

   Starting current ______ A.

8. Carefully wedge a piece of wood between the rotating shaft and the bench while holding the motor. This will provide a simple loading method for test purposes.

9. When the motor is loaded down, how does the running current respond?

____

10. If the motor is completely stalled, how does the running current respond?

____

11. Remove the load from the motor, turn off the clockwise switch, and turn on the counterclockwise switch. The motor should rotate equally well in the counterclockwise direction and have the same basic characteristics. Test these again to verify the theory.

12. Switch off the counterclockwise switch, then switch on the clockwise switch and notice the ease with which the direction change occurs.

13. Turn off the AC power source and disconnect the circuit. Return all materials.

Analysis

1. Discuss the operation of an AC synchronous motor.

Activity 47

Synchro Systems

Objectives

Synchro systems are commonly classified as two or more generator/motor units connected in such a way that they permit the transmission of angular shaft changes by electromagnetic field changes. When the shaft of the generator unit is turned, it automatically causes a corresponding change in the motor shaft position. Through systems of this type, it is possible to achieve accurate control over a distance.

Typical synchro systems contain two or more electromagnetic devices that are similar in appearance to a small electric motor. These devices are, however, connected in such a way that the angular position of the generator shaft is transmitted through interconnecting wires to the motor or receiving unit.

When electrical energy is first applied to a generator or motor unit, the unit will move to a specific location, then come to rest. This position is called electrical. Both synchro generators and motors are adjusted to zero degrees in the same manner.

In this activity, you will connect the motor and generator units of a simple synchro system together to test their operation. Through this experiment, you will become familiar with actual circuit terminal designations and be able to connect the units together to achieve clockwise and counterclockwise rotation.

Equipment and Materials

- Multimeter
- Oscilloscope

DOI: 10.1201/9788743810124-50

- Isolated variable AC power source
- Variable AC transformer
- Synchro generator/motor set

Safety

Safety is always a primary factor for laboratory activities. This lab should only be completed with supervision.

You may complete certain steps of the "Procedure" and the "Analysis" for better understanding without applying voltage.

Procedure

Section A: Electrical determination

1. Position the generator part of the synchro system so that you can readily observe the terminal connections at the rear of the unit. The stator terminals are usually labeled S_1, S_2 and S_3. The rotor terminals are generally labeled R_1 and R_2.

2. Connect the isolated variable AC power source to the rotor terminals as indicated in Figure 47.1.

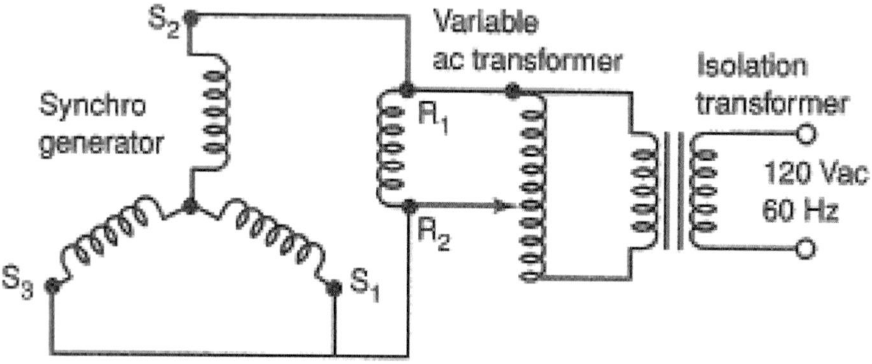

Figure 47.1 Circuit for determining the electrical position of a synchro generator unit.

3. Adjust the variable AC transformer to its lowest output voltage level. Then turn on the AC source and gradually increase the voltage until the rotor winding locks in place. This represents the electrical zero-degree position of the rotor.

4. Prepare the multimeter to measure AC voltage and connect it across R1 and R2. The supply voltage measures _______ volts AC.

5. With the rotor locked at zero degrees, adjust the indicating dial to the mechanical zero degree position.

6. Another way to determine the electrical zero-degree position of a synchro generator is using the voltmeter method. Turn off the power source and connect the generator unit so that S_1, S_2, and S_3 are open with respect to the AC source applied to the rotor. The stator windings in this case are energized entirely by induction from the rotor.

7. Turn on the power source and measure the voltage across S_1 and S_2. Turn the rotor while observing the voltage. Note that zero voltage occurs at two locations during one revolution. This represents electrical zero and 180 degrees.

8. To distinguish between zero degrees and 180 degrees, turn off the power source and connect a jumper wire between R_1 and S_2. Then turn on the power source and measure the voltage across R_2 and S_1. Turn the shaft or dial and note that the voltage has a high value at one position and a lower value at the opposite position. Electrical zero and the low point are the same. The 180-degree position is represented by the highest voltage value.

9. Turn off the power source and disconnect the generator unit.

10. Connect the motor and receiver unit in the same manner outlined for the generator, and set its electrical indicator.

11. Turn off the power source and disconnect the circuit.

Section B: Synchro position control

1. Connect the synchro generator/motor units together as indicated in Figure 47.2.

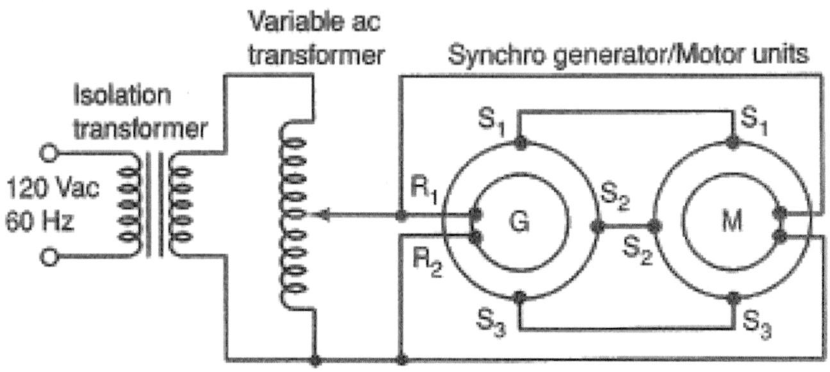

Figure 47.2 Circuit for observing the position control of a synchro generator/motor system.

2. Turn on the power source and increase the voltage of the variable AC transformer to the rated rotor voltage. As a general rule, this is usually 120 volts AC.

3. Move the generator rotor to the zero-degree position. The motor unit should follow or track the generator.

4. Rotate the generator 90 degrees in the clockwise direction. How does the motor respond?

   Return the generator to the zero-degree position and then rotate the generator counterclockwise. How does the motor respond?

5. You may want to try a variety of position changes to confirm your answer.

6. Prepare an oscilloscope for operation and connect the ground lead to R2 and the probe lead to R1. Turn on the power source and turn on the rotor through one complete revolution in each direction. How does the waveform respond to this physical change in the rotor?

7. Turn off the power source and connect the ground lead of the oscilloscope to S1 and the probe lead to R1. Turn on the AC power source and turn the rotor one revolution in each direction while observing the waveform. How does it respond?

8. Turn off the power source and reverse the R1 and R2 and rotor leads of the motor unit. Turn on the power source and test the operation. How has the circuit changed, compared with previous steps?
9. Turn off the power source and return R1 and R2 to they were originally connected in Figure 47.2. Connect stator leads S1 of the generator to S3 of the motor and S3 of the generator to S1 of the motor.
10. Turn on the power source and test the rotary control action of the generator/motor set. How does it respond when connected in this manner?
11. Turn off the power source and disconnect the circuit. Return all materials.

Analysis

1. Explain how the stator windings are energized by inductance as noted in Section A, Step 6.

2. What is meant by the term electrical zero?

3. What is the electrical difference between a synchro generator and a synchro motor of the same set?

4. What does the waveform of Section B, Step 6 show about the stator voltage?

5. Why does the motor rotor track the generator rotor of a synchro system?

UNIT IV

Electrical Machinery Control Basics

Activity 48

Industrial Symbols and Diagrams

Objectives

In Activity 4, basic electrical symbols were reviewed. There are several symbols that are unique to industrial applications. In this activity, you will examine the common industrial symbols and diagrams shown in Figure 48.1.

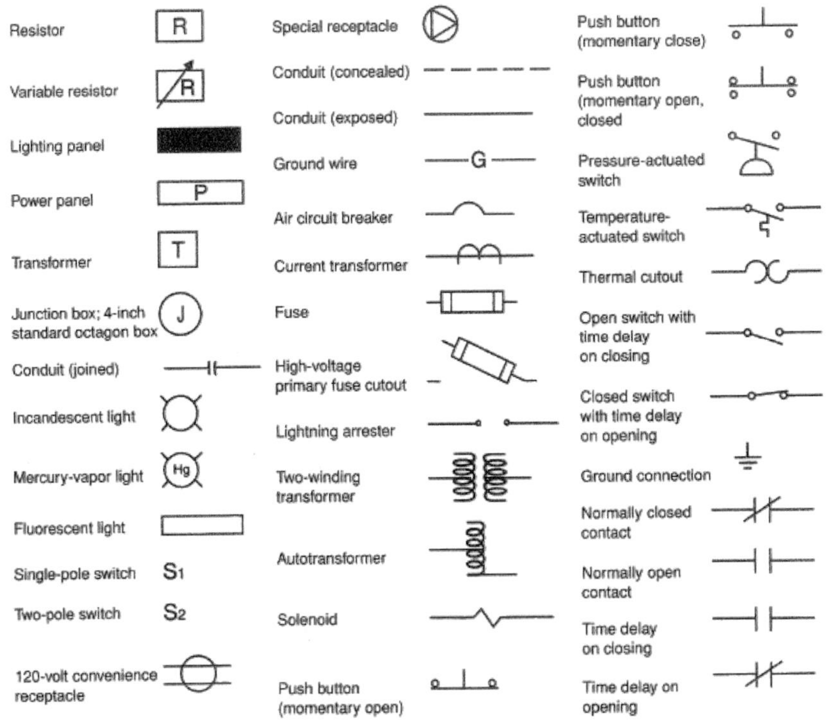

Figure 48.1 Industrial schematic symbols.

DOI: 10.1201/9788743810124-52

Industrial Diagrams

The most common diagrams used to illustrate the components, circuits, and subsystems of an industrial system are connection diagrams, block diagrams, schematic diagrams, and one-line diagrams.

In this activity, you will need to do some research to locate some simple industrial systems that can be represented by using the appropriate type of diagram.

Analysis

In the space next to each listed device, draw the proper symbols for that device.

1. Incandescent lamp:

2. Mercury vapor lamp:

3. Fluorescent lamp:

4. Circuit breaker:

5. Fuse:

6. Push-button switch (normally open):

7. Push-button switch (normally closed):

8. Pressure-actuated switch:

9. Temperature-actuated switch:

10. Thermal cutout:

11. Normally closed contact:

12. Normally open contact:

Connection diagram

The connection diagram shows the physical relationships of the parts of a system. It is used to identify various parts for installation or servicing. The wires that interconnect circuits are ordinarily shown in their correct location. In the following space, draw a simple connection diagram for an industrial system.

Block diagram

The block diagram is a very general method used to show how subsystems of any industrial system fit together to form a functional unit. Rectangles are used to represent subsystems, and arrows are used to connect the blocks and show the interrelationships of subsystems. In the following space, draw a block diagram of an industrial system.

Schematic diagram

The schematic diagram is used to show the electrical circuit relationships of the various components of an operational system. The electrical, but not necessarily the physical, layout of the circuit is shown. This type of industrial diagram is the most often used. In the following space, draw a schematic diagram of an industrial circuit.

One-line diagram

The one-line diagram is a simplified form of the schematic diagram. This type of diagram is used to show how the components of a system fit together. It is similar to the block diagram in several respects. However, actual symbols, rather than rectangular shapes, are used to show component parts. In the following space, draw a one-line diagram of an industrial system.

Activity 49

Types of Switches

Objectives

In this activity, you will learn to correctly identify the types of switches that are used in electrical control systems. It is important for a technician to be able to identify switch types since they are the most basic type of electrical control.

Procedure

1. Study the symbols of switches in **Chapter 13** of **Electrical Machinery Control Fundamentals**.
2. Learn the correct names of all types of switches.
3. Learn the uses of the switches as you study them.
4. The symbols of several common types of switches are shown in Figure 49.1 below. Fill in the correct names of the switches.

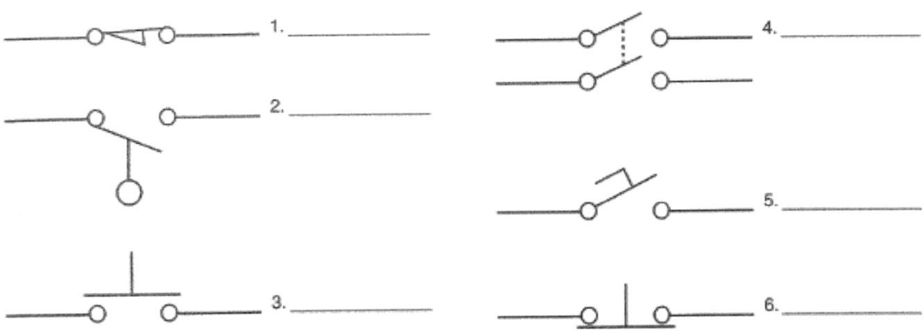

DOI: 10.1201/9788743810124-53

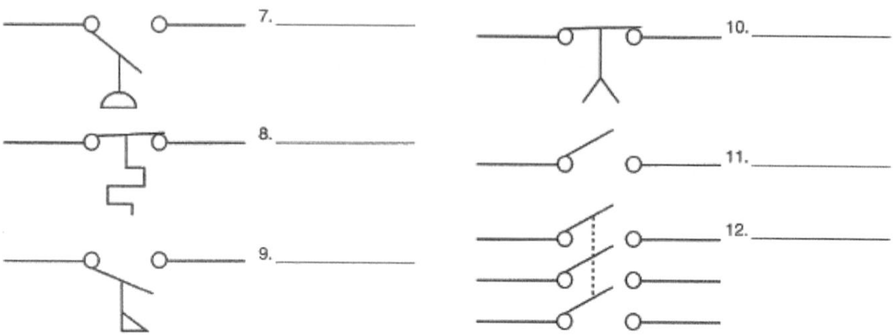

5. You will now be ready for testing to see how many of the types of switches available in your laboratory or shop you can identify.

Analysis

1. For each of the switches shown in Figure 49.1, write a brief description of the use of that type of switch.

Activity 50

Float Switches

Objectives

A float switch is an electromechanical control switch connected to a rod or chain that moves in response to the rise and fall of liquid in a tank or reservoir. Depending on its application, the normally open or normally closed contacts of the switch can be used.

Float switches are used on sump pumps, fill tanks, and flow lines. With the use of float switches, the automatic control of a pump motor, motor-driven gate valve, or other similar device can easily be arranged.

In this activity, a lever-type limit switch can be used in place of the regular float switch. The operation of the limit switch and the float switch is identical. The lever arm of the limit switch is usually activated by a moving object, while the float ball and lever arm of the float switch are activated by the level of liquid within the container.

The use of normally open or normally closed contacts is determined by whether the motor is to operate on the filling or emptying of a reservoir. In this activity, you will examine the operation of a float switch in both full-tank and sump operation.

Equipment and Materials

- Lever-type limit switch or float switch
- Small horsepower three-phase motor*
- Single-coil three-phase magnetic contactor
- Three-phase power supply*

*This activity can also be done with single-phase AC power.

DOI: 10.1201/9788743810124-54

Safety

Safety is always a primary factor for laboratory activities. This lab should only be completed with supervision.

You may complete certain steps of the "Procedure" and the "Analysis" for better understanding without applying voltage.

Be sure to wear eye protection while electric motors are in operation. Also, be very careful when working with high voltages.

Procedure

1. Connect the circuit of Figure 50.1. As connected, this is a full-tank operation.

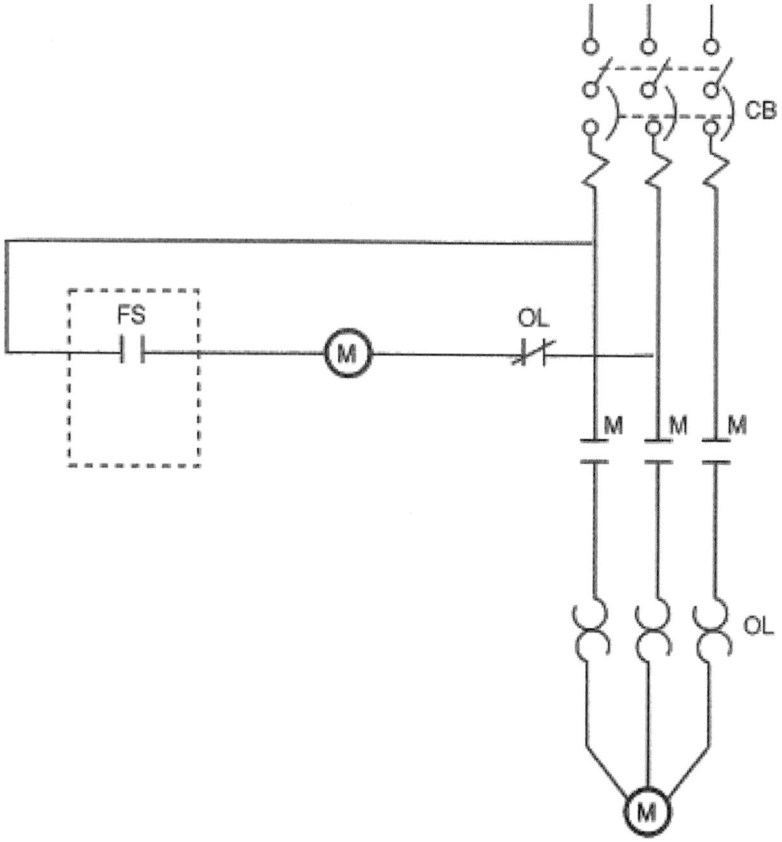

Figure 50.1 Float switch motor control.

2. Have the instructor/supervisor check your circuit before applying power.

3. Being a full-tank operation, the normally open contacts are used. Pull the lever of the switch to one side. As long as the lever is held, the motor should run.

4. Release the lever. What happens?

5. What condition of the liquid level does the depressed lever represent?

   What condition does the released lever represent? ____________________

6. Disconnect the power. Alter the circuit of Figure 50.1 so that the circuit operates as a sump pump operation. What must be changed to accomplish this?

7. Reconnect the power. Does the motor run now?

8. Pull the lever to one side. What happens?

9. Release the lever. What happens?

10. Disconnect the circuit and turn off the power. Return all materials.

Analysis

1. Write a brief explanation of the operation of both circuits in this activity. Also, explain what is meant by full-tank and sump operation. Include your own observations and conclusions about this activity in your discussion.

Activity 51

Motor Control with Magnetic Contactors

Objectives

A very common method used to turn electric motors on or off is a magnetic contactor used in conjunction with a set of push-button switches. A magnetic contactor is a type of relay. It is energized by applying a voltage to its electromagnetic coil. This control voltage causes a solenoid action that closes the contacts of the device. These contacts can be connected in series with the power line going to a motor to cause it to turn on. The coil of the contactor has a relatively high resistance, causing it to draw a small current from the power line. This small current is capable of energizing the relay so that the closing of its contacts will apply power to a motor. Thus, a small current is used to control the larger current of the motor. The surface area of the contacts is usually fairly large to accommodate the current level drawn by the motor.

In this activity, you will use a magnetic contactor and push-button switch assembly to control the operation of a motor. Be sure to note the difference in magnitude of the control current through the contactor coil and the current drawn by the motor.

Equipment and Materials

- Multimeter
- Single-phase AC power source
- Single-phase AC motor
- Start/stop push-button station
- Connecting wires

DOI: 10.1201/9788743810124-55

Safety

Safety is always a primary factor for laboratory activities. This lab should only be completed with supervision.

You may complete certain steps of the "Procedure" and the "Analysis" for better understanding without applying voltage.

Be sure to wear eye protection while electric motors are in operation. Also, be very careful when working with high voltages.

Procedure

1. Construct the circuit shown in Figure 51.1.

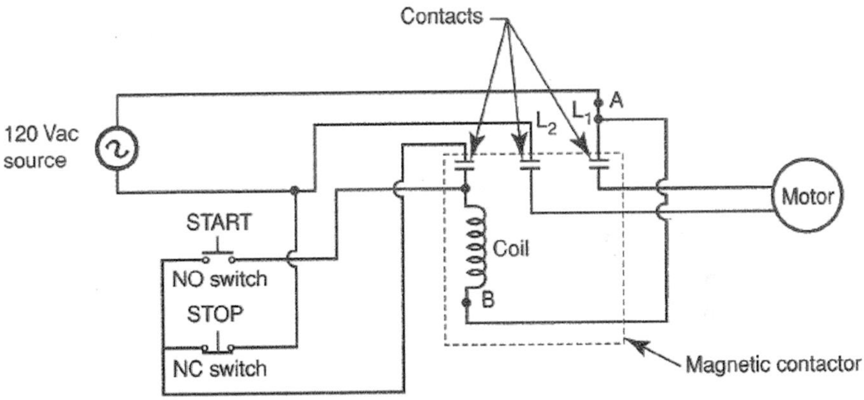

Figure 51.1 Magnetic contactor motor control circuit.

2. Press the normally open push-button switch (START switch) to test the circuit for operation. The relay coil should energize, causing the motor to rotate.

3. Press the normally closed push-button switch (STOP switch). The motor should stop running.

4. Turn off the 120 volts AC power source and connect an AC ammeter in series with the AC line at point A.

5. Apply power to the circuit, press the START switch, and measure the starting and running current of the motor.

   Starting current =_________ amperes AC.

   Running current = __________amperes AC.

6. Turn off the circuit and disconnect the AC power source. Reconnect the AC ammeter in series with the relay coil at point.

7. Apply power to the circuit, press the START switch, and measure the current through the coil. Coil current = _______ amperes AC.

8. With the AC line plugged into a variable 0–120 volt power source, perform the following steps (if variable AC is available).

9. Set the power source to zero volts and then turn it on. Keep the START button depressed while increasing the applied voltage. Measure and record the "pull-in" current and voltage of the contactor.

   Pull-in current = ______ amperes AC.

   Pull-in voltage = ______ volts AC.

10. Gradually reduce the applied voltage until the contacts open. Record the "drop-out" current and voltage.

    Drop-out current = ______ amperes AC.

    Drop-out voltage = _______ volts AC.

Analysis

1. Why are magnetic contactors used to control motors?

2. Explain the operation of a magnetic contactor circuit to control a single-phase motor.

3. Draw a schematic diagram showing how a magnetic contactor could be used to control a three-phase motor.

4. What is meant by “pull-in” current?

5. What is meant by “drop-out” current?

Activity 52

Magnetic Contactor Wiring

Objectives

During this activity, you will demonstrate your ability to correctly wire a magnetic contactor control.

Equipment and Materials

- Multimeter
- Single-phase AC power source
- Single-phase AC motor
- Start/stop push-button station
- Connecting wires

Safety

Safety is always a primary factor for laboratory activities. This lab should only be completed with supervision.

You may complete certain steps of the "Procedure" and the "Analysis" for better understanding without applying voltage.

Procedure

1. A control panel for a magnetic contactor used to control an electric motor is shown in Figure 52.1. This is a pictorial diagram of the laboratory activity circuit.

DOI: 10.1201/9788743810124-56

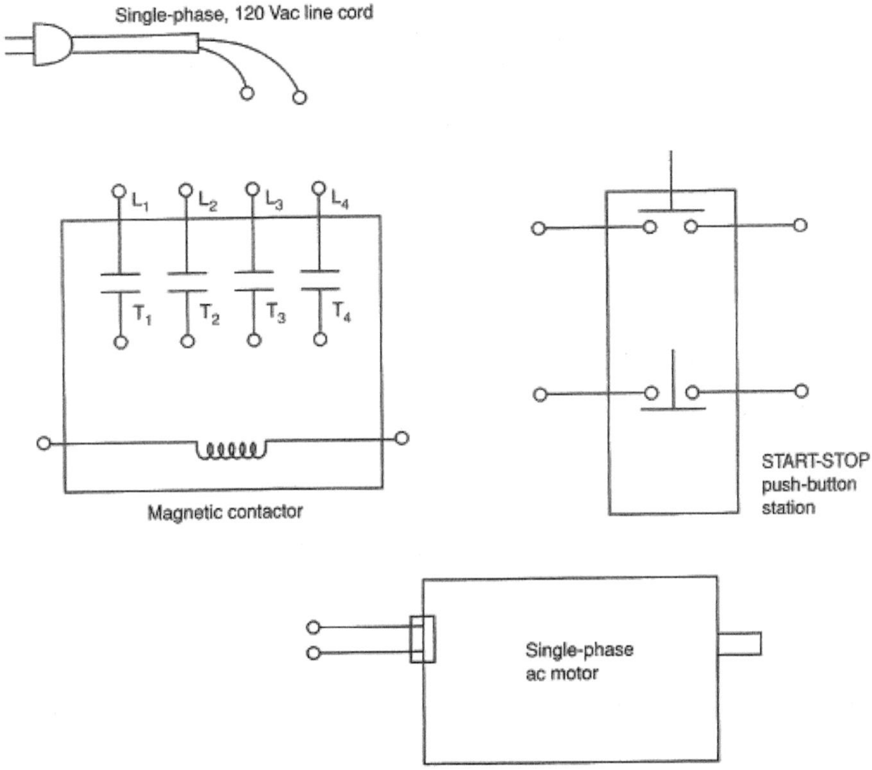

Figure 52.1 Control panel for a magnetic contactor.

2. Make a sketch of the control diagram for starting and stopping a single-phase, 120 Vac motor below.

3. Complete the wiring diagram of the pictorial below to correspond to your control diagram. Use Figure 52.1.

4. Wire a magnetic contactor, start push-button, and stop push-button to control a single-phase AC motor.

5. Have your completed circuit checked for proper operation.

Approval: ____________________________________

Activity 53

Electromagnetic Relays

Objectives

Relays are electromagnetic switches and are excellent examples of how a magnetic field attracts a magnetic material. These devices contain a coil that creates an electromagnetic field, an armature that is constructed of a magnetic material attracted by the coil, and a number of contacts or switches that open or close when the magnetic field attracts the armature.

In this activity, you will study the electromagnetic characteristics of a relay.

Equipment and Materials

- Multimeter
- Multi-contact relay
- Lamp with socket – 6 volt
- Variable DC power supply
- Resistor – 1 K Ohm
- Battery – 6 volt
- Connecting wires

Procedure

1. Prepare the multimeter to measure resistance. Measure and record the resistance of your relay coil. R = ________ Ohms.

DOI: 10.1201/9788743810124-57

2. Using the multimeter, determine how many normally open and normally closed contacts are used with your relay.

   Number of normally open contacts ________.

   Number of normally closed contacts _______.

3. Construct the circuit illustrated in Figure 53.1. Be sure that the variable DC power supply is adjusted to zero. The multimeter should be adjusted to measure DC current on the highest range.

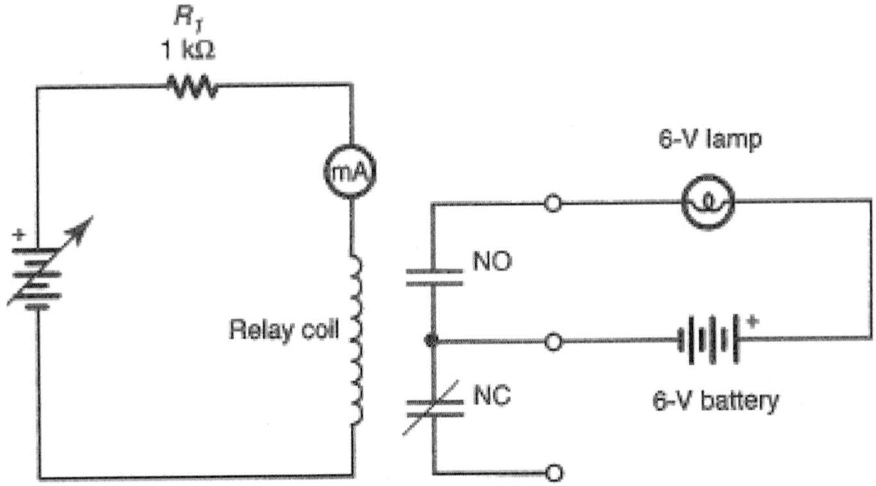

Figure 53.1 Circuit using an electromagnetic relay.

4. Slowly adjust the variable DC power supply from zero until the 6-volt lamp is turned on. Record the current measured by the multimeter when the relay is energized. This is the pickup current.

   Pickup current = _________ mA.

5. Slowly adjust the variable DC power supply toward zero until the 6-volt lamp is turned off. Record the current measured by the multimeter when the relay is de-energized. This is the dropout current.

   Dropout current = ________ mA.

6. Turn the variable power supply off.

7. Alter the circuit to match the circuit in Figure 53.2.

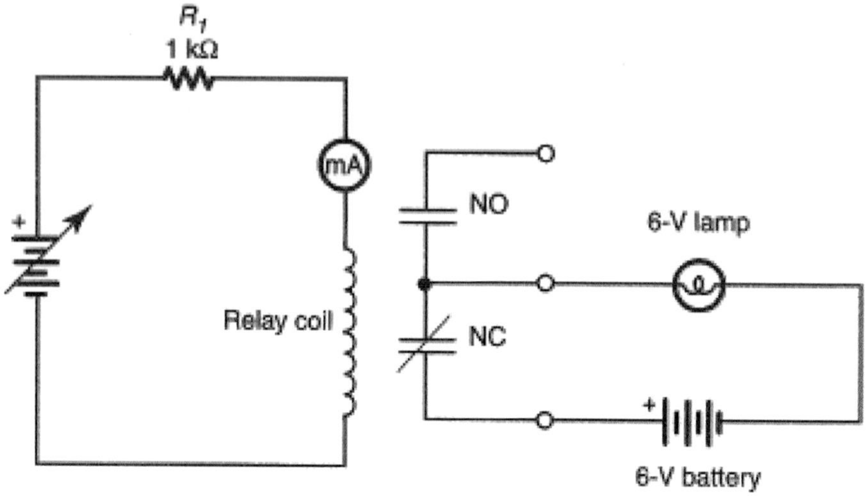

Figure 53.2 Circuit from Figure 53.1 with relay contact reversed.

8. You will notice that the only difference in the two circuits is the type of contacts used. In Step 3, the normally open contacts were used. In this procedure, the normally closed contacts are used, causing the lamp to remain on until the relay is energized.

9. Adjust the variable DC power supply and record the pickup and dropout currents as you did in Steps 4 and 5.

   Pickup current = ____ mA.

   Dropout current = ____ mA.

10. How do the currents recorded in Step 9 compare with the currents recorded in Steps 4 and 5?

11. How did the action of the 6-volt lamp in Steps 4 and 5 compare with the action of the lamp in Step 9?

Analysis

1. What are normally open contacts?

2. What are normally closed contacts?

3. What is meant by the term pickup current?

4. What is meant by the term dropout current?

5. Using Ohm's law, compute the voltage across the relay coil when the relay is energized (see Steps 1 and 4).

Activity 54

Load Centers

Objectives

Electrical power used to operate electrical devices is often generated from coal-fired steam systems, hydroelectric systems, or nuclear-powered systems. Power is carried across the country by high-voltage transmission lines. The voltage is then reduced to a usable value by means of a step-down transformer. The power enters a building through a service entrance that is connected to a metering device, through a main disconnect, and into the load center.

The load center is commonly known as a distribution panel, since this is where power is distributed to individual circuits. The load center is protected by the main circuit breaker, which breaks the hot power lines when current values are above the breaker rating. The neutral power line is not broken since it is connected to a grounding bar that is connected to earth ground. From the main breaker, the power is fed to bus bars from which the individual circuits are taken. Each circuit is protected by a circuit breaker with a current rating designated by the circuit.

In this activity, you will examine the circuit for a distribution panel, or load center.

Equipment and Materials

- Single-phase safety switch
- Single-phase load centers (2)
- Circuit breakers-15 amp, 20-amp and 30-amp, 120/240 volt
- Electrical conductors
- Wireway

DOI: 10.1201/9788743810124-58

Safety

Safety is always a primary factor for laboratory activities. This lab should only be completed with supervision.

You may complete certain steps of the "Procedure" and the "Analysis" for better understanding without applying voltage.

Be sure to wear eye protection while electric motors are in operation. Also, be very careful when working with high voltages.

Procedure

1. Properly wire the control panel shown in Figure 54.1.

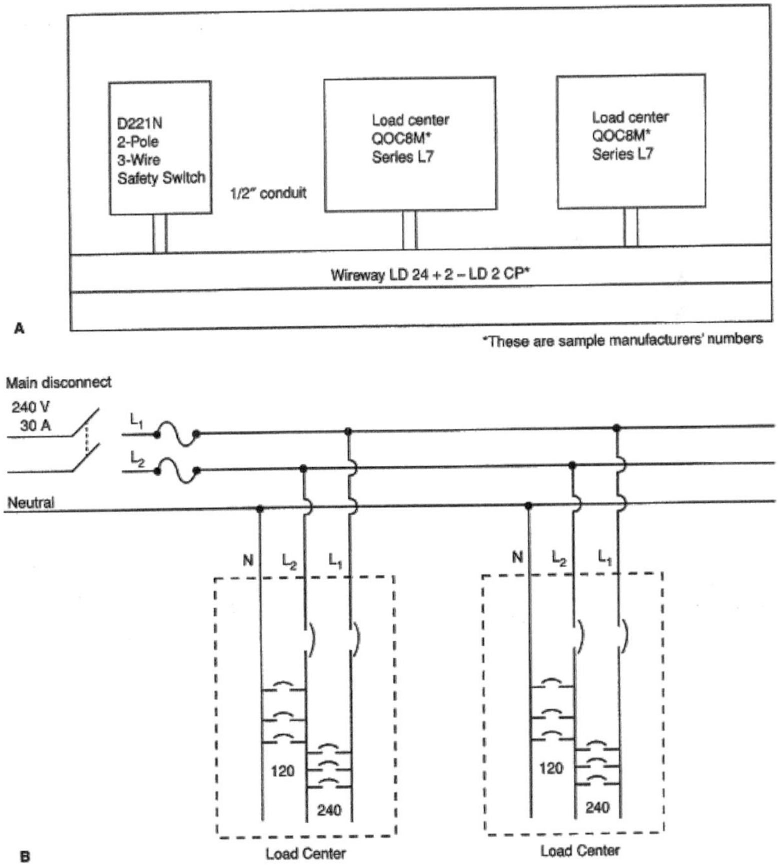

Figure 54.1 Load center controls. (A) Layout of suggested laboratory control panel. (B) Schematic of circuit wiring of the control panel.

2. Have your completed circuit checked for proper operation.

   Approval: ______________________________

Analysis

Discuss the following types of control equipment.

1. Load center.

2. Circuit breakers.

3. Single-phase safety switch.

4. Wireways (raceways).

Activity 55

Three-phase Motor Controls

Objectives

The panel of Figure 55.1 consists of a manual contactor, a magnetic contactor, and a combination starter. A manual starter is operated by human operators in most cases. The switch is spring-loaded to prevent excessive arcing of the contacts. This is the simplest contactor.

The magnetic contactor is operated by means of electrical energy, which is achieved by means of a start/stop control station. When the start button is pressed, current can flow from L2 through the closed stop contacts, through the momentarily closed start contacts, throughout the coil, and to L2. This completes the coil circuit and closes the line and holding contacts. When the start button is released, the coil stays energized through the holing contacts, which are in parallel with the start button contacts. A pilot light is connected in parallel with the coil, so anytime the coil is energized, the pilot light is energized. Since the stop button contacts are in series with the coil, the circuit can be opened by pressing the stop button.

The combination starter has the overload protection (which is a special fuse) and a circuit breaker. A current-limiting fuse interrupts the current before the short-circuit can occur. On low-value short-circuit currents, the breaker trips before the fuse opens. If the short-circuit current exceeds the capabilities of the circuit breaker, the fuse opens the circuit and protects the breaker from damage. All lines are opened when either device trips. This type of protection is valuable where extremely high currents flow under short-circuit conditions.

The combination starter has a start-stop station, overload devices, and magnetic contactors enclosed in a single enclosure. When the start button is pressed, the coil circuit is completed, which energizes the coil and closes the line and holding contacts. The stop button is in series with the coil, and when pressed, it opens the coil circuit. In this activity, you will examine a three-phase motor control panel.

DOI: 10.1201/9788743810124-59

Equipment and Materials

- Three-phase safety switch
- Three-phase manual starter
- Three-phase magnetic starter
- Combination starter
- Wireway
- Start/stop push-button station
- Melting-alloy thermal overloads

Safety

Safety is always a primary factor for laboratory activities. This lab should only be completed with supervision.

You may complete certain steps of the "Procedure" and the "Analysis" for better understanding without applying voltage.

Be sure to wear eye protection while electric motors are in operation. Also, be very careful when working with high voltages.

Procedure

1. Properly wire the control panel shown in Figure 55.1, using the equipment provided to study three-phase motor controllers.

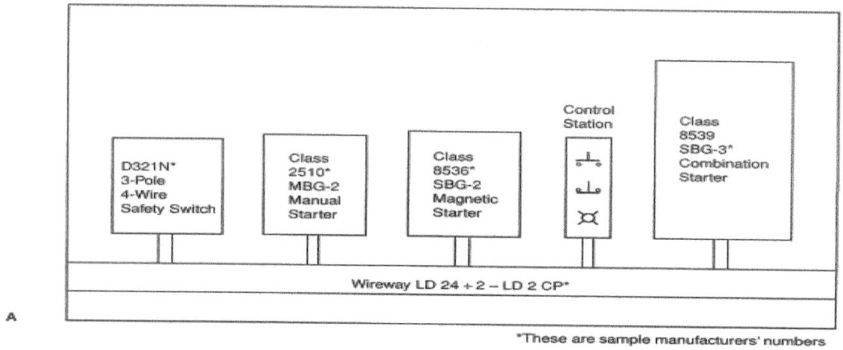

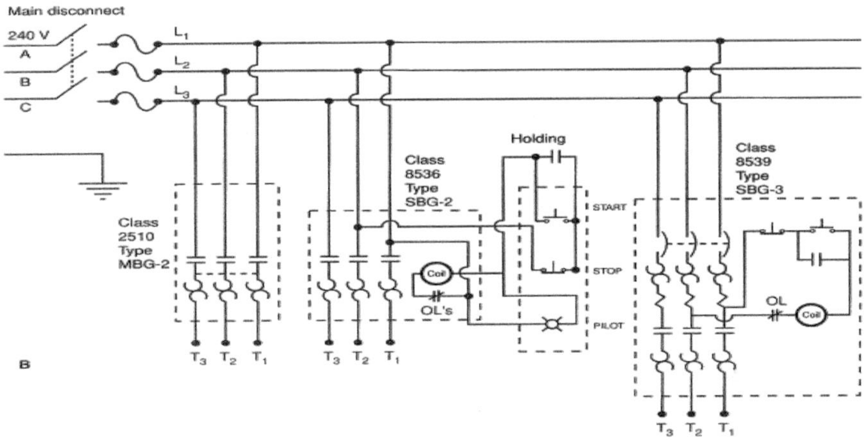

Figure 55.1 Three-phase motor controllers. (A) Layout of suggested lab control panel. (B) Schematic.

2. Have your completed circuit checked for proper operation.

   Approval: ______________________________

Analysis

Discuss the following types of control equipment.

1. Three-phase safety switch.

2. Three-phase manual starter.

3. Three-phase magnetic starter.

4. Combination starter.

5. Overload devices.
   a. Melting alloy.
   b. Bimetallic.

Activity 56

Three-phase Reversing Controls

Objectives

The panel of Figure 56.1 demonstrates the reversing controls for three-phase machines. These systems are used to reverse the direction of motor rotation by electrically interchanging the position of any two of the three conductors supplying the motor. A three-phase motor is the simplest of any motor to reverse. A direct short would result between phases if both the forward and reverse contactors of the starter were in the energized position at the same time. To prevent this, the contactors are mechanically interlocked so only one or the other can be closed. Three-phase motors can also be reversed by drum controllers. The drum controller is manually operated, and the mechanical interlocking is inherent in the construction.

The operation of this circuit is controlled by the forward/reverse/stop push-button station that can be located at any convenient location. The control circuits are operated from 120 volts, which is supplied from a step-down transformer. When the forward button is pressed, contacts are momentarily closed, allowing current to flow through the forward coil, which closes the magnetic contactors and the holding or auxiliary contacts that keep the circuit closed after the forward button is released. When the reverse button is pressed, the forward coil circuit is broken and the reverse coil circuit is closed, allowing current to flow through the reverse coil, which closes the reverse contactors and their auxiliary contacts. Either coil circuit can be opened by the stop button, which opens the circuit to the secondary side of the transformer. A maintain/stop button, which can be placed at any convenient place, is in series with the stop of the forward/reverse station. This button is a manual push-pull that has a lamp circuit that is closed when in the pressed position. Another way of opening the forward and reverse circuits is by means of limit switches. A limit switch is a device that is normally closed and opened by

DOI: 10.1201/9788743810124-60

mechanical means. In this activity, you will examine the circuit for three-phase reversing controls.

Equipment and Materials

- Three-phase safety switch
- Limit switches
- Emergency stop push-button switch
- Three-phase AC power source
- Three-phase reversing control
- Start/stop/reverse control station
- Transformer – 2:1 ratio
- Drum controller
- Wireway
- Conduit – 1/2 inch EMT
- Conductors – #12 AWG

Safety

Safety is always a primary factor for laboratory activities. This lab should only be completed with supervision.

You may complete certain steps of the "Procedure" and the "Analysis" for better understanding without applying voltage.

Be sure to wear eye protection while electric motors are in operation. Also, be very careful when working with high voltages.

Procedure

1. Properly wire the control panel shown in Figure 56.1.

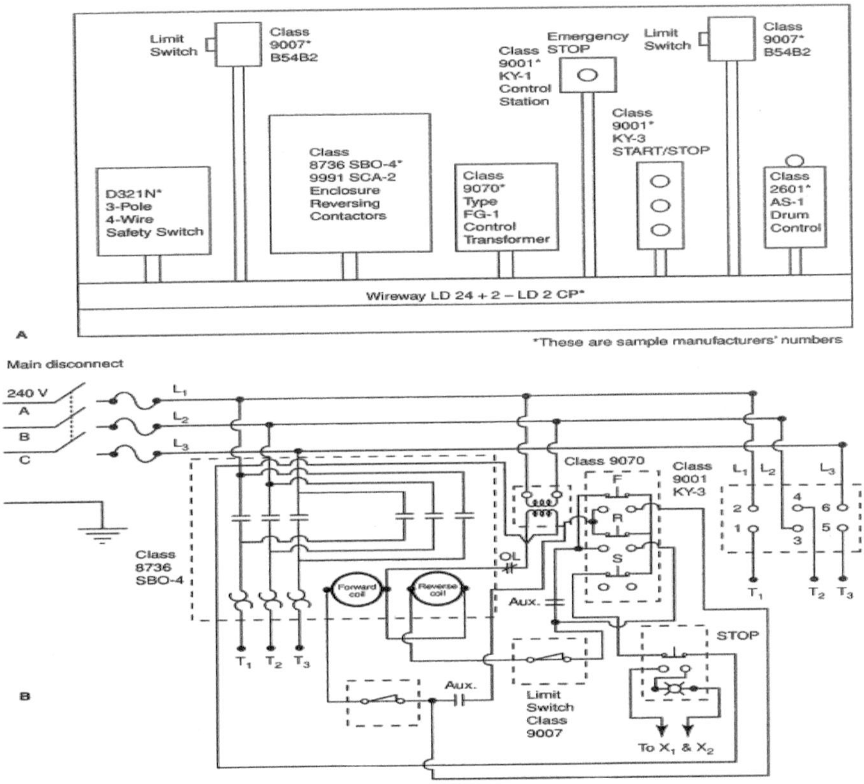

Figure 56.1 Three-phase reversing controls. (A) Layout of suggested laboratory control panel. (B) Schematic.

2. Have your completed circuit checked for proper operation.

   Approval: ______________________________

Analysis

Discuss the following types of control equipment:

1. Three-phase forward/reverse contactor

2. Control transformer

3. Push-button head designs

4. Drum controller

Activity 57

Single-phase Control Equipment

Objectives

The panel of Figure 57.1 demonstrates some of the many possibilities of single-phase control systems. On this panel, a motor can be delayed from the time the selector switch is activated until some time later. This type of motor control is used in sequencing control. When the selector switch is activated, a magnetic coil is energized, pulling the iron core in and allowing the dash-pot-type timer to time out. This closes the timer contacts, allowing the coil to energize the control relay, closing the control relay contacts that allow power to the motor.

Another circuit is the temperature control relays. When the temperature reaches a predetermined temperature, the temperature control contacts close by means of expanding liquid, causing the coil of a control relay to energize. This closes the control relay contacts, which allows power to flow to the motor.

A control relay can also be used as a magnetic contactor. This is done by using two push buttons in conjunction with a control relay. The coil is in series with normally open start contacts. When these contacts are closed, the coil circuit is completed through the normally closed stop contacts. At this time, the contacts are closed. Now there is a set of holding or maintaining contacts that are mechanically connected to the contactor and close at the same time. This contact parallels the normally open start contact and allows the coil to stay energized after the normally open contacts are released. The circuit can be turned off by depressing the stop button, which opens the circuit to the coil. In this activity, you will examine a single-phase control circuit.

DOI: 10.1201/9788743810124-61

Equipment and Materials

- Single-phase AC power source
- Single-phase safety switch
- Time delay relay
- Temperature switch
- Pressure switch
- Control relays
- Start/stop control station
- Conduit – ½ inch EMT
- Conductors – #12 AWG
- Wireway

Safety

Safety is always a primary factor for laboratory activities. This lab should only be completed with supervision.

You may complete certain steps of the "Procedure" and the "Analysis" for better understanding without applying voltage.

Be sure to wear eye protection while electric motors are in operation. Also, be very careful when working with high voltages.

Procedure

1. Properly wire the control panel shown in Figure 57.1, using the equipment provided to study single-phase control systems.
2. Have your circuit checked for proper operation.

   Approval: ______________________________________

Analysis

Discuss the following types of control equipment.

1. Time-delay relay.

2. Temperature switch.

3. Pressure switch.

4. Control relay.

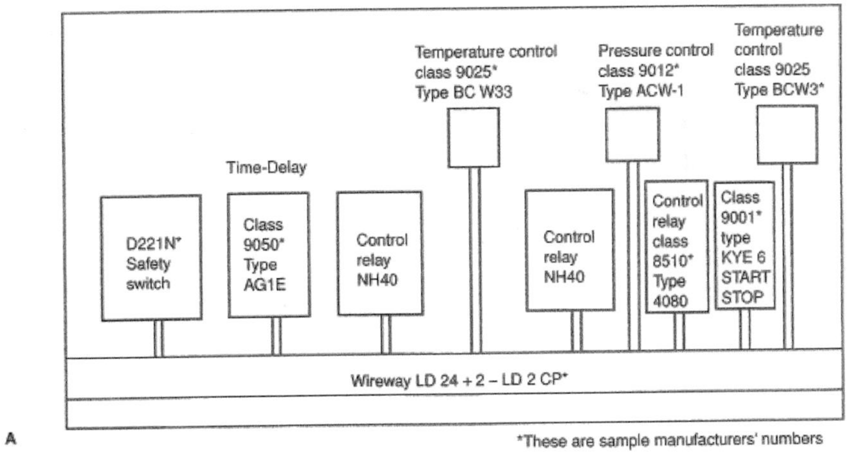

Figure 57.1 Single-phase control systems. (A) Layout of suggested laboratory control panel.

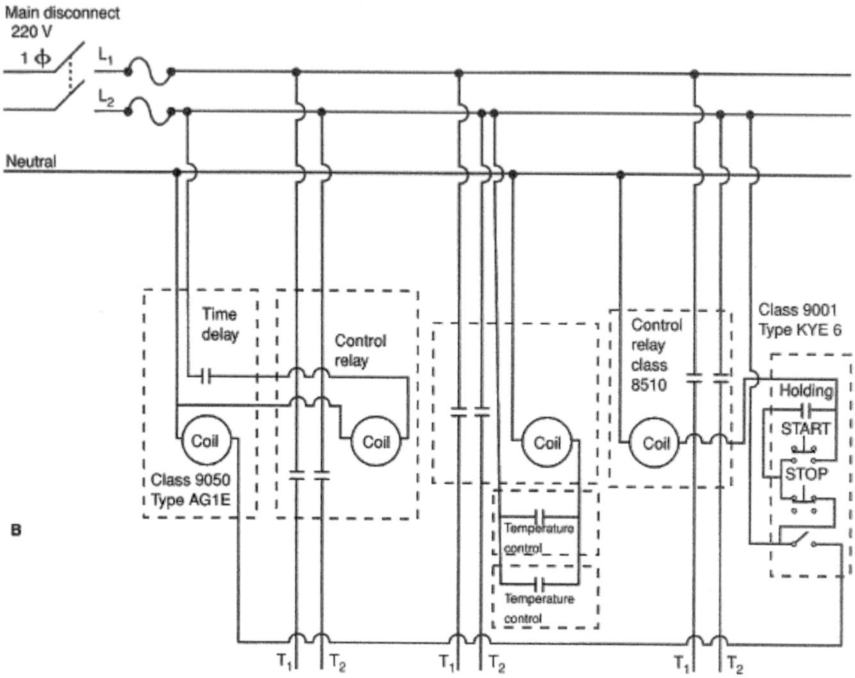

Figure 57.1 Single-phase control systems. (B) Schematic.

Activity 58

Forward and Reverse Control Circuits

Objectives

Most types of electrical motors can be made to rotate in either direction by simple modifications of their connections. Ordinarily, motors used in industry require two magnetic contactors to accomplish forward and reverse operation. These contactors are used in conjunction with a set of three push-button switches-FORWARD, REVERSE, and STOP. When the FORWARD push-button switch is depressed, the forward contactor will be energized. It is deactivated when the STOP push-button switch is depressed. A similar procedure takes place during the reverse operation.

In this activity, you will observe the forward and reverse operation of: (1) a DC shunt motor, and (2) a three-phase induction motor. The activity is designed for observation only. Therefore, no measurements will be taken. Keep in mind the function of the magnetic contactor as a low-current control device for controlling the large currents drawn by motors.

Equipment and Materials

- DC shunt motor
- Three-phase induction motor
- Push-button switches, normally open (2)
- Push-button switch, normally closed
- Magnetic contactors (2)
- DC power source
- Three-phase AC power source

DOI: 10.1201/9788743810124-62

Safety

Safety is always a primary factor for laboratory activities. This lab should only be completed with supervision.

You may complete certain steps of the "Procedure" and the "Analysis" for better understanding without applying voltage.

Be sure to wear eye protection while electric motors are in operation. Also, be very careful when working with high voltages.

Procedure

1. Construct the circuit shown in Figure 58.1 for forward and reverse operation of a DC shunt motor.

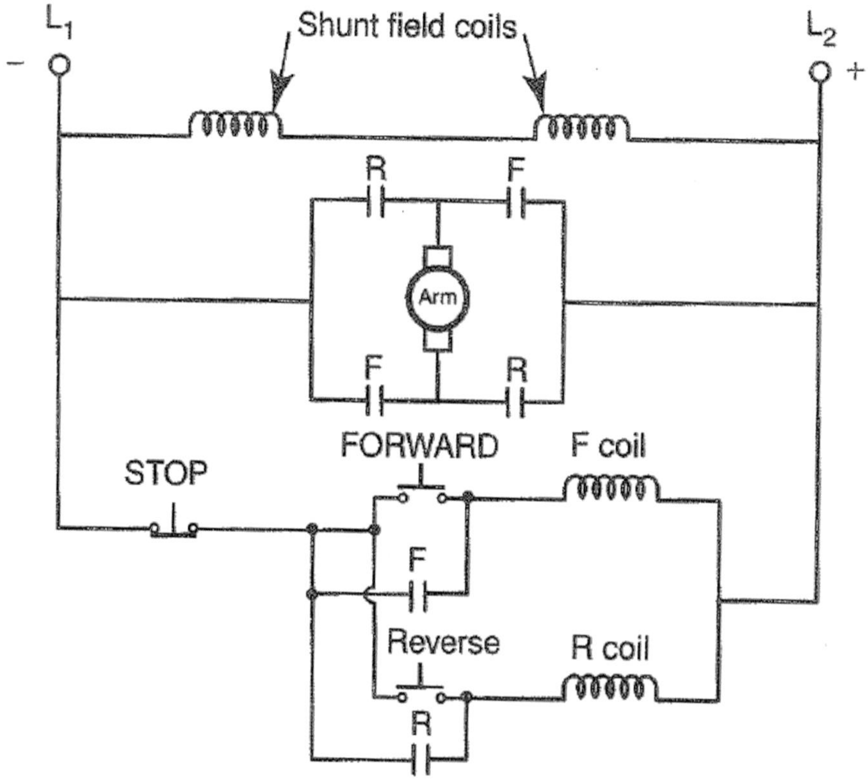

Figure 58.1 Control circuit for forward and reverse operation of a DC shunt motor.

2. Apply rated voltage to the DC shunt motor and check the circuit for operation. The FORWARD, REVERSE, and STOP switches should function. Have your circuit checked for proper operation.

   Approval: ______________________________

3. Disassemble the DC shunt motor control circuit and connect the circuit shown in Figure 58.2 for forward and reverse operation of a three-phase induction motor.

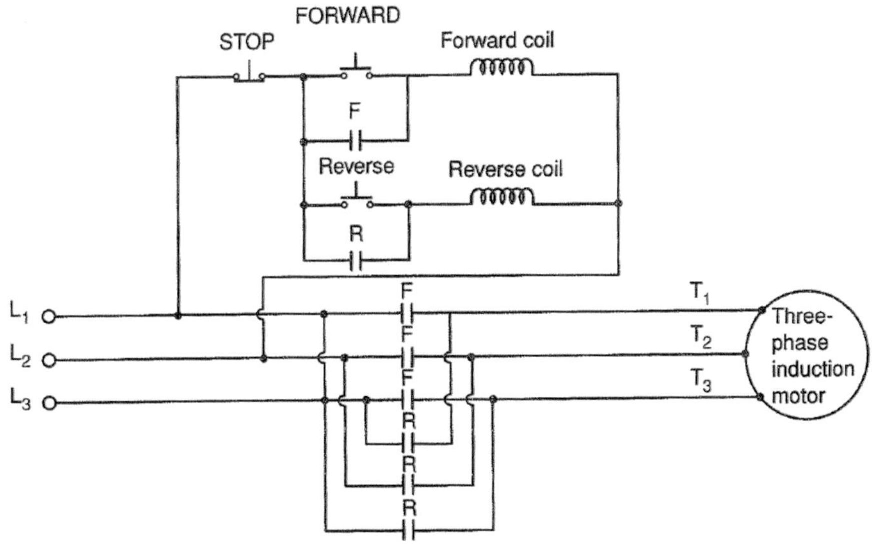

Figure 58.2 Control circuit for forward and reverse operation of a three-phase induction motor.

4. Apply the rated voltage to the three-phase motor and check the circuit for operation. The FORWARD, REVERSE, and STOP switches should function. Have your circuit checked for proper forward and reverse control.

   Approval: ______________________________

Analysis

1. How is the direction of rotation of a DC shunt motor changed?

2. How is the direction of rotation of a three-phase induction motor changed?

3. Explain the operation of the forward/reverse control of the DC shunt motor.

4. Explain the operation of the forward/reverse control of the three-phase induction motor.

Activity 59

Forward and Reverse Control Wiring for Three-phase Motors

Objectives

In this activity, you will create control and wiring diagrams for a three-phase AC motor. You will also observe the forward and reverse operation of a three-phase AC motor.

Equipment and Materials

- Three-phase AC power source
- Three-phase AC induction motor
- Magnetic contactors-three-phase (2)
- Start/stop push-button station
- Conductors-#12 AWG

Safety

Safety is always a primary factor for laboratory activities. This lab should only be completed with supervision.

You may complete certain steps of the "Procedure" and the "Analysis" for better understanding without applying voltage.

Be sure to wear eye protection while electric motors are in operation. Also, be very careful when working with high voltages.

DOI: 10.1201/9788743810124-63

Procedure

1. A control panel for forward and reverse control of a three-phase AC motor is shown in Figure 59.1. This is a pictorial diagram for this laboratory activity.

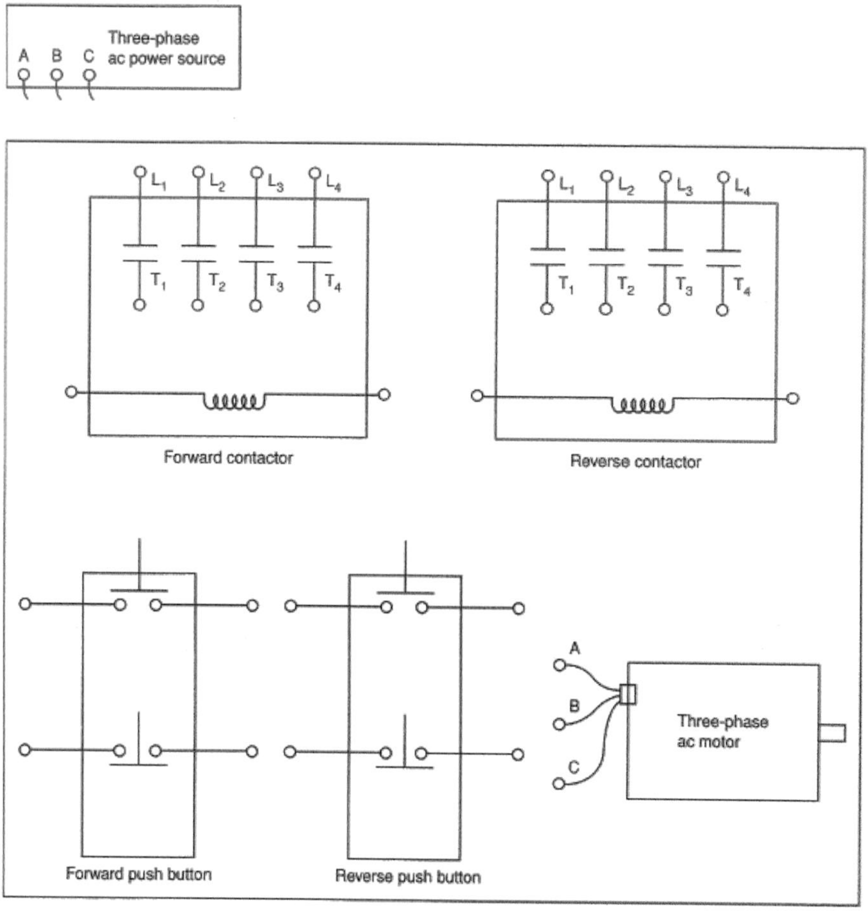

Figure 59.1 Control panel for forward and reverse control of a three-phase alternating current motor.

2. Make a sketch of the control diagram for forward/reverse/stop control of a three-phase AC motor.

3. Complete the wiring diagram of the pictorial above to correspond to your control diagram. Use Figure 59.1.

4. Wire the circuit of Figure 59.1 to accomplish forward and reverse control of a three-phase AC motor.

5. Have your completed circuit checked for proper operation.

   Approval: ______________________________

Activity 60

Forward and Reverse Control Wiring for Direct Current Motors

Objectives

In this activity, you will create control and wiring diagrams for a DC shunt-wound motor. You will also observe the forward and reverse operation of a DC motor.

Equipment and Materials

- DC power source
- DC shunt-wound motor
- Magnetic contactors (2)
- Start/stop push-button station
- Conductors-#12 AWG

Safety

Safety is always a primary factor for laboratory activities. This lab should only be completed with supervision.

You may complete certain steps of the "Procedure" and the "Analysis" for better understanding without applying voltage.

Be sure to wear eye protection while electric motors are in operation. Also, be very careful when working with high voltages.

DOI: 10.1201/9788743810124-64

Procedure

1. A control panel for forward and reverse control of a direct current shunt-wound motor is shown in Figure 60.1. This is a pictorial diagram for this laboratory activity.

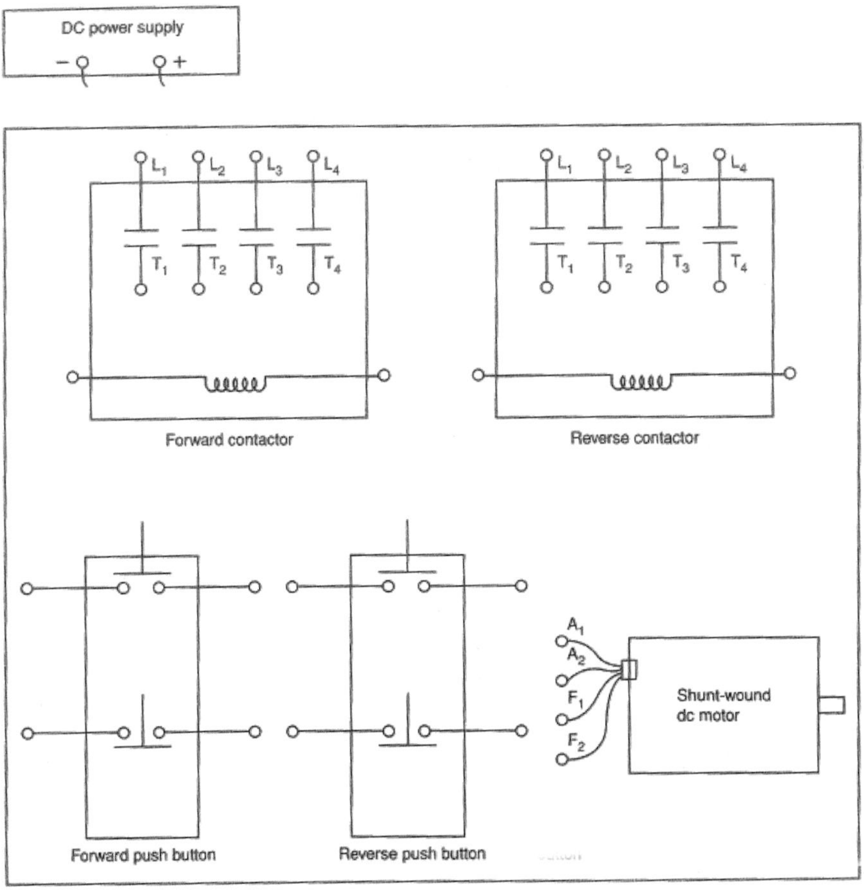

Figure 60.1 Control panel for forward and reverse control of a direct current motor.

2. Make a sketch of the control diagram for forward/reverse/stop control of a DC shunt-wound motor.

3. Complete the wiring diagram of the pictorial to correspond to your control diagram. Use Figure 60.1.

4. Wire the circuit of Figure 60.1 to accomplish forward and reverse control of a direct current (DC) motor.

   Have your completed circuit checked for proper operation.

   Approval: ______________________________

Activity 61

Dynamic Braking

Objectives

When a motor is turned off, its shaft will continue to rotate for a short period of time. This continued rotation is undesirable for many industrial applications. Dynamic braking is a method used to bring a motor to a quick stop when power is turned off. Motors with wound armatures utilize a resistance connected across the armature as a dynamic braking method. When power is turned off, the resistance is connected across the armature. This causes the armature to act as a loaded generator, making the motor slow down immediately.

In this activity, you will use simple dynamic braking methods to stop a single-phase AC induction motor and a DC shunt motor.

Equipment and Materials

- Single-phase AC induction motor (120 volts)
- DC shunt motor
- Double-pole, double-throw switch
- Resistor – 10 Ohm, 20 Watt

Safety

Safety is always a primary factor for laboratory activities. This lab should only be completed with supervision.

You may complete certain steps of the "Procedure" and the "Analysis" for better understanding without applying voltage.

DOI: 10.1201/9788743810124-65

Be sure to wear eye protection while electric motors are in operation. Also, be very careful when working with high voltages.

Procedure

1. Construct the dynamic brake for a single-phase AC induction motor as shown in Figure 61.1.

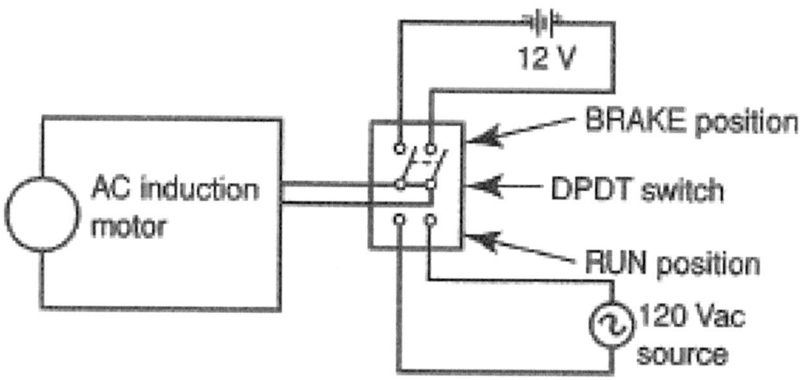

Figure 61.1 Dynamic braking circuit for a single-phase AC induction motor.

2. Place the DPDT switch in the RUN position, applying 120 volts AC to the motor.

3. Open the switch (center position) and observe the time required for the motor shaft to stop rotating.

   Estimated time for the motor to stop = _____ seconds.

4. Once again, place the DPDT switch in the RUN position. Now, place the switch in the BRAKE position while observing the time required to stop the motor.

   Estimated time for the motor to stop = _____ seconds.

5. This concludes the dynamic braking procedure for a single-phase induction motor. Turn off the power and disconnect the single-phase AC induction motor.

6. Construct the circuit shown in Figure 61.2 to demonstrate dynamic braking of a DC shunt motor.

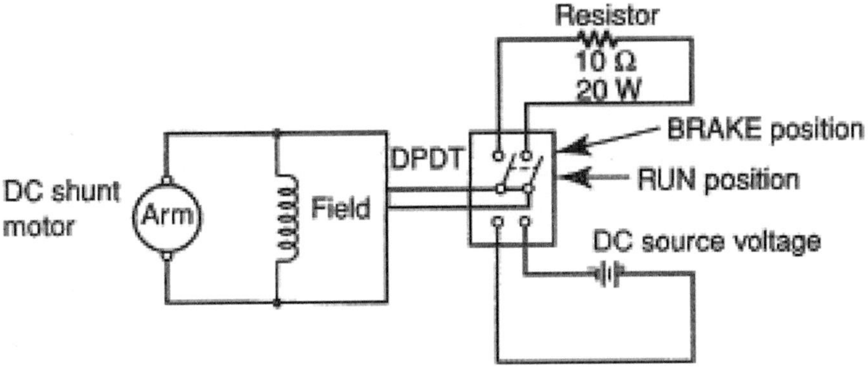

Figure 61.2 Dynamic braking circuit for a DC shunt motor.

7. Apply rated voltage to the DC motor by placing the DPDT switch in the RUN position.

8. Open the switch (center position) and observe the time required for the motor shaft to stop rotating.

   Time for motor to stop = _____ seconds.

9. Again, place the switch in the RUN position. Now, place the switch in the BRAKE position while observing the time required for the motor to stop.

   Time for motor to stop = _____ seconds.

10. Turn off the power and disassemble your circuit. Return all materials.

Analysis

1. What are some situations where dynamic braking of motors is needed?

2. Discuss the method used for dynamic braking of an induction motor.

3. Discuss the method used for the dynamic braking of a DC motor.

4. Draw a schematic diagram of a switching system that could be used for dynamic braking of a series-wound DC motor.

UNIT V

Electronic Control Basics

Activity 62

Industrial Electronic Symbols

Objectives

There are certain electronic symbols that are fundamental to industrial control systems. The majority of the symbols selected for this activity are used frequently in control circuits. You should be familiar with each symbol as a basis of understanding schematic diagrams.

In this activity, you should review the representative symbols, Figure 62.1, and demonstrate your understanding by completing the analysis.

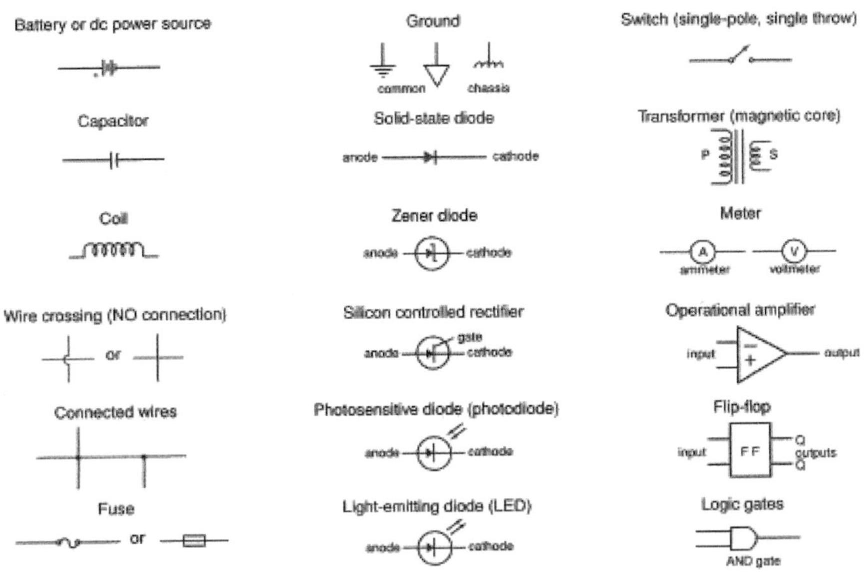

DOI: 10.1201/9788743810124-67

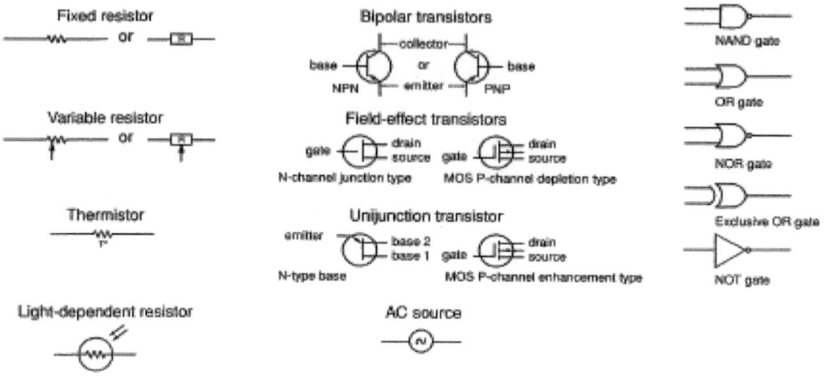

Figure 62.1 Frequently used industrial electronic symbols.

Analysis

Without looking at the symbols in Figure 62.1, identify each of the following symbols. Place the correct response in the space provided.

1. 

____

2.  ______________________________

3. 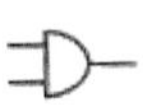 ______________________________

4.  ______________________________

5.  ______________________________

6.  ______________________________

7. 

____

8. 

9. 

10. 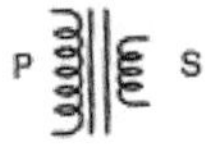

11. 

12. 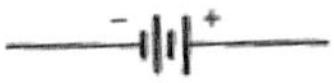

13. 

14. 

15. 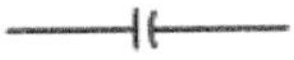

16. 

17. 

18. 

Activity 63

Characteristics of Diodes

Objectives

The I–V characteristics of a diode tell a great deal about its performance in an operating circuit. The data for an I–V curve can be determined by voltage and current measurements obtained from an operating circuit. This data can then be plotted on a graph and used to see how a diode will respond when it is placed in operation.

A number of important diode characteristics can be obtained from the data of an I–V curve. The characteristics are: the point where forward voltage causes an increase in forward current, the effect that reverse voltage has on reverse current, the influence that temperature has on forward and reverse conduction, and the conduction differences between germanium and silicon. The forward and reverse resistance of a diode can also be determined from I–V data. This information is very useful in predicting diode performance in a circuit.

In this activity, you will:

1. Determine the forward voltage, current, and resistance of a silicon and a germanium diode.
2. Determine the reverse voltage, current, and resistance of a silicon and a germanium diode.
3. Plot an I–V curve for a silicon and a germanium diode.

Equipment and Materials

- Multimeter
- Variable DC source

DOI: 10.1201/9788743810124-68

- Silicon diode, Texas Instruments – 1N4001
- Germanium diode, General Electric – 1N60
- Resistor – 1 K Ohm
- Circuit construction board

Procedure

1. Construct the circuit shown in Figure 63.1. The multimeter is used to measure different circuit voltages. It must be moved to different locations to observe these voltages.

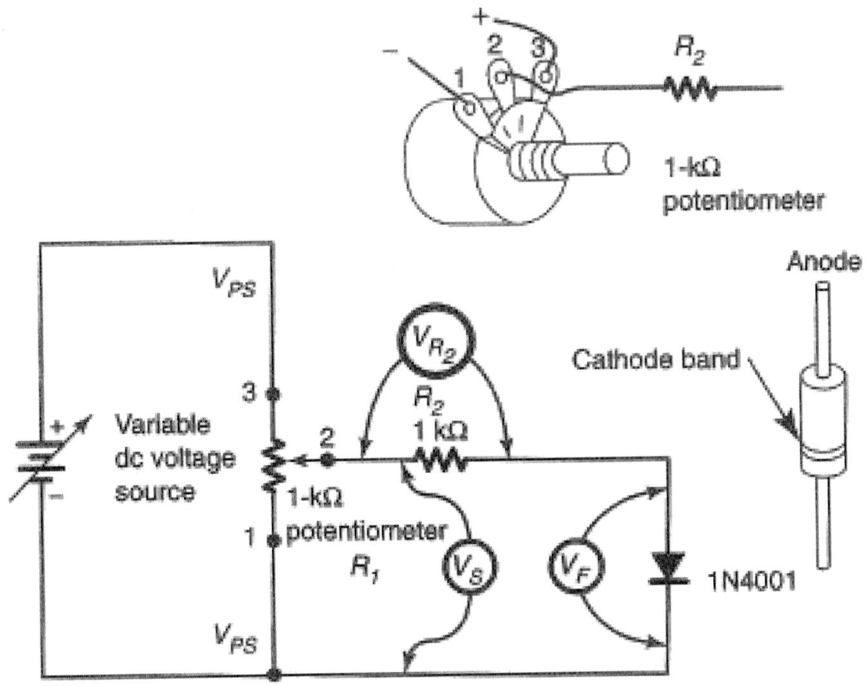

Figure 63.1 Diode characteristic testing circuit.

2. Turn on the DC voltage source and adjust it to produce 10 V, as measured across the 1 K Ohm potentiometer. This voltage will appear across the two outside terminals (1 and 3) of the potentiometer. The source voltage for the circuit is a variable value that is adjusted by the potentiometer and will appear across terminals 1 and 2 of the potentiometer. In this activity, you are directed to adjust the potentiometer to produce a specific voltage across the diode. You will then measure the voltage across the series resistor and circuit source voltage across terminals 1 and 2 of the potentiometer. Record these values in **Data Table 1** of Figure 63.2.

3. The first part of this experiment is designed to measure the forward characteristics of a diode. Note that only voltage values are called for in the procedure. Forward current and forward resistance are determined by calculation using Ohm's law.

4. Adjust source voltage to produce a V_F of 0 V across the diode. Measure and record the values of V_S and V_{R2}.

5. Adjust the value of V_S to produce a V_F of 0.1 V across the diode. Measure the values of V_S and V_{R2}. Record the measured values of V_S and V_{R2} in **Data Table 1.**

6. Repeat Step 3 for each of the values of the **Data Table 1**. Note that the voltage of a silicon diode is generally considered to be 0.7 V. In an operating circuit, this may vary from 0.5 to 0.7 V according to the amount of current flowing through the diode. As a rule, a large V_F will appear across a diode only when the value of I_F has been increased significantly. In this circuit, the value of I_F is kept at a rather low level. With some diodes, the upper value of V_F may not be obtained. If this occurs, use only the value that can be achieved by the circuit. Do not try to reach the upper values by increasing the source voltage.

7. Calculate the I_F and R_F values for each of the measured V_F values.

1N4001				
Data table 1				
Measured			Calculated	
V_F	V_{R_2}	V_S	I_F	R_F
0				
0.1				
0.2				
0.3				
0.4				
0.5				
0.6				
0.7				

1N4001				
Data table 2				
Measured			Calculated	
V_R	V_{R_2}	V_S	I_F	R_F
0				
2				
5				
7				
10				
12				
15				
20				

1N60				
Data table 3				
Measured			Calculated	
V_F	V_{R_2}	V_S	I_F	R_F
0				
0.1				
0.2				
0.3				
0.4				

1N60				
Data table 4				
Measured			Calculated	
V_R	V_{R_2}	V_S	I_F	R_F
0				
2				
5				
7				
10				
12				
15				
20				

Figure 63.2 Data tables.

8. Turn off the power source and remove the 1 N 4001 diode from the circuit. Reverse the two leads and return them to the circuit so that it is connected in the reverse-bias direction. The meter will be used to measure reverse voltage values across the diode, voltage across the resistor, and the voltage source. The reverse current and reverse resistance are calculated from these values.

9. Turn on the variable DC power source and adjust it to 20 V measured across terminals 1 and 3 of the potentiometer. Repeat Steps 2 through 5 for the reverse-connected diode. The values of V_R are indicated in **Data Table 2**. Note that these values are much larger than the previous values.

10. Turn off the power source.

11. Using the V_F and I_F data from **Data Table 1**, plot a forward conduction curve on the graph of Figure 63.3. The forward curve is located in the upper right quadrant of the graph.

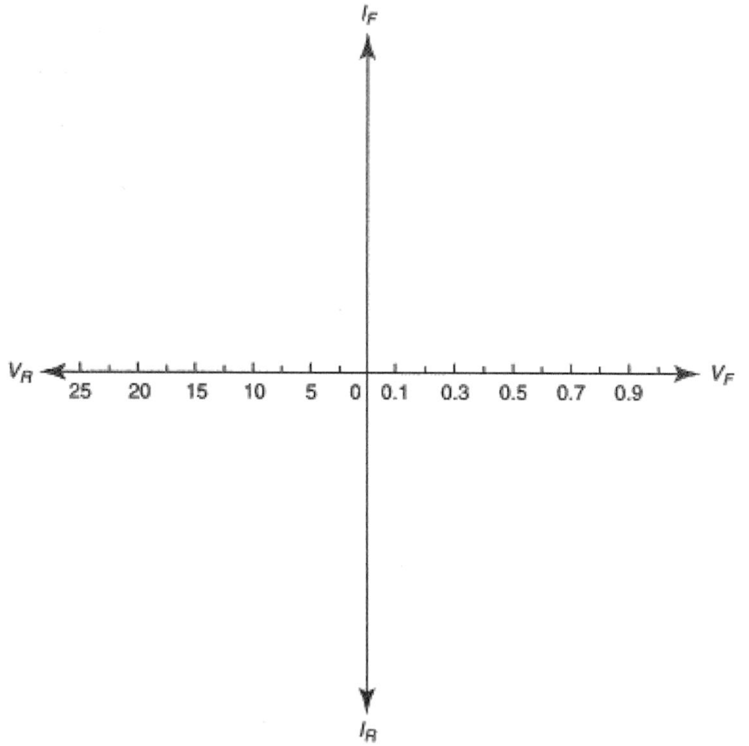

Figure 63.3 Chart of characteristics.

12. Using the V_R and I_R data from **Data Table 2**, plot a reverse conduction curve on the graph of Figure 63.3. The reverse curve is located in the lower left quadrant of the graph.

13. Repeat Step 1 through 12 for the 1 N 60 germanium diode. In this procedure, the value of the 1N60 diode should not be adjusted to values in excess of 0.4 V. Fill in the appropriate measurements and calculations into **Data Tables 3 and 4**. When plotting the forward and the reverse curves on the graph of Figure 63.3, use dots or dashed lines to distinguish 1N60 data from 1N4001.

14. Turn off the power to your circuit. Return all materials.

Analysis

1. Explain what the forward characteristics of a silicon diode tell about its operation.

2. Explain what the reverse characteristics of a silicon diode tell about its operation.

Activity 64

Zener Diode Testing

Objectives

An ohmmeter can be used to identify the leads of a Zener diode and to detect whether it is shorted or open. The ohmmeter, however, does not determine how a Zener diode will respond when voltage is applied. A simple test procedure can be performed to evaluate this condition of operation using a variable DC source, a series resistor, and a Zener diode. Its Zener voltage can be determined by altering the variable source voltage while observing the voltage across the Zener. The voltage will remain constant across the Zener when the V_Z value is reached. This will show the tolerance, the V_Z, and the knee voltage of the device. This is a significant characteristic when the Zener is reverse-biased.

In this activity, you will:

1. Demonstrate a practical method of testing a Zener diode with a voltage source.
2. Evaluate the V_Z of a Zener diode.
3. Connect a Zener diode in a circuit for normal operation.

Equipment and Materials

- Multimeter
- Variable DC source
- Zener diode – 5.1 Volt, 500 mW (Motorola 1N5231, 1N751 or equivalent)

DOI: 10.1201/9788743810124-69

- Resistor – 100 Ohm, ¼ Watt
- Potentiometer – 1K Ohm, 1 Watt
- Circuit construction board

Procedure

1. Prepare the multimeter to measure resistance on its R × 1 or lowest resistance range. For a digital meter, use the 2 K Ohm or an equivalent range.
2. Connect the two leads of the meter across the diode. A good Zener diode will show low resistance when it is forward biased and infinite resistance when reverse biased. Identify the anode and cathode of the diode with the meter.
3. Make a sketch of the identified leads of the device.

4. Construct the circuit shown in Figure 64.1. The potentiometer is adjusted to produce different values of input voltage. Adjust the DC power source to produce 12 V. Connect this voltage across the potentiometer as indicated.

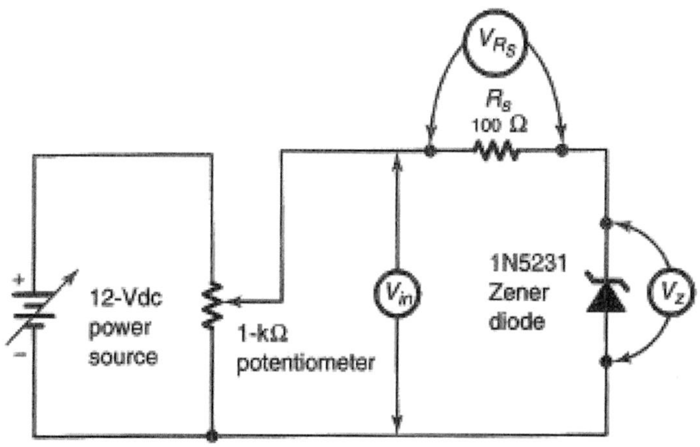

Figure 64.1 Zener diode test circuit.

5. Adjust the potentiometer to produce 0 V of input voltage for the circuit. This voltage is labeled DC input in the chart of Figure 64.2.

6. Measure the voltage that appears across the Zener diode. Record this value in the chart.

7. Adjust the potentiometer to produce 1 V of input voltage. Measure the V_Z and record its value in the chart.

8. Repeat this procedure for each of the indicated input voltage values.

9. The voltage across the series resistor of the circuit is the difference between the DC input voltage and V_Z. For a few of the input voltage values around 5 V, calculate the V_{RS}. These voltage values can also be measured with the multimeter for the appropriate input voltage value.

10. The Zener current of the circuit can be calculated using Ohm's Law.

11. Reverse the Zener diode in the circuit of Figure 64.1. Adjust the potentiometer to produce an input voltage of 0 V. Measure and record the diode forward voltage and the series resistor voltage. Adjust the input voltage across the potentiometer to 0.1 V. Again, measure and record the V_F and V_{RS}.

DC V_{in}	V_Z	V_{R_S}	I_Z
0			
1			
2			
3			
4			
5			
6			
7			
8			
9			
10			
11			
12			

DC V_{in}	V_F	V_{R_S}	I_F
0			
0.1			
0.2			
0.3			
0.4			
0.5			
0.6			
0.7			
0.8			
0.9			
1.0			

Figure 64.2 Data tables.

12. Repeat the procedure for each of the DC input values in the chart on the right side of Figure 64.2. Note that the input voltage does not exceed a value in excess of 1 V. This diode conducts very heavily in the forward direction. This should be noted by the observed voltage values. An increase in the value of V_F indicated a corresponding increase in I_F.

13. Calculate the forward current for the Zener diode for the measured values of the chart. Record the calculated values in the chart.

14. Turn off the power source and disconnect the circuit. Return all materials.

Analysis

1. Explain what causes a Zener diode to maintain its V_Z at a constant value when the source voltage is increased.

2. Explain why the power dissipation rating of a Zener diode is an important selection characteristic.

Activity 65

Characteristics of Transistors

Objectives

Any material can be electrically classified as a conductor, semiconductor, or insulator. Conductors are materials, such as copper or aluminum, through which an electric current will easily flow. Insulators are materials like rubber and glass, through which an electric current will not easily flow. Semiconductors are materials or devices classified somewhere between conductors and insulators because current through them can be controlled. This is achieved through controlling the resistance of semiconductors by using a bias current. When a semiconductor is allowed to conduct at its maximum, it exhibits the characteristics of a conductor. When a semiconductor is biased to not allow conduction, it exhibits the characteristics of an insulator. A semiconductor can be controlled to allow conduction between these two extremes.

The most common of all semiconductor devices is the transistor. Transistors are classified as NPN or PNP. These devices have three external leads: emitter, collector, and base. Figure 65.1 shows the electrical symbols, materials, and lead placement of the transistor.

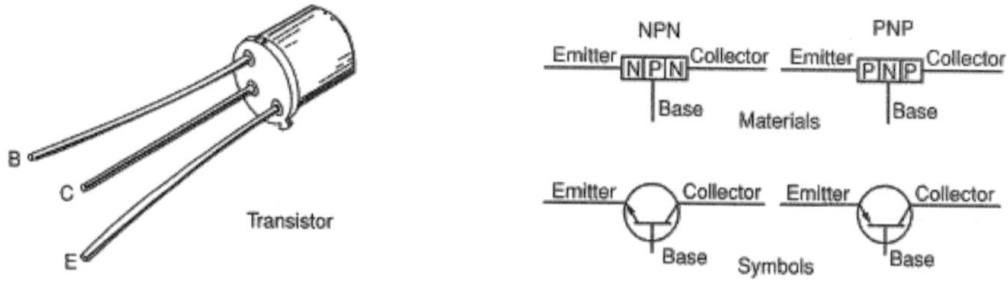

Figure 65.1 Transistor lead placement, symbols, and materials.

DOI: 10.1201/9788743810124-70

Transistors are used as AC or DC amplifiers and can be connected into circuits in one of three configurations: common-emitter, common-collector, or common-base. These three configurations are shown in Figure 65.2.

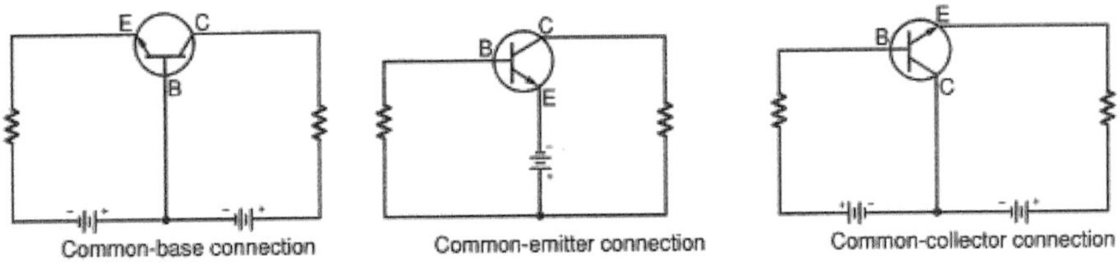

Figure 65.2 Three transistor circuits.

In this activity, you will observe the DC characteristics of an NPN transistor while it is connected in the common-emitter configuration.

Equipment and Materials

- Multimeter
- Variable DC power supply
- SPST switches (2)
- NPN transistor – 2N2405
- Resistors
- Circuit construction board and connecting wires

Procedure

1. With the multimeter adjusted to measure resistance in the R × 1 ohms range, measure and record the resistance between the leads of the 2N2405 NPN transistor. Note: Be sure to observe the meter polarity as indicated in the following illustrations.

a. When the emitter is negative and the base is positive, their forward resistance = ________ Ohms.

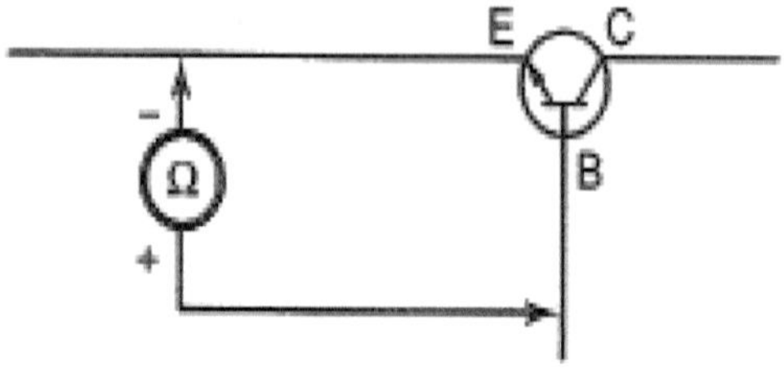

b. When the emitter is positive and the base is negative, their reverse resistance = ________ Ohms.

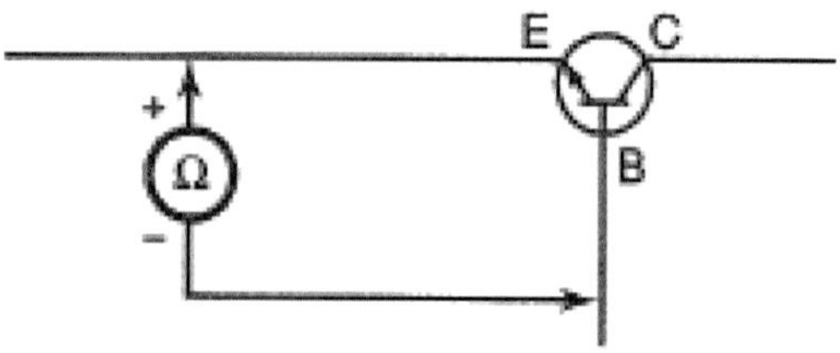

c. When the collector is negative and the base is (+), their forward resistance = ________ Ohms.

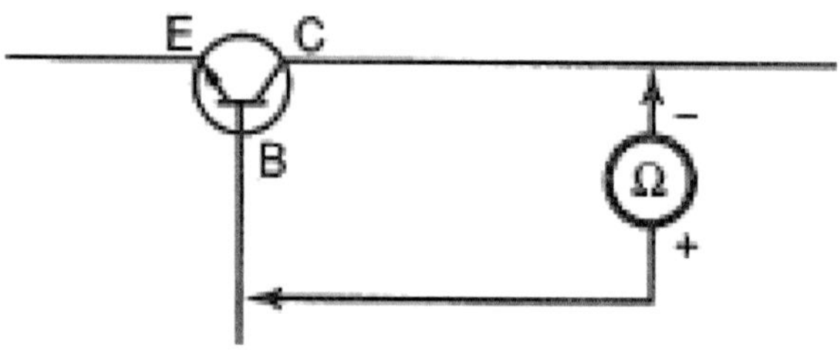

d. When the collector is positive and the base is negative, the reverse resistance = ________ Ohms.

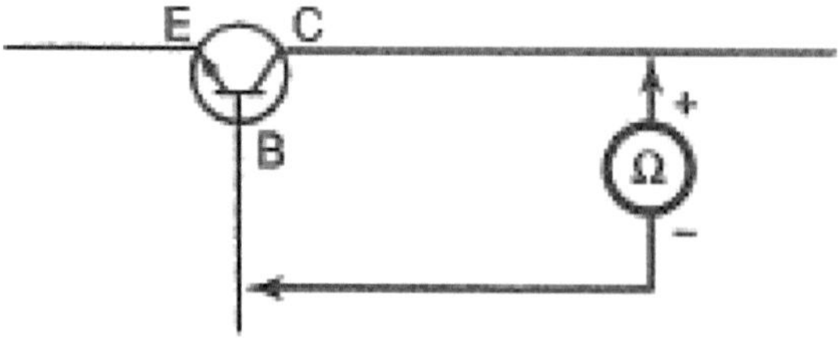

e. When the emitter is negative and the collector is positive, the resistance = ________ Ohms.

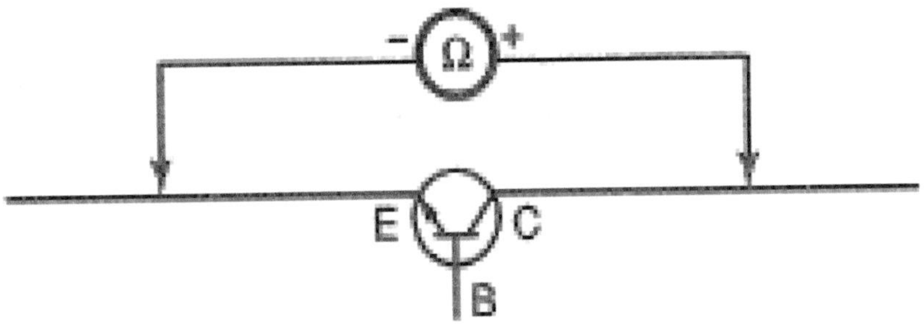

f. When the emitter is positive and the collector is negative, their resistance = _________ Ohms.

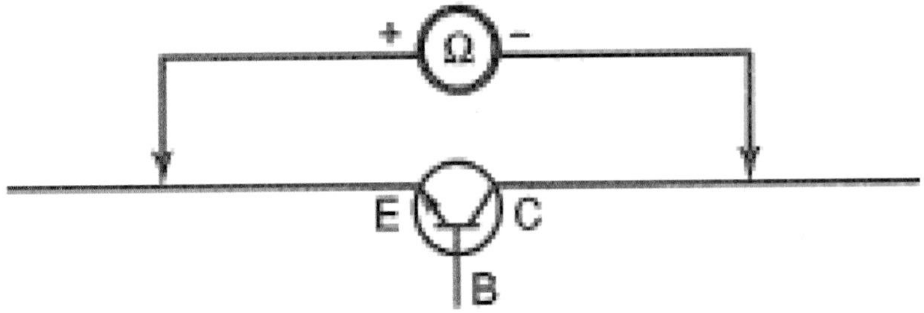

2. How did the forward resistance of the emitter-base in Step 1a compare with the reverse resistance in Step 1b?

3. How did the forward resistance of the collector-base in Step 1c compare with the reverse resistance in Step 1d?

4. How did the resistance of the emitter-collector in Step 1e compare with the resistance in Step 1f?

5. If the resistances of an identical PNP transistor had been measured, how would these compare with the resistances of the NPN transistor?

6. How would the ohmmeter polarity differ when measuring the resistance of the PNP transistor?

7. Construct the circuit in Figure 65.3. Note: The meter should be adjusted to measure direct current in the 100 mA range.

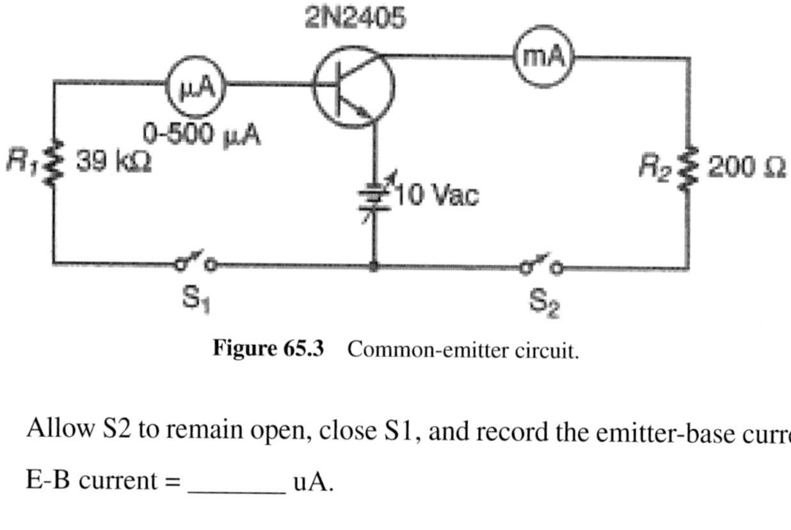

Figure 65.3 Common-emitter circuit.

8. Allow S2 to remain open, close S1, and record the emitter-base current.

   E-B current = ________ uA.

9. Open S1, close S2, and record the emitter-collector current.

   E-C current = ______ mA.

10. Close both S1 and S2. Record both the emitter-base and emitter-collector currents in the space below.

    E-B current = ________ uA.

    E-C current = ________ mA.

11. How did the E-B and E-C currents in Steps 8 and 9 compare with those in Step 10?

12. With both S_1 and S_2 closed, connect the proper values of R_1 into the circuit illus-trated in Step 7 and complete Figure 65.4.

Value of R_1	Measured base current	Measured collector current	Computed emitter current	Computed R_2 voltage	Computed emitter-collector voltage
27 kΩ					
33 kΩ					
39 kΩ					
51 kΩ					
68 kΩ					
100 kΩ					
200 kΩ					

Figure 65.4 Effects of changing.

13. Using the data from the chart completed in Step 12, complete the following statements:

   a. As the value of R1 increases, the base current ___________.

   b. As the base current decreases, the collector current ___________.

   c. As the base current decreases, the emitter current ___________.

   d. The _______ current controls the amount of ______ and ______ current.

   e. As the base current decreases, the voltage across R2 _________.

   f. As the base current decreases, the emitter-collector voltage ___________.

   g. As the voltage across R2 decreases, the emitter-collector voltage _________.

   h. The sum of the base and collector currents equals the ________ current.

   i. The ________ current controls the ________ current, which controls the __________ and _________ voltages.

14. Why is the transistor used as an amplifier?

15. Turn off the power and disassemble your circuit. Return all materials.

Analysis

1. What is a semiconductor?

2. What are two classifications of transistors?

3. In the space below, draw the electrical symbols for the PNP and NPN transistors.

4. What are the three external leads of a transistor?

5. How do the forward and reverse resistances of the emitter-base areas compare?

6. What three circuit configurations are most commonly associated with transistors?

7. What is the relation between base current and collector current when a transistor is connected in the common-emitter configuration?

8. What is the relation between base current and collector-emitter voltage when a transistor is connected in the common-emitter configuration?

Activity 66

Bipolar Transistor Testing

Objectives

The internal circuitry of an ohmmeter has a source voltage connected to its leads when measuring resistance. This voltage can be used to bias a transistor when it is connected to the ohmmeter. A bipolar transistor has two PN or NP junctions in its construction. When these junctions are connected together, an NPN or PNP type of transistor is formed. A very important step in transistor testing is the evaluation of the two junctions. They must respond like diodes in order for a transistor to be good. This means that forward biasing will cause low resistance, and reverse biasing will show infinite or high resistance. The two outside leads of a transistor will show infinite or high resistance to either method of biasing.

A good bipolar transistor must show low resistance when its emitter-base and base-collector are forward-biased. Reversing the polarity of the ohmmeter will cause the same two junctions to have an extremely high resistance. The emitter-collector junctions must show high resistance to either ohmmeter polarity if the device is good. If it does not, the transistor has leakage current. Silicon transistors generally have no leakage, while germanium transistors show some leakage. Leakage current is temperature-dependent in bipolar transistors.

In this activity, you will:

1. Test the status of a bipolar transistor with an ohmmeter.
2. Identify the polarity of the construction material of a bipolar transistor with an ohmmeter.
3. Identify the leads of a bipolar transistor with an ohmmeter.
4. Test the gain capabilities of a bipolar transistor with an ohmmeter.

DOI: 10.1201/9788743810124-71

Equipment and Materials

- Multimeter
- Transistor (Motorola MPSA20 or equivalent)
- Transistor (Motorola MPSA70, 2N2431 or equivalent)
- Resistor – 10 K Ohm

Procedure

1. Determine the lead polarity of the meter by placing it in the ohm position and measuring the voltage using a second voltmeter. If a second voltmeter is not available, determine the polarity of the meter from the instrument's operational manual.

   The polarity of the meter is (straight, reversed).

2. In the first testing operation, we will determine if the transistor is NPN or PNP. The polarity of the meter is very important in this procedure. Set the range switch of the meter to the R × 100 or R × 1K position. If a digital meter is used, set the range-select switch to a similar range.

3. Select the MPSA70 transistor or equivalent for this test procedure. Look at the three leads of the transistor. Notice if there are any identification marks or unusual lead placement for this transistor. Make a sketch of the lead placement of the transistor in the outer margin. Show the bottom view.

4. Pick up the center lead of the transistor and assume that it is the base. Connect the negative lead of the meter to it. If the lead is actually the base of the transistor, it will show a low resistance reading between each of the other two leads for a PNP transistor.

5. If the resistance in Step 4 is high in both directions, reverse the meter leads so that the positive lead goes to the assumed base. Once again, test the transistor by connecting the negative lead alternately between the

two outermost leads. If a low resistance reading is obtained between the center and outside leads, the transistor is an NPN type.

6. If using the center lead does not produce a low resistance between the two outside leads, it is not the base terminal. Select one of the outside leads and assume that it is the base. Try the same test procedure again. Note that a shorted transistor will more than likely show low resistance in either polarity direction. It is rather difficult to determine the polarity of a shorted transistor.

7. After you have determined the polarity of a transistor to be either NPN or PNP by this procedure, determine if the transistor has any leakage. This test is applied to the two outside leads or to the two leads that do not include the assumed base. These leads will be identified as the emitter and collector. In general, there should be no leakage current or resistance in either direction where the meter is connected. Test the emitter to collector for leakage. Record the leakage resistance.

   Leakage resistance is __________.

8. We have assigned the names emitter, base, and collector to the leads of the transistor. We must now verify if the lead assignment is correct. To perform this test, assume that the base has been identified and that the transistor is of the PNP type. Select one lead and assume that it is the emitter and that the other lead is the collector. Connect the positive lead of the meter to the assumed emitter and the negative lead to the assumed collector. Then connect a resistor between the collector and the base. If the lead assignments are correct, the meter will show a substantial change in resistance. This is an indication of transistor gain or beta. If there is no change in resistance, the lead assignments for the emitter and collector have been reversed. Switch the meter leads and again connect the resistor between the newly assigned collector (negative meter lead) and the base. This should produce a change in resistance, indicating current gain or beta. If the meter responds in this manner, the lead assignments of the emitter, base, and collector are correct.

9. Make a bottom-view drawing of the transistor in the outer margin and identify the location of the leads.

10. Repeat the test procedure for the MPSA20 or equivalent transistor. The drawing, type of transistor, and lead identification should be placed in the space that follows.

11. Turn off your meter and return all materials.

Analysis

1. Explain how an ohmmeter can be used to determine if a transistor is an NPN or a PNP type.

2. How is the leakage of a transistor evaluated with an ohmmeter?

Activity 67

Field-effect Transistor Testing

Objectives

The internal circuitry of an ohmmeter has a voltage source connected to its lead when it is used to measure resistance. This voltage can be used to bias a field-effect transistor. The status of the terminals, the condition of the device as good or bad, and to some extent, the leads can be identified with an ohmmeter. An ohmmeter is a quick and easy method of field-effect transistor evaluation. This method of evaluation is rather commonly used in electronic troubleshooting procedures.

In this activity, you will:

1. Test the status of a field-effect transistor with an ohmmeter.
2. Identify the polarity of the construction materials of a field-effect transistor.
3. Identify the leads of a field-effect transistor.

Equipment and Materials

- Multimeter
- N-channel JFET (Motorola 2N5459 or equivalent)
- N-channel MOSFET (Motorola 2N3796 or equivalent)
- Resistor – 100 K Ohm

DOI: 10.1201/9788743810124-72

Procedure

1. Determine the lead polarity of the meter by placing it in the ohms position and measuring the probe voltage with a voltmeter. Measure the voltage on the R × 1K or R × 2K Ohm or equivalent range. If the common lead is negative and the probe lead is positive, the meter has forward polarity. An opposite polarity indicates reverse meter polarity.

   Record the voltage of the meter = __________ V.

   The meter used for this test has (forward, reverse) polarity.

2. Select the 2N5459 JFET or equivalent. Make a sketch of the bottom view of the transistor lead layout.

3. First, determine the source (S) and the drain (D) leads of the transistor by connecting the multimeter, set to measure resistance, between two randomly selected leads. If the two selected leads indicate a resistance, reverse the meter probes. Two leads that show the same resistance in either direction of polarity are the source and drain. It may be necessary to try several different two lead combinations in order to determine the source and drain of the JFET. When the leads have been identified, indicate your findings on the JFET sketch of Step 2.

   The measured source-drain resistance is _________ Ohms.

4. The gate-channel of a JFET responds as a diode. This can be observed by connecting the meter between two leads and then reversing the meter leads. The gate-source or gate-drain will respond as a diode by showing low forward resistance and high or infinite reverse resistance.

5. If the gate shows low resistance when the positive probe is connected to the gate and the negative to either the source or drain, the JFET is an N-channel device. A P-channel device will produce low resistance when the gate is negative and the source or drain is positive. The resistance ratio between forward and reverse connections should be 1:100 for a device to be good.

6. An approximate test of the JFET's ability to amplify is displayed by connecting the meter to the source and drain. The polarity of the meter is not significant. Touch the gate of the JFET with your finger. It should cause a decrease in resistance.

7. A more significant resistance change occurs when a 100 K Ohm resistor is connected between the gate and either the source or drain. Very little or no change in resistance during this test indicates an open device or improper lead identification.

8. Using the same procedure outlined in Steps 2 through 6, test the 2N3796 MOSFET or equivalent. This device has a fourth lead that is connected to the substrate. The substrate and the source or drain leads will respond as a diode. The gate of a MOSFET will show an infinite resistance between all leads in either direction of meter polarity.

9. Turn off your meter and return all materials.

Analysis

1. Explain how an ohmmeter can be used to evaluate the status of a JFET.

2. How would an ohmmeter test distinguish between an N-channel and a P-channel JFET?

3. What are some of the differences that an ohmmeter will show between a JFET and a MOSFET?

Activity 68

MOSFET Characteristics

Objectives

In this experiment, you will develop a partial family of characteristic curves for a MOSFET. The curves are derived by setting the gate voltage to a specific value and adjusting the source-drain voltage to specific values. The drain resistor voltage is then measured for each value and used to calculate the drain current. These values are then used to plot a characteristic curve for the MOSFET and to plot representative curves. The polarity of the gate voltage is then reversed to show how the MOSFET will respond with a different polarity of gate voltage. This curve is also plotted.

The developed family of curves of the MOSFET is then used to show the operational characteristics of the device in a representative circuit. A load line is plotted for the circuit. An operating point is established according to the circuit's data, and characteristic values of drain current and source-drain voltage are determined from the curves. These values are used to calculate transconductance and voltage gain. This procedure shows how the amplifier circuit responds graphically.

In this activity, you will:

1. Demonstrate the operating characteristics of a MOSFET by measuring some representative source-drain voltages and calculating the drain current.
2. Plot a partial family of characteristic curves for a MOSFET.
3. Determine a load line for a representative circuit and plot it on the partial characteristic curves.
4. Evaluate the operation of a MOSFET circuit with a family of curves and determine voltage gain, transconductance, and amplifier phase relationships.

DOI: 10.1201/9788743810124-73

Equipment and Materials

- Multimeter
- Dual-section variable power supply – (or two 9 V batteries)
- N-channel MOSFET (Motorola 2N3796 or equivalent)
- Resistors – 100, 1 K, and 4.7 K Ohm
- Potentiometers – 1 K Ohm and 100 K Ohm
- Circuit construction board and connecting wires

Procedure

Part A: Characteristic curve data collection

1. Construct the MOSFET characteristic curve test circuit of Figure 68.1. The leads or pin-out of the MOSFET are shown in the diagram.

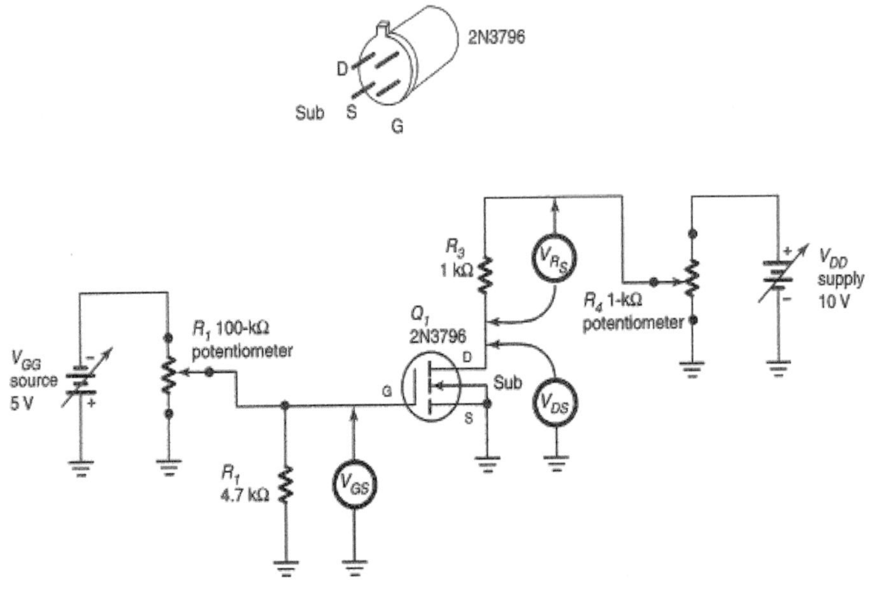

Figure 68.1 MOSFET characteristic curve test circuit.

2. With circuit construction complete, remove the MOSFET shorting ring or conducting foam.

3. Adjust the two power supply voltage controls to the 0 V position and turn on the power supply. Adjust the supply to produce 10 V. This is measured across the two outside leads of the potentiometer. Adjust the supply to produce 5 V. This is measured across the two outside leads of the potentiometer. Potentiometers R1 and R4 are used to change the value of the gate-source voltage and the source-drain voltage of the MOSFET.

4. Voltage measured across the drain resistor is used to determine the drain current by calculation. This calculation will be made for each voltage value and recorded in the data column of the chart as I_D.

5. Voltage is measured across the gate-source resistor. This voltage is adjusted to a specified value and then adjusted to the values indicated in the chart of Figure 68.2.

	$V_{GS} = 0$ V		$V_{GS} = -1$ V		$V_{GS} = +1$ V	
V_{DS}	Measured V_{R_3}	Calculated I_D	Measured V_{R_3}	Calculated I_D	Measured V_{R_3}	Calculated I_D
0 V						
1 V						
2 V						
4 V						
6 V						
8 V						
10 V						

Figure 68.2 Chart of MOSFET values.

6. Connect the meter across the gate-source of the MOSFET and adjust R1 to produce zero volts. Connect the meter across the source-drain and adjust potentiometer R4 to produce zero volts. Measure and record the voltage across resistor R3. Calculate the drain current for this value. Adjust potentiometer to produce 1 V. Measure and record the voltage across resistor R3 for this setting. Calculate the drain current and record its value in the chart. Repeat this procedure for values of 2, 4, 6, 8, and 10 V.

7. Return the value of V_{DS} to 0 V.

8. Increase the value of $V_{G\text{-}S}$ to −1 V. Repeat the procedure outlined in Step 6. Record your data values in the appropriate column of the chart.

9. Return the two potentiometers to 0 V. Disconnect the negative lead of the power supply connected to potentiometer. Connect this potentiometer lead to the positive side of the $V_{D\text{-}D}$ power supply. Adjust the power supply to 10 V.

10. Adjust potentiometer R1 to produce + 1 $V_{G\text{-}S}$. Repeat the procedure outlined in Step 6 for different values of $V_{D\text{-}S}$ for 0 to 10 V. Record your data in the column of the chart of Figure 68.2.

11. Using the data collected in the chart of Figure 68.2, plot a partial set of curves for the MOSFET in Figure 68.3.

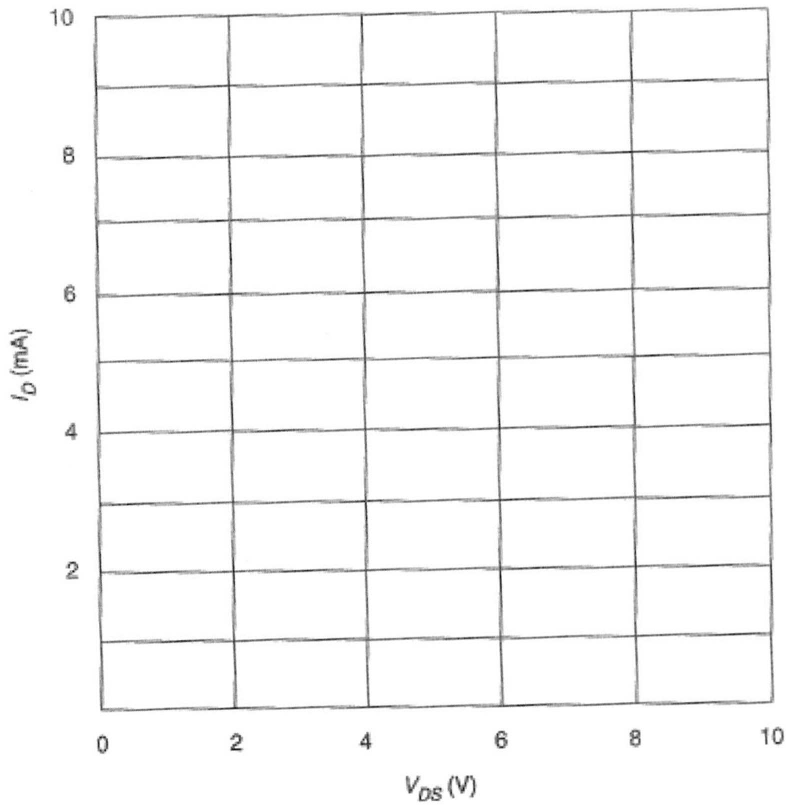

Figure 68.3 Graph of MOSFET values.

Part B: Load line analysis

1. Plot a load line on the partial set of curves of Figure 68.3 for the circuit of Figure 68.4.

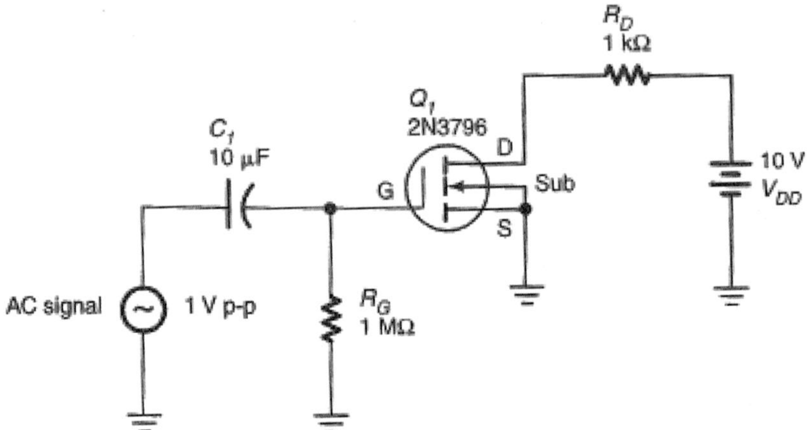

Figure 68.4 MOSFET circuit for load line analysis.

2. Resistor R_D serves as the load, and $V_{D\text{-}D}$ serves as the source. The two conditions of operation are full conduction and non-conduction.

3. The operating point of this circuit is zero, based on the value of $V_{G\text{-}S}$. Mark the Q point on the load line.

4. Project lines from the Q point to the I_D scale and to the $V_{D\text{-}S}$ scale of the graph.

5. With the indicated AC signal, the total change in $V_{G\text{-}S}$ is ________ peak to peak. The resulting output from this input is ________ Vp-p. Note the polarity of $V_{D\text{-}S}$ compared with $V_{G\text{-}S}$. What does this indicate?

6. The resulting I_D that occurs for this input is ________ mA p-p. How does the polarity of I_D compare with $V_{D\text{-}S}$?

7. Using the value of the indicated input and the output voltage, calculate the voltage gain for the circuit.

8. Transconductance is a measure of the effectiveness of this device to amplify a signal. Calculate the transconductance of the amplifier circuit using your data.

   Transconductance = ________ S.

9. Turn off the power and disassemble your circuit. Return all materials.

Analysis

1. Explain how an N-channel E-type MOSFET achieves amplification.

2. How would an increase in the resistance of the drain resistor or load change the operation of a MOSFET?

3. What are the operational differences between an E-type MOSFET and a JFET amplifier?

Activity 69

SCR Testing

Objectives

The internal circuitry of an ohmmeter has a voltage source connected to its leads when it is used to measure resistance. This voltage can be used to bias the junctions of an SCR. The good/bad status of the device and its leads can be identified through this process. The ohmmeter is a quick and easy method of evaluating an SCR. This method of evaluation is used widely in electronic troubleshooting procedures.

As a rule, an analog multimeter or hand-deflection meter works very well for this method of evaluation. A digital multimeter does not work as well for some of the measurements. Digital instruments generally have a very small current flow when measuring resistance. This can restrict the test procedure by not permitting the meter to trigger an SCR and have it latch or hold in its conduction range. All other test procedures are essentially the same.

In this activity, you will:

1. Test the status of a silicon-controlled rectifier with an ohmmeter.
2. Identify the leads of a silicon-controlled rectifier.

Equipment and Materials

- Multimeter
- Silicon-controlled rectifier (General Electric C106F or equivalent)
- Circuit construction board and connecting wires

DOI: 10.1201/9788743810124-74

Procedure

1. Prepare the multimeter to measure resistance in the low resistance range. Use

   R × 1 for analog meters or equivalent for a digital instrument.

2. Select any two SCR leads at random and test the forward and reverse resistance ratio. A two-lead combination that responds as a diode (low-resistance in one direction and infinite resistance in the other direction) is the cathode-gate. A two-lead combination that shows infinite or high resistance in both directions is either the anode–cathode or the gate-anode. Find the two lead combinations that respond as a diode. The third or alternate lead of this combination is the anode. For the two-lead diode combination, connect the meter to produce a low-resistance reading. This will show the cathode connected to the negative and the gate to the positive lead (this assumes that your meter has straight polarity).
3. Make a sketch of the SCR, indicating the lead assignments you have selected.

4. To verify the correctness of the lead assignments, connect the meter as shown in Figure 69.1. This should initially make an infinite resistance indication on the meter.

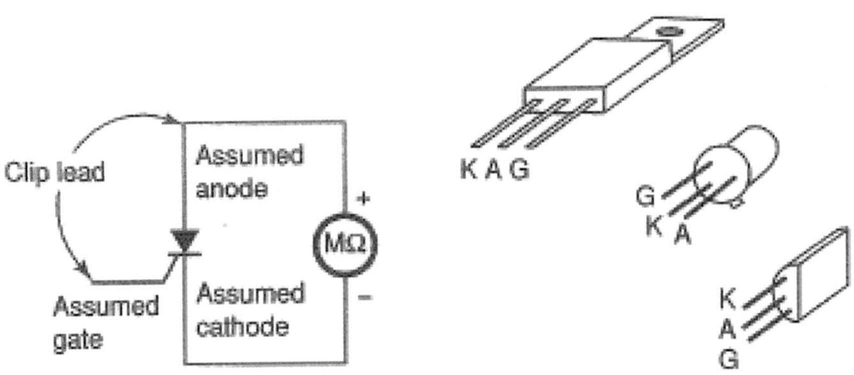

Figure 69.1 Lead assignments for testing an SCR.

5. Connect a wire between the anode and the gate. If the lead assignment is correct, the meter will change to a low resistance value. If the wire is disconnected, the meter should continue to show low resistance. This shows the latching or holding function of the SCR. If a digital meter is used, the SCR may not latch or hold in its conduction state. A standard analog meter will have enough current in the ohmmeter range to latch and hold the SCR into conduction. To alter the latching function, momentarily disconnect one side of the meter, then return the meter leads to the anode–cathode. This should show an infinite resistance with the gate lead disconnected. Connect the gate lead again to the anode to see the latching function. Try this test procedure several times to be certain of how it is performed.

6. This test procedure applies only to good SCRs. If a two-lead combination cannot be found that responds as a diode, the gate-cathode junction is open. If any two-lead combination shows low resistance in either direction of polarity, the junction or combined junctions are leaky or shorted. SCRs can become open or shortened in normal usage. As a rule, the SCR is rather free of trouble and responds as a very reliable device.

7. Turn off your meter and return all materials.

Analysis

1. Why does an analog multimeter work best in evaluating the status of an SCR?

2. Why does the gate-cathode of an SCR show low resistance in one direction and high resistance in the other direction of ohmmeter polarity?

3. Explain why the anode–cathode of a good SCR shows infinite resistance in either direction of ohmmeter polarity.

Activity 70

Triac Testing

Objectives

The internal circuitry of an ohmmeter has a voltage source connected to its leads when it is used to measure resistance. This voltage source can be used to bias a triac into conduction. The good/bad status of a triac can be tested, and the leads can be identified. An ohmmeter is a quick and easy method of evaluating the status of a triac.

An analog ohmmeter works best for this method of evaluation. This type of instrument generally has a larger value of current in its ohmmeter. The current value is somewhat critical when triggering a triac into conduction and getting it to latch. Digital meters, as a rule, have a very nominal current flow when the meter is used as an ohmmeter. This current value is generally not large enough to cause a triac to latch. All other tests being performed will respond equally with either type of meter.

The test procedure of a triac is primarily the same as that of an equivalent SCR. The triac, however, can be triggered in four different modes or conditions. Each lead has a dual-crystal polarity that is voltage selective. The polarity of the ohmmeter is used to select the appropriate combination of the test. Identification of the respective leads is somewhat more complex than the procedure used in the evaluation of the SCR.

In this activity, you will:

1. Test the status of a triac with an ohmmeter.
2. Identify the leads of a triac with an ohmmeter.

DOI: 10.1201/9788743810124-75

Equipment and Materials

- Multimeter
- Triac (RCA 2N5724 or equivalent)
- Circuit construction board and connecting wires

Procedure

1. If an analog meter is being used, prepare it to measure ohms in the R × 1 range. If a digital meter is used, prepare it to measure resistance in an equivalent range. The R × 1 range (lowest) is preferred because it usually provides the necessary holding current to latch this device into conduction. We will assume that the instrument being used has straight polarity with the negative connected to the black or common lead and the positive connected to the red or ohms lead.
2. Select any two of the three leads at random and determine the forward-to-reverse resistance ratio. Terminal 1 to Terminal 2 should indicate an infinite resistance or extremely high resistance in both directions. The gate and terminal 1 will show a rather low resistance in both directions. With the two leads that show low resistance in either direction identified, the third lead is terminal 2.
3. Make a sketch of the triac leads. Identify Terminal 2 in the sketch. Terminal 1 and the gate have not been identified at this time.
4. The next test procedure is based on a right or wrong selection opportunity of determining the correct lead assignment for Terminal 1 and the Gate. Connect the positive side of the meter to Terminal 2, then select one of the two remaining leads and connect the negative side of the meter to it. We will assume in this case that the negative meter lead is connected to Terminal 1. Now attach a connecting wire to the third lead, which we have assumed to be the Gate. Connect the other end

of this wire to Terminal 2. If the assumed leads are correct, the meter will show a low resistance, which indicates triggering. Momentarily disconnecting the Gate from Terminal 2 should cause the triac to stay in its conduction state. If this procedure does not produce a change in resistance, the assumed leads are reversed. Switch the negative meter lead from the first assumed lead to the assumed gate lead. Place the connecting wire to the newly assumed leads and try the triggering procedure again. If triggering is produced, the correct lead assignment has been made. Label these two leads on the diagram. If the triac cannot be triggered by this procedure, it may be faulty. Only a good triac will respond to this test.

5. With the lead assignments made, once again try the triggering procedure. Connect the negative lead to Terminal 1 and the positive lead to Terminal 2. Then attach a connecting wire between the Gate and Terminal 2. This should cause a resistance change in the meter and trigger the triac into conduction.
6. If the lead assignments are correct, the triac should also trigger in the reverse polarity direction. In this case, connect the negative side of the meter to Terminal 2 and the positive side to Terminal 1. Attach a connecting wire to the gate and momentarily connect it to Terminal 2. This action should also trigger the triac into conduction. If the triac is functioning properly, it can be triggered in either direction of polarity. Test the triac to see if it has this capability.
7. A digital meter will work for all test procedures except the latching function. The analog meter will latch the triac only when it is in the range of operation.
8. Turn off your meter and return all materials.

Analysis

1. What are some of the common things to look for in testing an SCR to see if it is faulty?

2. Why does the gate and T_1 of a triac show low resistance in either direction of ohmmeter polarity?

3. Why does and of a good triac show infinite resistance in either direction of ohmmeter polarity?

Activity 71

Universal Motor Speed Control

Objectives

An important advantage of a universal (AC/DC) motor is the ease of speed control. Since this type of motor is similar to a series-wound DC motor, it has many of the same characteristics. The universal motor has a brush/commutator assembly, with the armature circuit connected in series with the field windings. By varying the voltage applied to universal motors, their speed can be varied from zero to maximum.

In the activity, you will observe the speed control characteristics of a universal motor. The circuit used for this purpose will be a gate-controlled triac. The triac is a semiconductor device whose conduction can be varied by a trigger voltage applied to its gate. A silicon-controlled rectifier (SCR) could also be used as a speed control device for a universal motor. Speed control circuits like the one you will construct are used for many applications, such as electric drills, sewing machines, electric mixers, and several industrial applications.

A commercial light dimmer control can be substituted for the control circuit of this activity, if desired.

Equipment and Materials

- Multimeter
- Universal motor
- Resistors – 4.7 K, 47 K, 68 K Ohm
- Potentiometer – 500 K Ohm
- Capacitors (2) – 0.1 uF

DOI: 10.1201/9788743810124-76

- Diac-GE-X13 or equivalent
- Triac-GE-X12 or equivalent

Safety

Safety is always a primary factor for laboratory activities. This lab should only be completed with supervision.

You may complete certain steps of the "Procedure" and the "Analysis" for better understanding without applying voltage.

Procedure

1. Construct the universal motor speed control circuit shown in Figure 71.1.

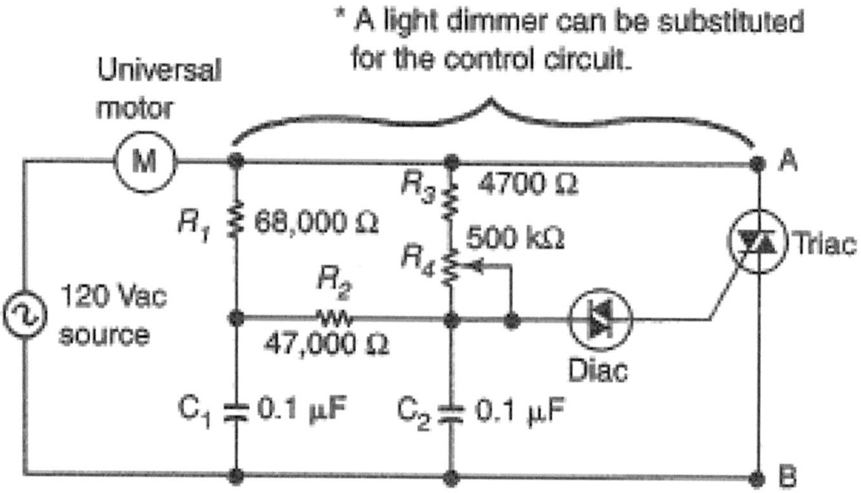

Figure 71.1 Speed control circuit for a universal motor.

2. Connect your multimeter, set to measure voltage, across the triac (point A to point B) to observe the voltage across the device.
3. Adjust the speed control potentiometer through its range to check the circuit for operation. The motor speed should vary from zero to rated speed.

4. Monitor the voltage across the triac as the potentiometer is adjusted. Observe the voltages across the triac and the motor at the voltage levels indicated in Figure 71.2. Record these voltages.

Voltage across motor	Voltage across triac
20 Vac	
40 Vac	
60 Vac	
80 Vac	
100 Vac	
120 Vac	

Figure 71.2 Voltage comparison of motor and triac circuit.

5. Monitor the voltage across the potentiometer. Record the maximum voltage as the potentiometer is varied and observe its effect on the speed of the motor.

   Maximum voltage __________ volts AC.

6. Monitor the voltage across capacitor 2. Record the minimum and maximum voltage across as the potentiometer is varied.

   Minimum voltage = __________ volts AC.

   Maximum voltage = __________ volts AC.

7. Turn off the power and disassemble your circuit. Return all materials.

Analysis

1. From the data in Figure 71.2, what is the relationship between the voltage across the motor and the voltage across the triac?

2. At what voltage across the motor does it begin to rotate?

3. How does the triac functionally control the speed of the universal motor?

4. From the data of Step 5, as potentiometer resistance increases, what happens to motor speed?

5. From the data of Step 6, as the voltage across capacitor 2 increases, what happens to motor speed?

6. Why does this type of control allow the universal motor to produce more torque through its entire operational range than a rheostat control that reduces the applied AC voltage?

Activity 72

Inverting Op-amps

Objectives

Applications of the op-amp are very common in control circuits. Integrated circuits of this type are used primarily as high-gain amplifiers. The versatility of this amplifier makes it very useful in automatic control systems and instrumentation applications. With no external feedback circuit, an op-amp can achieve an open-loop gain that is around 100,000. With a feedback circuit, gains can be achieved from 1 to 100,000 according to the selected resistance values.

In this activity, you will:

1. Construct a simple inverting op-amp, including both AC and DC signals.
2. Employ calculations that show voltage gain, current gain, and resistance value selection.
3. Design a simple operational amplifier that will achieve a specific amount of gain.

Equipment and Materials

- Multimeter
- Oscilloscope
- Split power supply (or two 9 V batteries)
- Variable low-voltage DC source
- AC signal generator

DOI: 10.1201/9788743810124-77

- Op-amp – uA741C or equivalent
- Resistors – 2K, 10K, 200K Ohm
- Capacitor – 4 uF, 50 Vdc

Procedure

1. Construct the inverting DC op-amp of Figure 72.1. The power source of the uA741C is derived from a split power supply of two 9 V batteries, Figure 72.2, or variable DC supplies. Figure 72.3 shows the top view of the IC package.

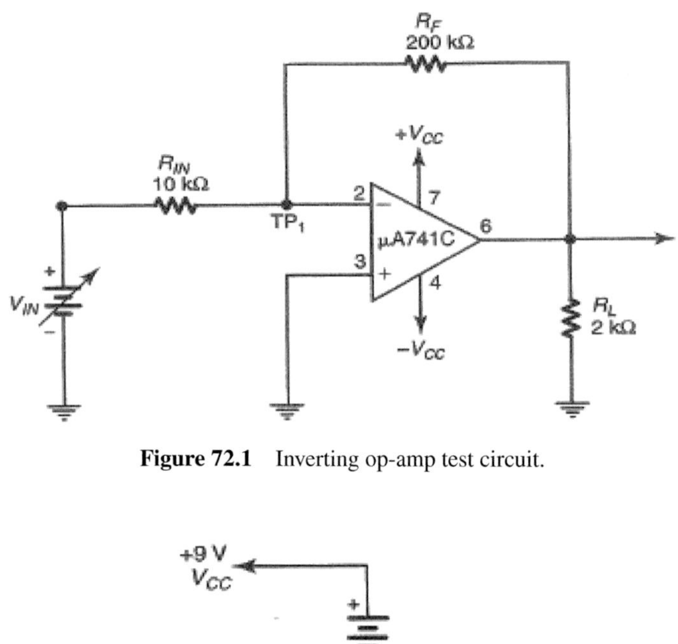

Figure 72.1 Inverting op-amp test circuit.

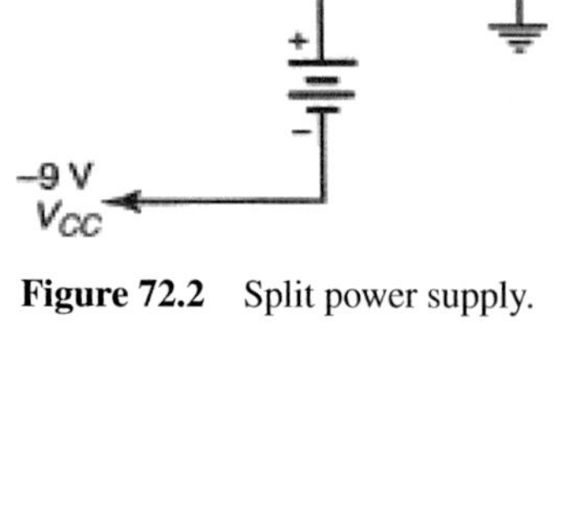

Figure 72.2 Split power supply.

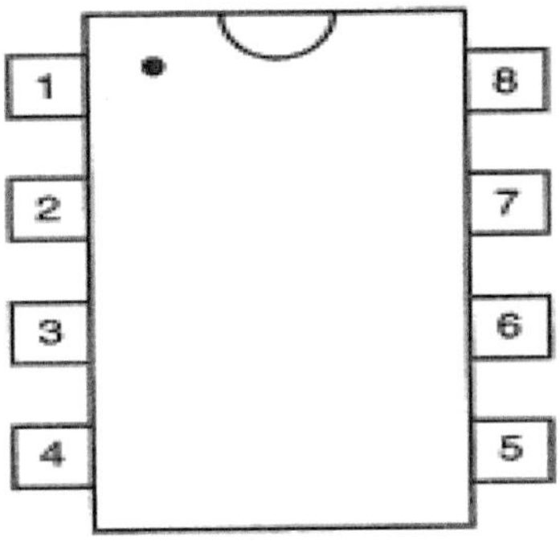

Figure 72.3 Pin layout of the op-amp IC package.

2. Adjust the DC input source to produce 0.1 Vdc.
3. Energize the IC by turning on or connecting the split power supply. Measure and record the output voltage across R_L.

   V_{RL} = ________ Vdc.

4. How does the polarity of the output compare with that of the input?
5. Carefully increase the voltage to 0.3 V. Measure and record the output.

   Output = ________ Vdc.

6. The DC voltage gain of an operational amplifier is determined by calculation. Use a negative sign of denotes the inversion function.
7. Calculate the gain factor for 0.3 V input.

   Voltage gain =________.

8. Test the measured value of Steps 3 and 5 with calculated values. How do they compare? How do you account for any differences between calculated and measured values?

9. Using the measured voltage across R_L in Step 5, calculate the load current.

   Calculated load current = _______ mA.

10. Measure the load current and record your findings.

    Measured load current = _______ mA.

11. How do the measured and calculated values compare?

12. Measure the input current at test point 1. Then calculate the input current. We will assume that the resistance of a regulated power supply is quite small and not applicable in this case. How does the measured value compare with the calculated value?

13. The current gain of an inverting DC amplifier is determined by calculation.

    Calculate this value. Current gain = ______.

14. Turn off the power supply and disconnect the DC input source. Connect a capacitor in series with R_{IN} and an AC signal source of 400 Hz. Adjust the output of the signal source to its lowest value. Connect an oscilloscope across the output resistor.

15. Turn on the power supply and signal source. Carefully adjust the signal source input until it produces its maximum undistorted output. Record the peak-to-peak output voltage.

    Output voltage = _______ V p-p.

16. Measure and record the input voltage at test point 1.

    Input voltage ____V p-p.

17. Calculate the AC voltage gain of the amplifier.

    Voltage gain = _______

18. Turn off the power and disassemble your circuit. Return all materials.

Analysis

1. If the source resistance is changed to 500 Ohms, how would it alter the load current in Step 10?

2. If R_F is increased to 250 K Ohms, how would it alter the voltage gain in Step 7?

3. If R_F is changed to , how would it alter the voltage gain in Step 7?

4. How does the current gain compare with the voltage gain in Steps 13 and 7?

Activity 73

Noninverting Op-amps

Objectives

Circuits frequently need amplifiers that develop high voltage and current gains without signal inversion. The non-inverting op-amp is specifically designed for this type of application. Both AC and DC signals can be amplified by this circuit, with the input and output remaining in phase. The input resistance or impedance of this amplifier is quite high, with typical values exceeding 100 M Ohms. With the input signal applied to the non-inverting input terminal, the voltage gain is dependent upon the values of R_{IN} and R_F. The output signal of this amplifier is developed across the load resistor. Typical R_L values are 35 to 50 Ohms.

In this activity, you will:

1. Connect a simple non-inverting op-amp and observe its operation.
2. Calculate amplifier gain using both the resistance method and the voltage method.
3. Design a simple non-inverting op-amp and alter its gain by changing the resistance values.

Equipment and Materials

- Multimeter
- Oscilloscope
- Split power supply (or two 9 V batteries)

DOI: 10.1201/9788743810124-78

- Variable DC source (or a 1.5 Volt C cell with a potentiometer)
- AC signal generator
- Op-amp – uA741C or equivalent
- Resistors – 2 K, 10 K, 200 K Ohm (2)
- Capacitor – 0.01 mF, 100 V_{DC}

Procedure

1. Construct the non-inverting op-amp of Figure 73.1.

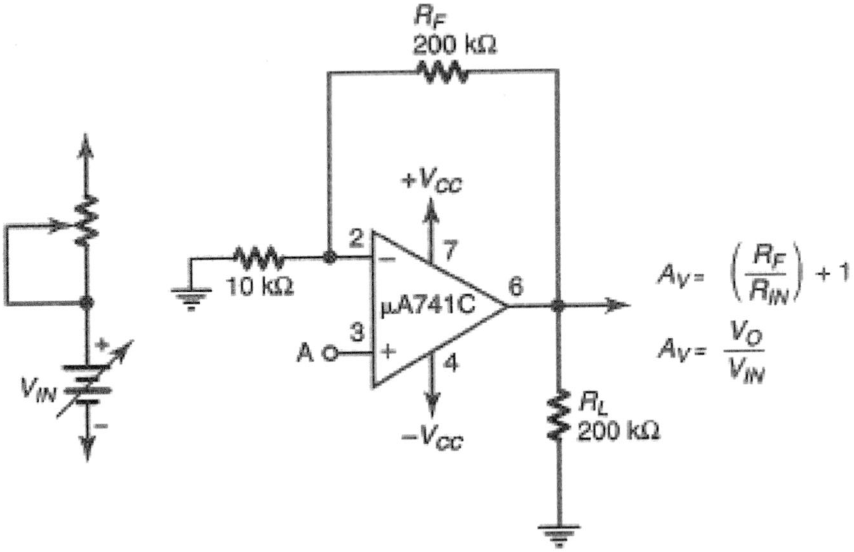

Figure 73.1 DC non-inverting op-amp test circuit.

2. Adjust the voltage input source to 0.1 V and connect it to point A. Turn on the power source, and measure and record the output voltage developed across R_L.

   Output voltage = ____ _____ Vdc.

3. Using the voltage amplification formula, calculate the A_V for the circuit using measured voltage values.

   A_V = ________.

4. Using the formula for resistance, calculate the gain using the circuit values for R_{IN} and R_F.

   A_V = ________.

5. If there is a significant difference in the values of these two voltage calculations, measure the actual value of the resistances and calculate using these actual values.

6. Connect two 220 K Ohm resistors in series to make a 400 K Ohm R_F. Repeat Steps 2, 3, and 4 for the new value of R_F. Record the measured value and calculated values.

   V_{OUT} = ________ V.

   A_V (by voltage) = ________.

   A_V (by resistance) = ________.

7. Connect the two 220 K Ohm resistors in parallel and repeat Step 6 for a V_{IN} of 0.3 V. Record the measured and calculated values.

   V_{OUT} = ________ V.

   A_V (by voltage) = ________.

   A_V (by resistance) = ________.

8. Alter the DC amplifier of Figure 73.1 to conform with the AC amplifier of Figure 73.2. Resistors R_{in} and R_1 should be of equal value.

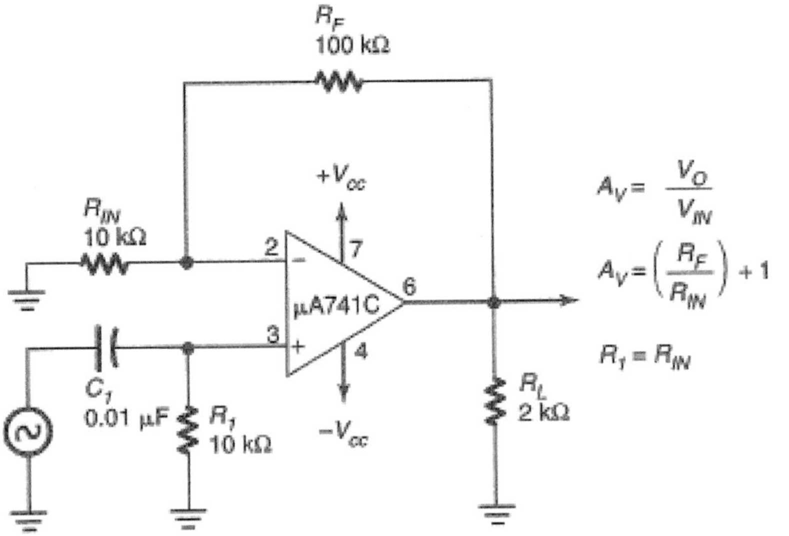

Figure 73.2 AC non-inverting op-amp test circuit.

9. Turn on the AC signal generator and adjust it to produce a signal of 0.1 V p-p. Turn on the IC power supply.

10. Measure and record the output voltage appearing across R_L.

    V_{OUT} = ________ $V_{P\text{-}P}$.

11. Using the voltage formula, calculate the voltage gain.

    A_V = ________.

12. Calculate the voltage gain with the resistor formula.

    A_V = ________.

13. Observe the phase relationship of the input signal while observing the output signal across R_L. At what input value does the output become distorted?

14. Turn off the power and disassemble the circuit. Return all materials.

Analysis

1. What is the primary difference between an AC and a DC non-inverting op-amp?

2. Why is a capacitor used on the input of the AC amplifier?

3. How would the value of the capacitor limit the amplifying frequency of the amplifier?

4. Why does R_F influence voltage gain?

Activity 74

Proximity Detectors

Objectives

Proximity detectors are used to trigger an alarm or turn on a circuit by detecting a physical change within a designated area. These circuits are designed to respond in some way to a change in capacitance. This change usually alters the frequency of an oscillator so that the output can be amplified and used to trigger a load control device. Typically, the circuit contains an oscillator, a tuned circuit, an amplifier, and a trigger device.

One of the most popular proximity detectors in operation today utilizes the loaded oscillator principle. In this type of circuit, the LC components of the oscillator are shunted to ground by an external capacitance. This capacitance is usually called the sensor plate. Moving a hand or producing a physical change near the sensing plate upsets the electrostatic field of the capacitance. This, in turn, alters the frequency of the oscillator or may load down the circuit in such a way as to quench the oscillation process. Liquid level detection, automatic thickness gauge testers, intruder alarm systems, counters, and switching circuits employ this type of detector.

In this activity, you will construct a loaded oscillator proximity detector. The oscillator is a Colpitts type with split capacitors in the collector circuit. The output of the oscillator is rectified and applied to a voltage comparator circuit, where it is amplified by an op-amp. The amplified output is used to trigger a load control device. The 5 K Ohm resistor in the comparator circuit is used to establish a balanced condition for detection. When the oscillator senses a change, it shuts down and stops oscillating. This creates an imbalance, which is then detected by the comparator circuit.

Through this activity, you will test the operation of a loaded oscillator circuit and observe typical waveforms. The test procedure used to analyze circuit operation is also used to troubleshoot the circuit. Through this testing

DOI: 10.1201/9788743810124-79

experience, you will become familiar with the oscillator loading principle and be able to follow a logical troubleshooting procedure.

Equipment and Materials

- Multimeter
- Oscilloscope
- DC power supply
- Op-amp – LM3900 or equivalent
- Transistor – 2N3397 or equivalent
- Coil – 10 mH
- Diode – 1N4004
- Light-emitting diode
- Resistors – 470, 6.8 K, 10 K (2), 68 K, 1 M Ohm (2),
- Potentiometers – 5 K, 500 K Ohm
- Capacitors – 47 pF, 100 pF, 0.01 uF
- SPST toggle switches (2)
- Sensor plate – (6 in. × 6 in. metal)

Optional according to the selected load control circuit

- Transistor – 2N3397 or equivalent
- Relay – 12V, (Guardian 1335-2C-120D or equivalent)
- No. 47 lamp with socket or equivalent
- Resistors – 47, 220 Ohm, ½ Watt
- SCR (General Electric C-122D or equivalent)
- AC source – 6 Volt

Procedure

1. Construct the proximity detector of Figure 74.1.

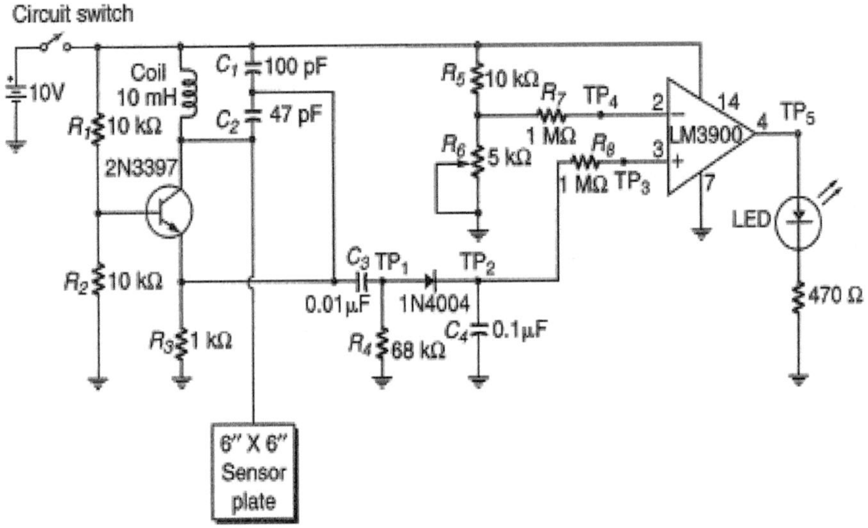

Figure 74.1 Proximity detector

2. Close the circuit switch and adjust the 5 K Ohm potentiometer for illumination of the LED. There should be a specific point where the LED turns on. Increasing resistance will cause it to remain on, while decreasing the resistance causes it to turn off. The point just before turn-on represents the most sensitive detection setting of the potentiometer. Adjust the potentiometer several times to locate the most sensitive point.

3. If the circuit is not operating properly, perform Steps 4, 8, 9, 10, and 11 first to determine which part of the circuit is malfunctioning. If the circuit is operating satisfactorily, follow the procedure steps in order and record the indicated data.

4. Prepare an oscilloscope for operation and connect it to test point 1. Make a sketch of the observed waveform, Figure 74.2.

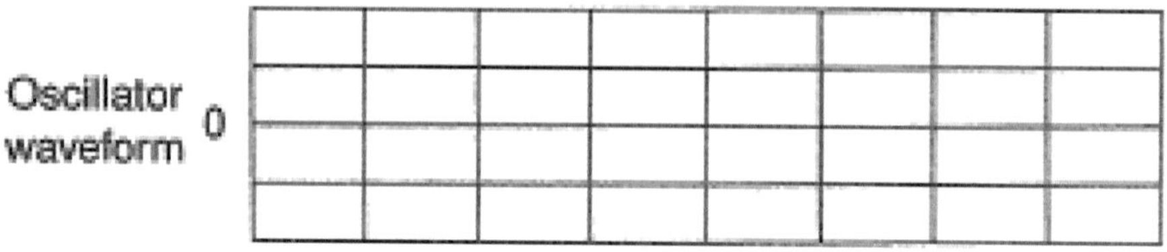

Figure 74.2 Observed waveform at test point.

5. While observing the oscilloscope, touch the sensor plate. What influence does this have on the waveform?

6. Move your hand over the sensor plate while observing the oscilloscope. What influence does this have on the waveform?

7. Determine the frequency of the oscillator with your oscilloscope.

   Oscillator frequency ____ Hz.

8. Measure and record the voltages at the emitter, base, and collector of the oscillator transistor Q1. Compare your measurements with those indicated on the schematic diagram.

   Emitter voltage ____V.

   Base voltage ____V.

   Collector voltage ____V.

9. Measure and record the DC output voltage at test point.

   Unloaded oscillator output voltage = _______ V.

10. Touch the sensor plate and record the measured voltage at test point 2. This represents the loaded output voltage of the oscillator.

    Loaded output voltage ____ V.

11. The output of the oscillator will change approximately 2 V between the loaded and unloaded conditions of operation.

12. Measure and record input voltages at TP3 and TP4 when the LED is on. This indicates a balanced condition.

    TP3 = ________ V.

    TP4 = ________ V.

13. Measure and record the voltages at which the circuit is unbalanced.

    TP3 = ________ V.

    TP4 = ________ V.

14. Measure and record the output voltage of the circuit at test point 5 for the balanced and unbalanced conditions.

    The total change in voltage is ____ V.

15. Select one of the alternate load control circuits in Figure 74.3 and construct it. Connect it to the output of the proximity circuit.

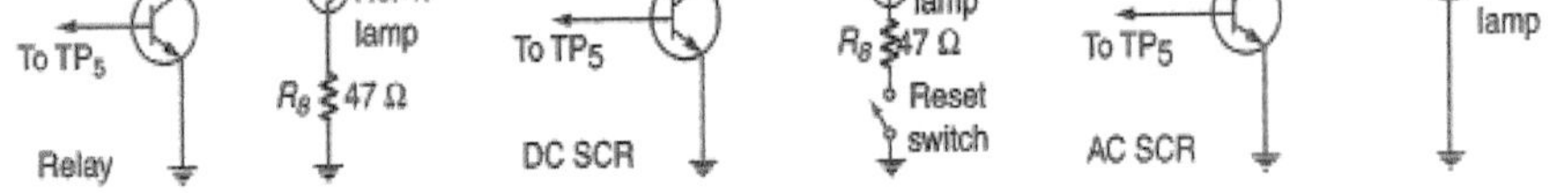

Figure 74.3 Alternate load control circuit.

16. Close the circuit switch, adjust the balance potentiometer, and test the control capacity of the completed load control circuit.

17. Open the circuit switch. Disconnect the circuit and return all parts.

Analysis

1. Describe what is meant by loaded and unloaded oscillator conditions.

2. What is the function of the IC op-amp in the proximity circuit of Figure 74.1?

3. Draw a block diagram representing the major parts of the proximity circuit constructed in Step 15.

Activity 75

Reed Switches

Objectives

Reed switches are devices that respond to a controlled magnetic field. They are designed to close or open when exposed to either a permanent magnetic field or to an electromagnetic field. The contacts of a reed switch are housed inside a hermetically sealed glass tube. When actuated, contact sparks are isolated from the outside environment. Industrial applications of this device are quite numerous in explosive areas and dirt-prone environments. They are used to verify Z-axis movement on some robots.

A reed switch contains two flat metal strips, or reeds, housed in a hollow glass tube filled with an inert gas. When the reeds are exposed to a magnetic field, they are forced together, making or breaking contact depending on their design. The normally closed switch breaks contact when exposed to a magnetic field. Normally open contacts, by comparison, are forced closed when exposed to a magnetic field.

In this activity, you will construct a simple electrical circuit and control its action by changes in magnetic field strength. Through this experience, you will gain some insight into the operation, sensitivity, and control action of a reed switch.

Equipment and Materials

- Multimeter
- DC power supply – 5 Volt
- AC power source – 6 Volt
- Reed switch (GE-X7 or equivalent)

DOI: 10.1201/9788743810124-80

- Bobbin-wound, reed switch coil
- Reed switch magnet
- Lamp – No. 47 or equivalent
- Switch – SPST
- Circuit construction board and connecting wires

Procedure

1. Construct the reed-switch circuit of Figure 75.1.

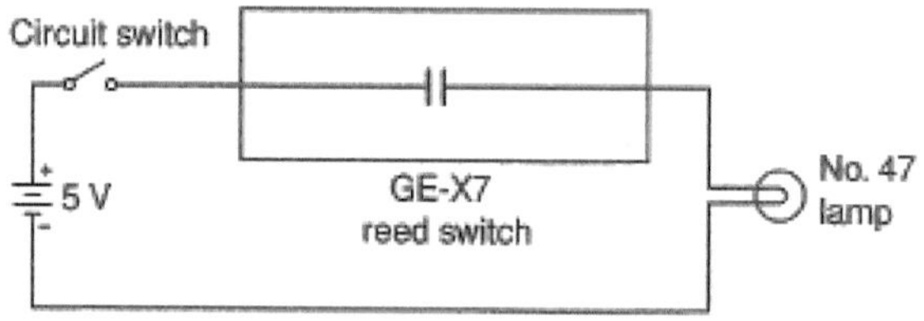

Figure 75.1 Reed switch test circuit for observing permanent magnet actuation.

2. Turn on the circuit switch and move the small permanent magnet near the reed switch. Try the magnet at a right angle with reeds, parallel with the reeds, etc. Which type of magnet orientation produces the best control capability?

3. Place the reed switch in the center of the bobbin-wound coil. Connect the circuit of Figure 75.2 to test the electromagnetic action of the reed switch.

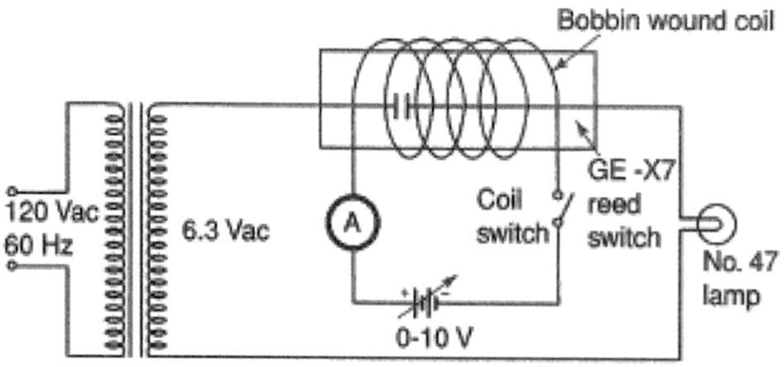

Figure 75.2 Reed switch test circuit for observing electromagnetic actuation.

4. Turn on the AC power source and the DC coil-control circuit.

5. Starting at zero volts DC, gradually increase the DC source voltage until the reed switch is actuated. The meter should be set at a high current range until an approximate actuating value is determined. Run at least two trial tests to determine the actuating current needed to energize the switch.

   DC actuating current ____ mA.

6. In some applications, the actuating coil is used to increase the sensitivity of the reed switch by producing a partial field. Increase the DC coil current close to the actuating value. Then place the permanent magnet near the coil to actuate the reed. In this case, the switch can have some degree of variable sensitivity.

7. Turn off the AC and DC power sources for the circuit of Figure 75.2. Use 5 volts DC to supply the coil. The primary side of the coil should be connected to a variable AC transformer.

8. Turn on both the AC and DC sources, then increase the AC voltage applied to the coil. Measure the applied AC voltage. Do not increase the voltage to a value that will cause the reed coil to overheat. How does the reed switch respond to AC compared with DC?

9. Turn off the AC and DC power sources and disconnect the circuit. Return all components.

Analysis

1. How could a reed switch be used to control a pump motor in a liquid sump tank?

2. Make a sketch of this circuit.

3. How could a reed switch be used to control a circuit in an explosive area without danger of an explosion?

Activity 76

Touch Switch Circuit

Objectives

A number of switches used to control industrial machinery do not employ any moving mechanical parts. This classification of a switch is commonly called a touch switch. It achieves control by someone simply touching a seemingly solid button or plastic cover plate. The actual circuit is a capacitance switch that responds to a physical change in circuit capacitance caused by touching a finger to a certain area. In many cases, the circuit can be adjusted so that switching is achieved by simply placing a finger in close proximity to the actual contact area. Switches of this type are very reliable and, in many cases, are used for safety reasons in explosion-proof installations.

In this activity, you will construct a FET/op-amp touch switch that triggers an SCR. The load control of a lamp is achieved without moving parts. When the source of the load control circuit employs DC, manual resetting is required. Alarm circuits frequently employ this type of control. When the source of the SCR is AC, the touch switch circuit has an automatic reset after each half cycle of operation.

Through this activity, you will test the signal path from the FET through the op-amp and ultimately to the SCR. This signal tracing process is often used to troubleshoot circuits of a similar nature. The touch switch employs a number of components that you have studied in other applications. The combination of FET, op-amp, and SCR does, however, represent a rather unusual circuit application.

Equipment and Materials

- Multimeter
- Split power supply – 0–5 V (2) or two 9 V batteries

 DOI: 10.1201/9788743810124-81

- AC source – 6 V
- SCR (General Electric C-122D or equivalent)
- Op-amp – A741C or equivalent
- FET (General Electric FET-1 or equivalent)
- Potentiometer – 5 K Ohm, 2 Watt
- Resistors – 120, 4.7 K, 10 K, 56 K, 220 K, 1 M Ohm
- No. 47 lamp with socket or equivalent
- SPST toggle switch
- IC circuit construction board and connecting wires
- A piece of metal of any gauge (1 inch square)

Procedure

1. Construct the touch switch of Figure 76.1.

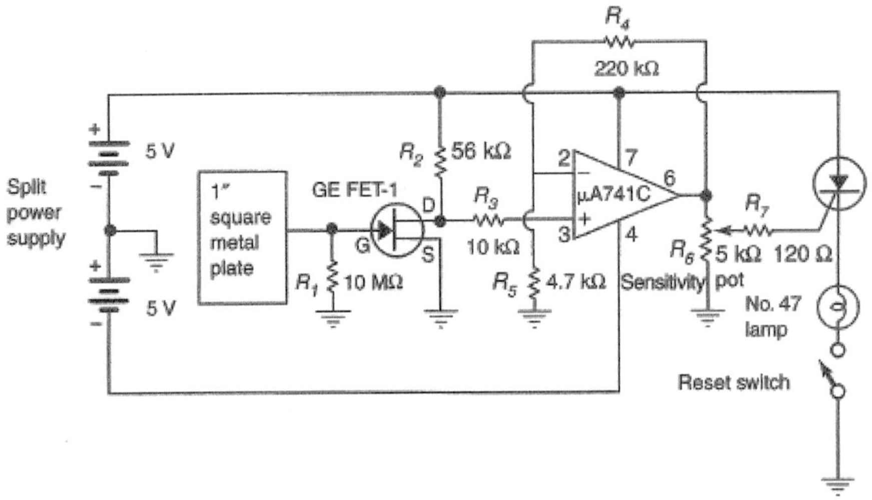

Figure 76.1 Touch switch circuit.

2. Note that a split power supply is used to supply power to the uA741C and other parts of the circuit. Two batteries can be used in place of the variable DC power supplies.

3. Close the reset switch and touch the square metal plate. Adjust the 5 K Ohm sensitivity control while touching the plate. The lamp should turn on at a certain setting of the potentiometer.

4. Remove your finger from the touch plate and open the reset switch momentarily, then turn on the switch. If the sensitivity is properly set, the lamp should remain off until the metal plate is touched. You may need to try several attempts at setting the sensitivity to its best range. The circuit must be reset each time.

5. If the circuit does not work properly, you may want to make the following test as a troubleshooting procedure. If the circuit is working properly, the following tests are made to analyze the operation of the circuit.

6. With power applied to the circuit, connect a multimeter between the drain and ground of the FET. Touch the metal plate while observing this voltage. As a general rule, the voltage change is very small.

   Change in drain voltage is _______ to _______V.

7. The change in drain voltage is then applied to the positive or non-in-¬verting input of the uA741C. This voltage is amplified and appears as the output voltage at pin 6. Measure and record the IC output voltage change.

   Output voltage ranges from________ to _______V.

8. The output voltage from the op-amp is applied to the trigger sensitivity control of the SCR. This voltage controls the gate current needed to trigger the SCR into conduction. Measure and record the change in gate voltage.

   The range of gate voltage is ________ to _______V.

9. Measure the voltage across the gate resistor when the SCR is being triggered.

   Gate triggering voltage is _______V.

10. Calculate the gate current needed to trigger the SCR.

    Gate current = ________ mA.

11. Carefully disconnect the +5 volt DC line from the anode of the SCR. In its place, connect a 6.3-volt AC source between the anode and ground. Turn on the reset switch.

12. Using the procedure outlined in Steps 3 and 4, adjust the sensitivity so that the SCR triggers when the metal plate is touched. How does the operation of the circuit differ from that of the DC SCR circuit?

13. Turn off the circuit power supplies. Disconnect the circuit and return all components.

Analysis

1. Why does the SCR need to be reset after it has been triggered?

2. What does touching the metal plate actually change in the FET?

3. Why does an AC supply to the SCR have automatic resetting?

Activity 77

Temperature Sensors

Objectives

Temperature sensing circuits are widely used in industry as system control elements. Heat sensors, which are by far the most popular of all sensors, are commonly found in alarm circuits that detect changes in temperature at remote locations. Machinery, electronics equipment, and measuring instruments also employ temperature sensors to detect unusual operating conditions. The thermistor is a very popular sensor element for temperature detection equipment.

Thermistors are primarily classified as temperature-sensitive resistors with an operating range of from 230 to 650 °C. When used in bridge circuits with high-gain amplification, thermistors can detect temperature changes as small as 0.001 degrees.

The physical makeup of a thermistor includes a mixture of nickel, manganese, and cobalt oxides formed into a piece of semiconductor material. These oxides are mixed together and fired to form a coherent nonporous material. The mixture is then formed into a variety of shapes. In this activity, a tiny piece of the ceramic material is formed into a bead and enclosed in glass. Changes in temperature cause the resistance of the thermistor to produce a wide range of values. Typically, thermistors have a negative temperature coefficient. This means that an increase in temperature causes a decrease in resistance. Metal, by comparison, has a positive temperature coefficient. This means that a rise in temperature causes a corresponding increase in resistance.

In this activity, you will build a temperature-detecting bridge circuit. The output of the bridge is then fed into an op-amp for high-gain amplification and increased sensitivity. Any imbalance in the bridge is detected by the amplifier and is used to trigger an output circuit. Circuits of this type are typically found in alarm circuits that function as heat detectors.

DOI: 10.1201/9788743810124-82

Equipment and Materials

- Multimeter
- Split DC power supply – 9 volt or two 9 volt batteries
- Resistors – 47, 1K, 4.7 K, 10 K (2), 200 K Ohm
- Potentiometer – 5 K Ohm
- Op-amp – uA 741 C or equivalent
- Thermistor – Fenwal GB 32J2 or equivalent
- SCR – GE C-122D or equivalent
- No. 47 lamp with socket or equivalent
- IC circuit construction board and connecting wires
- SPST toggle switch
- Decade resistance box (optional)

Procedure

1. Construct the thermistor bridge circuit of Figure 77.1. Note that a voltmeter is used to detect the amplified output of the circuit.

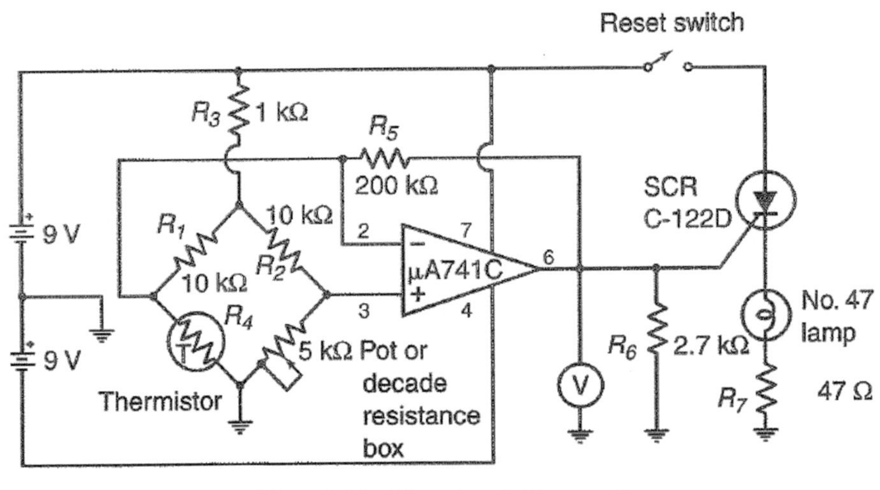

Figure 77.1 Thermistor bridge circuit.

2. Before connecting the thermistor to the circuit, measure its resistance. Try to avoid touching the glass part of the thermistor when attaching the test leads. With the meter connected, the resistance should stabilize after a few seconds.

   The stabilized resistance is _______ Ohms.

3. While observing the meter, grasp the glass bead between your index finger and thumb. How does body temperature influence the resistance of the thermistor?

4. Connect the thermistor to the bridge circuit as indicated. The reset switch should be in the off position, and the meter should be adjusted to the zero center position. It should be in the 50 V or equivalent range.

5. Turn on the split power supplies and energize the circuit. Balance the bridge by adjusting the potentiometer to produce a zero indication of the meter. A decade resistance box can be used in place of the potentiometer.

6. After the bridge has been nearly balanced, you can switch the meter to a lower voltage range to improve the balancing accuracy.

7. To test the sensitivity of the circuit, place your finger near the glass bead of the thermistor while observing the meter. If the circuit is working properly, the meter should deflect upscale or in the positive direction. Within a few seconds, it should return to the balanced indication when the thermistor has reached its stabilized resistance. If you touch the glass bead with your finger, it usually takes longer for it to return to the balanced state.

8. Try blowing on the thermistor or placing it near some heat-producing source. Avoid temperatures over 150 °C (302 °F).

9. Close the reset switch of the SCR. Place your finger near the thermistor. While observing the meter, see how much voltage is needed to trigger the SCR into conduction. You must wait a few seconds for the thermistor to stabilize before resetting the SCR.

10. Try several sources of heat to turn on the second circuit.

11. Turn off the power supply and disconnect the circuit. Return all components.

Analysis

1. What type of temperature coefficient does the thermistor used in this activity have?

2. Explain how a change in thermistor resistance causes the bridge circuit to be imbalanced.

3. Why is it advantageous to use an op-amp with the bridge circuit of this activity?

Activity 78

Thermocouple Applications

Objectives

Thermocouples are frequently used to measure temperatures in an industrial setting. Due to the relatively low voltage output associated with most thermocouples, amplification circuits are used to increase this output as well as to increase sensitivity. The resulting output of the amplification circuit is used to drive or activate the readout device.

In this activity, you will see the thermocouple used to control the conductivity of an FET. The conductivity of the FET, in turn, controls the action of a single-stage transistor amplifier and thus the current through a multimeter that is used as the readout device.

Equipment and Materials

- Digital multimeter
- Variable DC power supply – 0–5 volts
- DC power supply – 10 volts
- AC power supply – 120 V
- Thermocouple-type J or equivalent
- Resistors – 5.6, 560 Ohm
- Field-effect transistor-GE-FET-1 or equivalent
- NPN transistor – 2N2405 or equivalent
- Connecting wires
- Heat cone – 660 W or equivalent

 DOI: 10.1201/9788743810124-83

Safety

Safety is always a primary factor for laboratory activities. This lab should only be completed with supervision.

You may complete certain steps of the "Procedure" and the "Analysis" for better understanding without applying voltage.

Procedure

1. Construct the circuit shown in Figure 78.1.

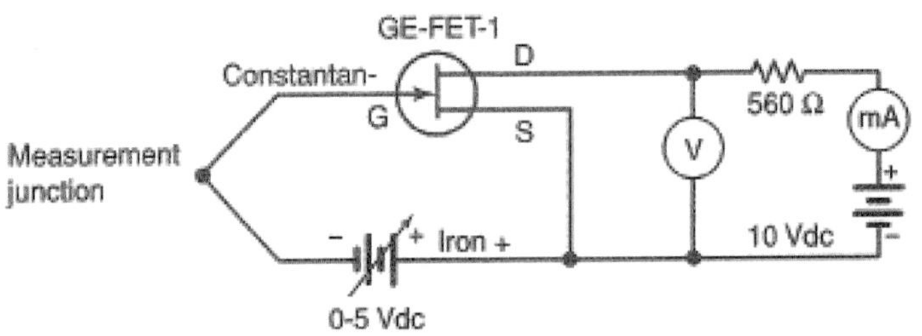

Figure 78.1 Type J thermocouple FET circuit.

2. Allow the measurement junction of the thermocouple to remain at room temperature. Alter the gate voltage of the FET to equal those listed in Figure 78.2. Record the source-drain current and voltage for each FET gate voltage value.

Gate voltage (V)	Source-drain current (mA)	Source-drain voltage (V)
0.2		
0.4		
0.6		
0.8		
1.0		
1.5		
2.0		

3. Adjust the gate voltage to 0.2 V.

4. Grasp the measurement junction of the thermocouple between your thumb and forefinger. Describe how this action affects the source-drain current and voltage of the FET as compared to the data gathered in Step 2 when the gate voltage was 0.2 V.

5. Connect the 660 W heat cone to 120 Vac and allow it to warm up for about three minutes.

6. Position the measurement junction of the thermocouple inside the heat cone for a period of three minutes and record the source-drain current and voltage of the FET.

7. How do the values of source-drain current and voltage recorded in Step 6 compare with the data gathered in Step 2, when the FET gate voltage was 0.2 Vdc?

8. Disconnect the heat cone from the 120 Vac power supply. What effect does the removal of the heat cone have on the source-drain current and voltage?

9. Construct the circuit illustrated in Figure 78.2.

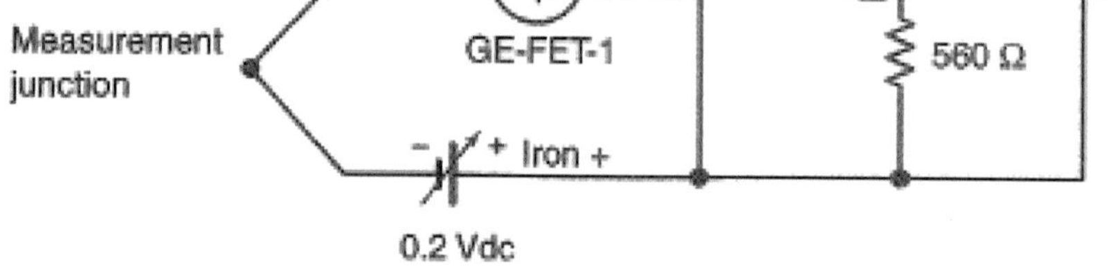

Figure 78.2 Modified Type J thermocouple FET circuit.

10. Record the collector current of Q2 as displayed by the digital multimeter with the measurement junction of the thermocouple at room temperature.

I_C = ________ mA.

11. Grasp the measurement junction of the thermocouple between your thumb and forefinger. Describe how this action affects the collector current as displayed on the digital multimeter.

12. Place the measurement junction of the thermocouple inside the cone of the cool 660 W heater. Connect the 120 Vac to the heat cone and allow it to warm up for five minutes.

13. Record the collector current as displayed on the digital multimeter after the heat cone has warmed up.

    I_C = _______ mA.

14. How does the current recorded in Step 10 compare with the current recorded in Step 13?

15. How do you explain the difference?

16. If the digital multimeter used in the circuit shown in Step 9 was calibrated in degrees Fahrenheit, how could the circuit be used to measure temperature?

17. Disassemble your circuit and return all components.

Analysis

1. Why is it sometimes necessary to use an amplifier when a thermocouple is used to measure temperature?

2. What was the readout device used in the circuit constructed in Step 9?

3. What would determine the maximum temperature that could be measured by the circuit in Step 9?

4. How did the action of the circuit used in Step 9 differ from that used in Step 2?

5. Explain how the circuit in Step 2 could be used to measure temperature.

Activity 79

Linear Variable Differential Transformer (LVDT) Detector

Objectives

Frequently, it is not enough to measure only the amount of the displacement of an object. Often, the direction of displacement, as well as the amount of displacement, must be indicated. When this becomes the case, a detector circuit similar to that shown in Figure 79.1 can be used in conjunction with the output of the linear variable differential transformer (LVDT).

In this detector circuit, the diodes conduct equally when the movable core is in the null position. This results in an equal and positive voltage drop across R_1 and R_2, causing the readout device (voltmeter) to indicate zero displace¬ment. When displacement occurs, one diode will conduct more than the other, causing an upset in the balanced voltage across R_1 and R_2. The result is an output voltage, as measured by the voltmeter, whose polarity indicates the direction of displacement and whose magnitude indicates the amount of displacement. In this activity, you will examine the action of an LVDT detector circuit.

Equipment and Materials

- Digital multimeter
- Power supply – 6 Vac
- Air-core coils – 100 turns, A.S. No. 16 wire (2)
- Air-core coil – 200 turns, A.S. No. 24 wire
- Steel core – 6 inch length, ¾ inch diameter
- Diodes-1N4004 (2) or equivalent

DOI: 10.1201/9788743810124-84

- Resistors – 1K Ohm
- Capacitor – 1 uF, 25 VDC
- Connecting wires
- SPST switch

Safety

Safety is always a primary factor for laboratory activities. This lab should only be completed with supervision.

You may complete certain steps of the "Procedure" and the "Analysis" for better understanding without applying voltage.

Procedure

1. Construct the circuit illustrated in Figure 79.1.

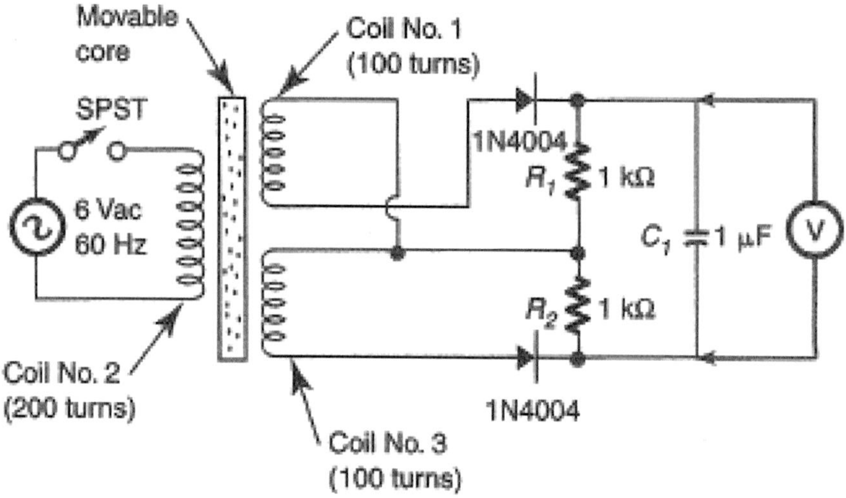

Figure 79.1 LVDT detector test circuit.

2. Close the SPST switch and slide the movable core until the meter indicates zero volts.

3. Moving the core in the direction and by the amount indicated in Figure 79.2, complete the tale by recording the voltage output of the detector circuit along with the proper voltage polarity.

Direction of movement from null position	Amount of movement Inches	Centimeters	Voltage output	Polarity (+ or –)
Left	1/16	0.159		
Left	1/8	0.318		
Left	3/16	0.477		
Left	1/4	0.635		
Left	5/16	0.793		
Left	3/8	0.953		
Return to null position				
Right	1/16	0.159		
Right	1/8	0.318		
Right	3/16	0.477		
Right	1/4	0.635		
Right	5/16	0.793		
Right	3/8	0.953		

Figure 79.2 Output polarity and magnitude of the LVDT test circuit.

4. How do the voltages generated by moving the core to the left compare with the voltages generated when the core is moved to the right?

5. Ideally, what is the minimum displacement of the movable core that would result in a voltage output?

6. Disassemble your circuit and return all components.

Analysis

1. What is meant by coils being connected in series to oppose?

2. Explain how the detector circuit used in conjunction with the LVDT enables the direction of displacement to be measured.

3. When is it necessary to measure direction, as well as the amount of displacement?

4. How would the null position of the core be affected if R1 of the detector circuit were a value different from R2?

Activity 80

Strain Gauges

Objectives

The strain gauge is a commonly used resistive transducer. It is used in the measurement of displacement, torque, weight, pressure, and other force parameters. A strain gauge has a fine-wire element looped back and forth on a mounting plate, which is usually cemented to the member undergoing stress. The extra length due to the looping increases the effect of any stress applied lengthwise. A tensile stress stretches the wire, increasing its length and, therefore, its resistance. The measure of the sensitivity of a material to strain is called the gauge factor (GF), and is the ratio of the change in resistance to the change in length. The initial resistance is typically around 120 Ohms, and the gauge factor may run from −12 to +6. A gauge factor of 2 is common for wire strain gauges.

The more recent semiconductor strain gauges have greater sensitivity and can be used exactly like a metallic strain gauge. Semiconductor strain gauges have a gauge factor roughly 50 times greater (around 100) than the metallic types. This large gauge factor is accompanied by a thermal rate of change of resistance, also approximately 50 times higher than in older gauges. Thus, the semiconductor strain gauge is as stable as the metallic type, but has a much higher output. Simpler temperature compensation methods can be applied to the semiconductor strain gauge, making it more useful and extending its range to the measurement of small values of micro-strain (micro-inches/inch of length), which could not be measured with earlier gauges.

In this activity, a strain gauge used in a bridge network will be observed. This method can be used for most strain gauge measurements.

DOI: 10.1201/9788743810124-85

Equipment and Materials

- Multimeter
- Galvanometer
- Resistors – 1K, 100 K Ohm
- Strain gauge
- Variable resistor (calibrated) – 20 K Ohm
- DC power source

Procedure

1. Construct the strain gauge balance detector circuit shown in Figure 80.1.

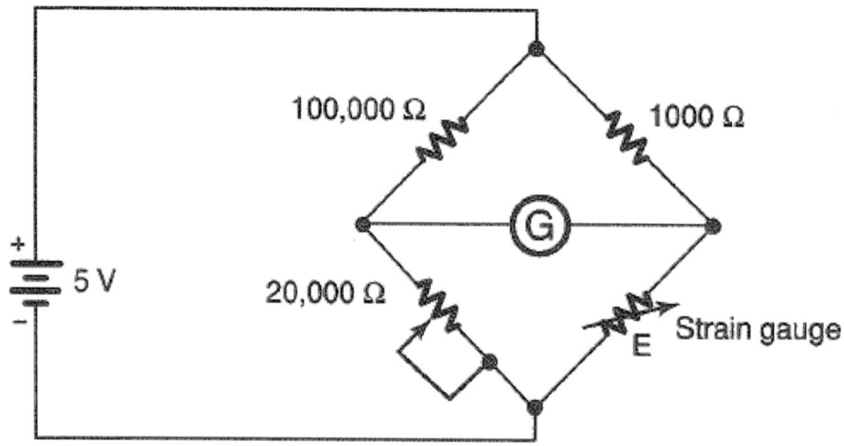

Figure 80.1 Strain gauge balance detector test circuit.

2. Measure the resistance of the strain gauge with the meter.

   Resistance = _______ Ohms.

3. While gently bending the metal mounting plate, monitor the galvanometer null detector and observe the effect on the meter reading. What occurs?

4. Disassemble your circuit and return all components.

Analysis

1. Write a brief analysis of the operation of the strain gauge. Include in your analysis your observation of the bridge network as a measuring method using a strain gauge.

Activity 81

Thermistors

Objectives

Thermistors are commonly used as temperature-sensitive transducers. Their most important property is that their resistance decreases with increases in temperature. In this activity, you will observe the characteristics of one type of thermistor.

Equipment and Materials

- Multimeter
- Variable DC power source
- Thermistor – JA41J1 or equivalent
- Resistor – 1 K Ohm
- Lamp – 60 W

Procedure

1. Construct the circuit shown in Figure 81.1.

DOI: 10.1201/9788743810124-86

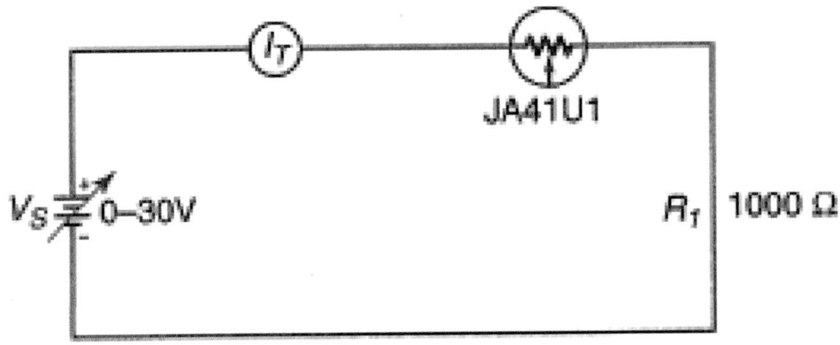

Figure 81.1 Thermistor characteristics test circuit.

2. Complete the data for Figure 81.2 relative to the circuit of Figure 81.1 with the thermistor at room temperature (note: allow about two minutes between source voltage changes).

Source Voltage (V_S)	I_T	Voltage Across Thermistor (V_T)	Calculated Resistance of Thermistor (V_T/I_T)	(V_{R_1})	Calculated Circuit Resistance (V_S/I_T)	Resistance Change of Thermistor
0.5						
1.0						
2.0						
3.0						
4.0						
5.0						
10						
15						
20						
25						
30						

Figure 81.2 Thermistor characteristics (room temperature).

3. Repeat Step 2 using a 60-watt lamp placed about one inch from the thermistor. This will increase the temperature and change the characteristics of the thermistor (again, allow about two minutes between voltage changes). Record your data in Figure 81.3.

(V_S)	I_T	(V_T)	Calculated Resistance of Thermistor	(V_{R_1})	Calculated Circuit Resistance	Resistance Change of Thermistor
0.5						
1.0						
2.0						
3.0						
4.0						
5.0						
10						
15						
20						
25						
30						

Figure 81.3 Thermistor characteristics (heated).

4. Allow the thermistor to return to room temperature.
5. Apply 0.5 volts DC to the circuit and record the current caused by the circuit with the thermistor at room temperature.

   Current = _______ mA.

6. Grasp the thermistor with your hand for three minutes and observe what happens to the current in the circuit.
7. Disassemble your circuit and return all components.

Analysis

1. Draw a current versus voltage graph to illustrate the characteristics of the thermistor from the data of Figures 81.2 and 81.3.

2. Briefly discuss the characteristics of a thermistor.

3. Compare the data of Figures 81.2 and 81.3 by noting whether each of the following measurements increased, decreased, or remained the same with an increase in temperature.

   a. Voltage across R1:______________________________

   b. Circuit current: ______________________________

   c. Voltage across thermistor: ______________________

   d. Resistance of thermistor: _________________________

   e. Circuit resistance: ______________________________

   f. Resistance changes of thermistor: __________________

4. In Step 2, why is it necessary to allow two minutes between source voltage changes?

5. In Step 6, what happened to the circuit current when the thermistor was heated?

UNIT VI

Industrial Electronic Control Systems

Activity 82

Motor-driven Timers

Objectives

The motor-driven timer provides a wide variety of timing actions for industrial circuit applications. In its simplest form, this timer has an electric drive motor, a ratchet release coil, and a ratchet dial that is held stationary until released. When the timing cycle reaches its set time, the ratchet is released and the dial resets itself by spring action. Both on-delay and off-delay reset timers are available. More sophisticated reset timers permit a wide range of timing operations in a single unit.

In this activity, a simple rest timer will be used to build a load control circuit that produces either interval or delay timing operations. Initially, when the timer is energized by the control switch, load A is turned on and load B is turned off. After the expired time setting or the time-out condition has been reached, the load is turned off and then turned on.

Equipment and Materials

- AC power supply – 120 V
- Reset timer – Eagle Signal HD-50 Series or equivalent
- Incandescent lamps with sockets – 7.5 W (2)
- SPST toggle switch

Safety

Safety is always a primary factor for laboratory activities. This lab should only be completed with supervision.

DOI: 10.1201/9788743810124-88

You may complete certain steps of the "Procedure" and the "Analysis" for better understanding without applying voltage.

Procedure

1. Refer to the timer manufacturer's product manual before attempting to complete this activity.
2. Wire the reset timer for the on-delay operation of Figure 82.1.

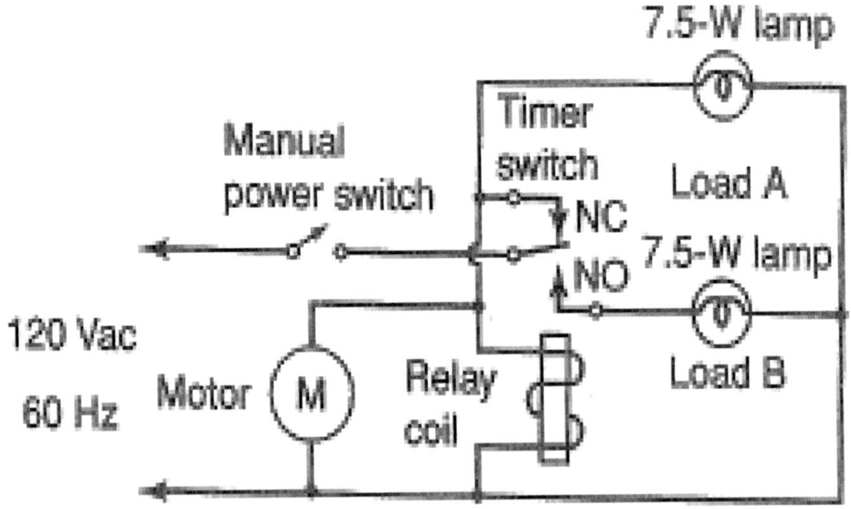

Figure 82.1 Reset the timer with on delay operation.

3. Adjust the time setting dial by pulling it out and turning it to a desired setting. Releasing the dial automatically locks it in the new setting.
4. Turn on the manual power switch and describe the operating condition of loads A and B.
5. When the dial setting trips, describe the operating condition of the load and the motor.
6. Start a new condition cycle by opening the control switch momentarily, then turning it on again. After the cycle has been in operation for a few seconds, momentarily open the control switch, then close it. What action does this initiate?

7. Disassemble your circuit and return all components.

Analysis

1. What are some industrial applications of a reset timer?

2. Could this timer be modified to achieve a different function of some type? Explain.

3. Explain the difference between interval and delay timers.

Activity 83

Motor-driven Repeat Cycle Timers

Objectives

Industrial applications of motor-driven repeat cycle timers are very common. With this timing device, it is possible to control a number of specific operations in certain periods in a desired sequence. The operational time of this device is primarily based upon the rotational speed of a series of cams that actuate snap-action switches. The switching position of the cam can usually be altered within a range of from 2 % to 5 % of a single revolution of the shaft. The timer motor rotates the shaft through a gear train. Each cam-actuated switch can be used to control a separate load device.

In this activity, you will connect up a simple repeat cycle timer and test its operational cycle. Sequence changes and cycle timing alterations will also be made. As a general rule, repeat cycle timer cam adjustments vary to some extent with each manufacturer. The timer used in this activity is only representative of one adjustment method. Specific adjustment procedures for timer devices must conform to the directions supplied by the manufacturer.

Equipment and Materials

- AC power source – 120 V
- Motor-driven repeat cycle timer – Automatic Timing & Controls Co. Series 324C 02-B2D-R-1-A-01-X, General Time Corp. RJ Series Unit or equivalent
- Incandescent lamp and socket
- SPST switch

DOI: 10.1201/9788743810124-89

Safety

Safety is always a primary factor for laboratory activities. This lab should only be completed with supervision.

You may complete certain steps of the "Procedure" and the "Analysis" for better understanding without applying voltage.

Be sure to wear eye protection while electric motors are in operation. Also, be very careful when working with high voltages.

Procedure

1. Refer to the manufacturer's data sheet for the repeat cycle timer, Figure 83.1, before proceeding.

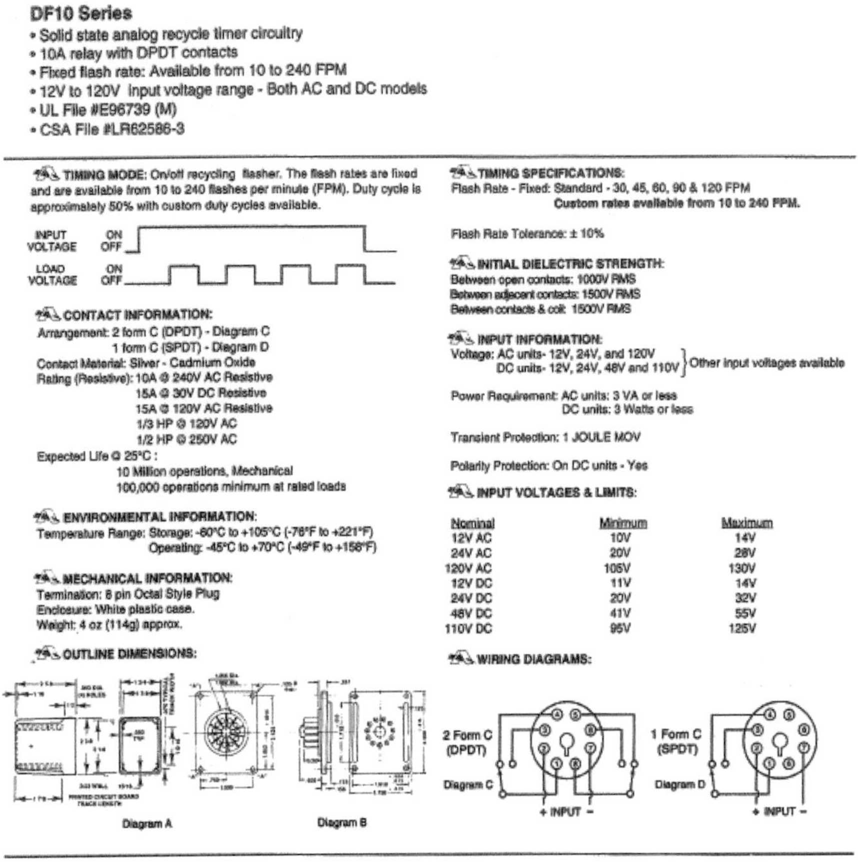

DF10 Series

- Solid state analog recycle timer circuitry
- 10A relay with DPDT contacts
- Fixed flash rate: Available from 10 to 240 FPM
- 12V to 120V input voltage range - Both AC and DC models
- UL File #E96739 (M)
- CSA File #LR62586-3

TIMING MODE: On/off recycling flasher. The flash rates are fixed and are available from 10 to 240 flashes per minute (FPM). Duty cycle is approximately 50% with custom duty cycles available.

INPUT VOLTAGE ON OFF
LOAD VOLTAGE ON OFF

CONTACT INFORMATION:
Arrangement: 2 form C (DPDT) - Diagram C
1 form C (SPDT) - Diagram D
Contact Material: Silver - Cadmium Oxide
Rating (Resistive): 10A @ 240V AC Resistive
15A @ 30V DC Resistive
15A @ 120V AC Resistive
1/3 HP @ 120V AC
1/2 HP @ 250V AC
Expected Life @ 25°C :
10 Million operations, Mechanical
100,000 operations minimum at rated loads

ENVIRONMENTAL INFORMATION:
Temperature Range: Storage: -60°C to +105°C (-76°F to +221°F)
Operating: -45°C to +70°C (-49°F to +158°F)

MECHANICAL INFORMATION:
Termination: 8 pin Octal Style Plug
Enclosure: White plastic case.
Weight: 4 oz (114g) approx.

OUTLINE DIMENSIONS:

Diagram A

Diagram B

TIMING SPECIFICATIONS:
Flash Rate - Fixed: Standard - 30, 45, 60, 90 & 120 FPM
Custom rates available from 10 to 240 FPM.

Flash Rate Tolerance: ± 10%

INITIAL DIELECTRIC STRENGTH:
Between open contacts: 1000V RMS
Between adjacent contacts: 1500V RMS
Between contacts & coil: 1500V RMS

INPUT INFORMATION:
Voltage: AC units- 12V, 24V, and 120V
DC units- 12V, 24V, 48V and 110V } Other input voltages available

Power Requirement: AC units: 3 VA or less
DC units: 3 Watts or less

Transient Protection: 1 JOULE MOV

Polarity Protection: On DC units - Yes

INPUT VOLTAGES & LIMITS:

Nominal	Minimum	Maximum
12V AC	10V	14V
24V AC	20V	28V
120V AC	105V	130V
12V DC	11V	14V
24V DC	20V	32V
48V DC	41V	55V
110V DC	95V	125V

WIRING DIAGRAMS:

2 Form C (DPDT) Diagram C + INPUT −

1 Form C (SPDT) Diagram D + INPUT −

Ordering Information:
Definition of a part number for the Amperite DF10 Series Flasher.
Example:

120 A F 60 DF10
A B C D E

A: Denotes nominal input voltage. Voltages Available: 12, 24, & 120V AC
12, 24, 48 & 110V DC
Custom Voltages are available.

B: Denotes type of current required for operation: A= AC - Alternating Current
D= DC - Direct Current

C: Denotes flasher configuration.

D: Denotes flash rate. Standard rates are 30, 45, 60, 90 & 120 FPM.
Custom rates are also available from 10 to 240 FPM

E: Denotes 10A DPDT 2 form C flasher - DF10 Series.

Figure 83.1 Repeat cycle timer used for timing applications. Study this specification sheet (Amperite Company, Inc.).

2. Construct the repeat cycle timer of Figure 83.2.

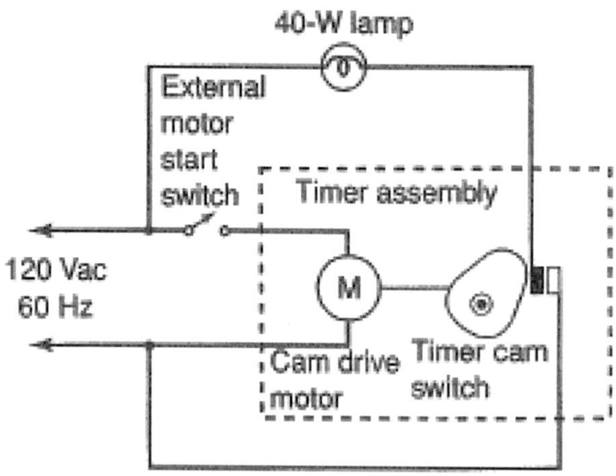

Figure 83.2 Repeat cycle timer.

3. Adjust the timer cam for an on and off operational cycle. Then turn on the external motor start switch and test the operational cycle. The circuit should repeat itself as long as the external motor start switch is in the on position.

4. How long is the lamp on and off during a cycle of operation?

   On = ________ seconds; Off = __________ seconds.

5. Modify the repeat cycle timer to form a stop cycle operation as indicated in Figure 83.3.

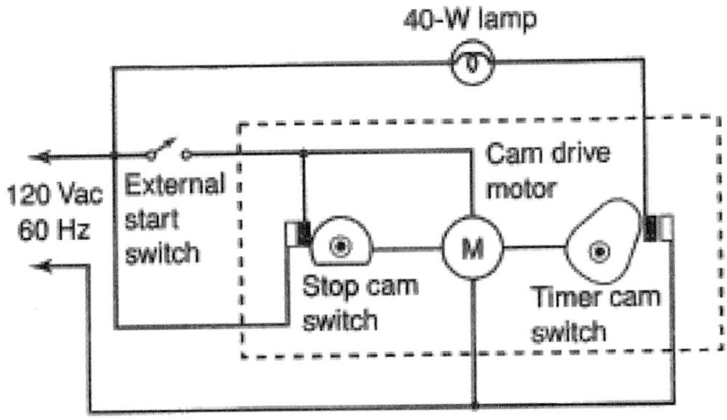

Figure 83.3 Timer with a stop cycle operation.

6. To test the operation of this circuit, the external motor start switch must be turned on for at least 1% of the cycle time. It can then be opened. At the end of one operational cycle, the stop cam switch turns off the motor. To repeat the cycle, the external start switch must be turned on again.
7. If time permits, you might want to change the cam setting for a different range of control. If multiple cams are available on your timer, you can connect the unit to achieve control of the alternate load.
8. Disassemble your circuit and return all components.

Analysis

1. Describe the basic action of the cam switch of a motor-driven cam timer.

2. How many control operations can the timer used in this experiment achieve during an operational cycle?

3. Describe an application for a repeat cycle timer.

Activity 84

Thermal Time-delay Relays

Objectives

Thermal relay devices are quite common in industrial equipment, and they have been used for industrial applications for a number of years, and continue to be rather popular. Thermal timers are discrete devices designed to control circuit operation through the thermal expansion principle. When certain metals are heated, their expansion characteristic commonly causes a bending of the metal. Through this principle, it is possible to force electrical contacts to come together or to open after a prolonged period. Devices that do this are called thermal time-delay relays because they are housed in a self-contained glass or metal package. Hermetically sealed units are not affected by atmospheric conditions, moisture, dust, or altitude.

The heat source of a typical thermal timer is provided by a resistive element mounted inside the enclosure. When the element is energized by an outside source, it causes normally open (NO) contacts to close and the normally closed (NC) contacts to become open. When the element source is removed, the contacts change back to their original state. As a general rule, the element can be energized by either AC or DC. Most timers of this type are designed to have a fixed timing action at the rated heater element voltage. At reduced voltage, however, timing action can be altered somewhat.

In this activity, you will construct an electrical circuit that utilizes a thermal relay. You will be able to test the actual expansion of a control element when it is energized. You will also alter the timing range of the device by decreasing the heating element energy source voltage.

DOI: 10.1201/9788743810124-90

Equipment and Materials

- Multimeter
- AC power source – 6 V
- Variable DC source – 0 to 6 V
- Thermal time-delay relay – Amperite 6 NO 30 T or equivalent
- 9-pin miniature socket
- Circuit construction board and connecting wires
- Incandescent lamp and socket
- SPST toggle switch

Procedure

1. Study the thermal time-delay relay through the glass enclosure. See if you can identify the two heater pins. After deciding upon the two pins, measure the resistance with your multimeter.

   Coil resistance = _______ Ohms.

2. Examine the specification sheets in Figure 84.1.

(Figure A)

(Figure B)

(Figure C)

(Figure D)

(Figure E)

(Figure F)

(Figure G)

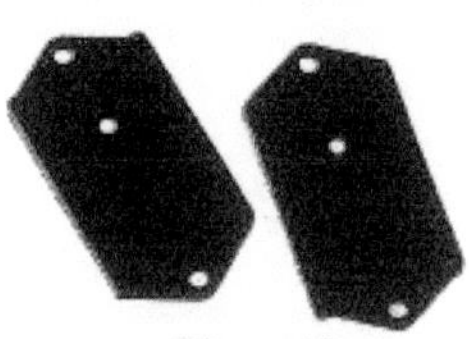
(Figure H)

SOCKETS

TYPE	AMPERITE PART #	USED WITH THESE AMPERITE SERIES
Chassis Mount Octal Socket (Figure A)	Octal	C10, CI, DC10, DF10, G, GF
Surface Mount Octal Socket (Figure B)	8-Pin-SM	C10, CI, DC10, DF10, G, GF
Surface Mount 11-Pin Octal Style Socket (Figure C)	11-Pin-SM	CR10, CIR
Chassis Mount 9-Pin Socket (Figure D)	9-Pin-SM	G, GF
P.C. Mount 9-Pin Socket (Figure E)	9-Pin-PC	G, GF

QUICK CONNECTS

.110 Female Quick Connect Terminals (Figure F)	110 Female	B, BF, BR, C, CR, D
.250 Female Quick Connect Terminals (Figure G)	250 Female	B, BF, BR, C, CR, D, DF, DFW

BRACKETS

2-Screw Panel Mount Bracket (Figure H)	Panel Mount	B, BF, BR, C, CR, D, DF

A

Figure 84.1 Specification sheets. (A) Sockets and mountings used with some types of thermal timers. (B) Description of a G Series timer (Amperite Company, Inc.).

G Series

- Hermetically sealed
- Delay on Make or Delay on Break timing modes
- Thermal device
- 3 AMP rating
- 1 - 115V input voltage range - works on AC or DC
- Isolated output contacts
- Fixed delay times only
- Initial and reset (release) delay device
- Long life
- UL File #E96739 (M)

TIMING MODE: Timing cycle begins upon application of power to the heater terminals. At the end of the initial delay time the relay contacts transfer and remain in a transfered state until input power is removed . When the heater input power is removed, the contacts transfer back to their original state at the end of a reset (release) delay period.

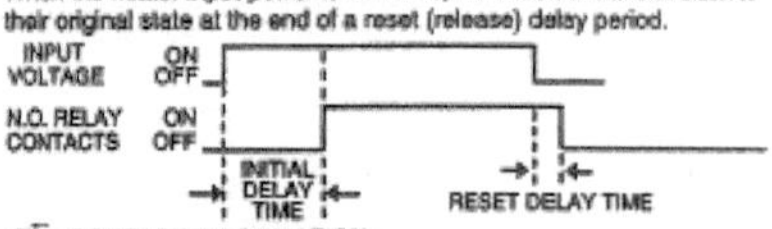

CONTACT INFORMATION:

Arrangement: 1 form A (SPST - Normally open) - Delay on Make
1 form B (SPST - Normally closed) - Delay on Break

Contact Material: Silver - Cadmium Oxide
Rating (Resistive): 3A @ 115V AC
Expected Life @ 25°C :
500,000 operations minimum at rated loads

ENVIRONMENTAL INFORMATION:
Temperature Range: Operating & storage: -55°C to +80°C, (-67°F to +176°F)

TIMING SPECIFICATIONS:
Timing - Fixed: 1 through 300 secs - (octal style) or
1 - 120 secs. (9-pin miniature style)
Timing Tolerance: ± 20% - Tighter tolerances are available.
Repeatability: ± 5%
Release Time: Contact factory
Timing Cycle Interrupt Transfer: none

INITIAL DIELECTRIC STRENGTH:

1 - 10 Second Type:	15 - 300 Second Type:
Between open contacts: 250V RMS	Between open contacts: 800V RMS
Between contacts & coil: 500V RMS	Between contacts & coil: 500V RMS

INPUT INFORMATION:
Voltage: AC or DC - 6V, 12V, 26V, 50V and 115V
(Other voltages are available)

Power Requirement: 2.0 Watts approx.

Transient Protection: impervious to transients

Polarity Protection: None required

INPUT VOLTAGES & LIMITS:

Nominal	Minimum	Maximum

Temperature Range: Operating & Storage:

Nominal	Minimum	Maximum
6V AC/DC	4V	8V
12V AC/DC	10V	14V
26V AC/DC	22V	30V
50V AC/DC	42V	58V
115V AC/DC	90V	130V

MECHANICAL INFORMATION:
Termination & Enclosure: Octal style, or 9-pin miniature style glass envelope. See Diagram A & B.

Weight: 1 oz (28g)

OUTLINE DIMENSIONS:

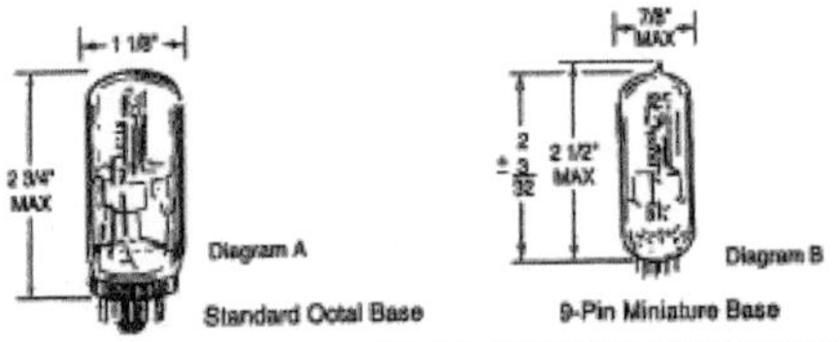

Diagram A — Standard Octal Base

Diagram B — 9-Pin Miniature Base

WIRING DIAGRAMS:

Base Wiring 9-Pin Miniature
Pins 1 & 5 - Heater
Pins 3 & 4 - First Contact
Pins 8 & 9 - Second Contact

Base Wiring Standard Octal
Pins 2 & 3 - Heater
Pin 5 - First Contact
Pin 7 - Second Contact

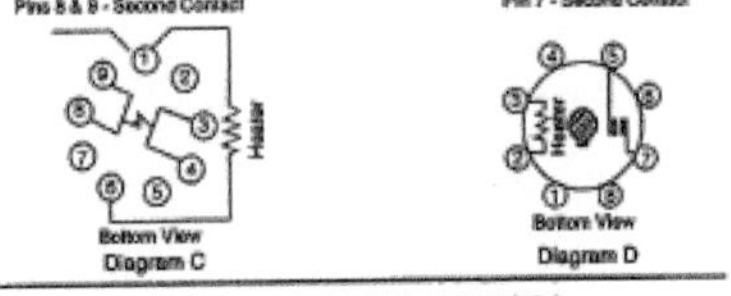

Diagram C

Diagram D

Ordering Information:
Definition of a part number for the Amperite G Series Time Delay Relay.
Example:

115 NO 60 T
A B C D

A: Denotes nominal input voltage. Voltages Available: 6, 12, 26, 50 & 115V AC/DC
Custom Voltages are available.

B: Denotes contact form: NO = normally open (Delay on Make) - 1 form A - SPST
C = normally closed (Delay on Break) - 1 form B - SPST

C: Denotes timing value: Factory preset time delays from 1 - 300 secs. are available (octal style) and 1 - 120 secs. (9-pin miniature style).

D: Denotes type of glass envelope: Blank = octal style. T = 9-pin miniature style.

3. Study the base diagram of the thermal time-delay device used in this activity. Make a sketch of the base diagram of this device in the space that follows.

4. If the thermal time-delay device used has a number, record it. _______.
The number of this device often reveals a great deal about its operating characteristics. An Amperite type 24 NO 120 T, for example, denotes a 24 V heater, normally open contacts, and a 120-second time delay.

5. Identify the operating characteristics of the thermal time-delay in Step 4.

6. In the space that follows, sketch a wiring diagram showing the thermal time-delay relay controlling a 7 W incandescent lamp. Energize the relay with 6.3 Vac and the lamp with 120 V at 60 Hz. Connect an SPST switch in series with the heater coil so that it can be easily turned on and off.

7. Have your supervisor approve your circuit before applying power to the delay device or lamp.

8. Apply energy to the lamp part of the circuit first. Then turn on the heater coil switch and determine the delay time of the circuit.

Delay time = ________ s.

9. Turn off the circuit and disconnect the source from the heater coil. Connect a variable DC source in its place. Adjust the DC source to its zero volts position before connecting it to the heater coil part of the circuit.

10. Prepare the multimeter to measure DC voltage, and connect it to the variable DC source. Adjust the DC to 6 V. Then, turn on the heater switch and record the delay time.

    Delay time = __________s.

11. Prepare the multimeter to measure the DC current in the heater coil part of the circuit.

    Heater coil current = ________A.

12. Using the applied heater voltage and current, calculate the operating coil resistance.

    Operating coil resistance = ________.

13. How does this resistance compare with the cold coil resistance of Step 1?

14. Turn off the heater coil switch and adjust the source to 5 Vdc. Turn on the switch

    Delay time = ________ s.

15. Measure and record the heater current when supplied with a source.

    Heater current = ________ A.

16. Plot a time delay versus heater coil voltage curve on the graph of Figure 84.2.

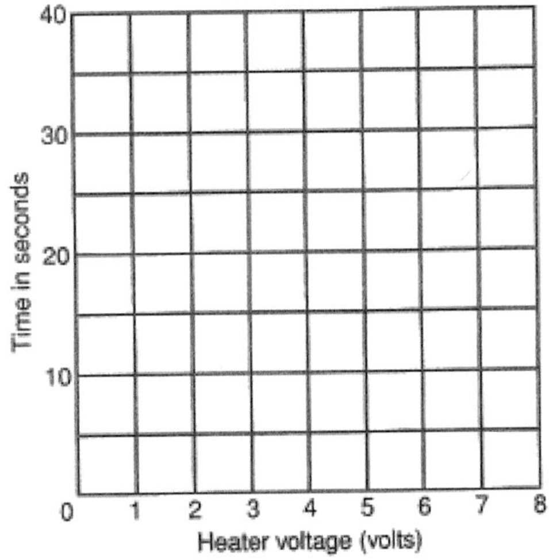

Figure 84.2 Characteristics of thermal time-delay relay.

17. Disassemble your circuit and return all components.

Analysis

1. What would be an application of the thermal time-delay device in an industrial installation?

2. What would be the advantages and disadvantages of using a thermal time-delay device in a delay circuit?

3. Since the DC heater coil voltage influences the time delay, how could the delay time of this device be altered when a fixed DC source is used?

Activity 85

Time Constant Circuits

Objectives

Inductance opposes any change in current in a circuit, whereas capacitance opposes any change in voltage. In both instances, the reaction time associated with inductance and capacitance in opposing changes to current and voltage is dependent upon resistance. The time, in seconds, required for a capacitor or an inductor to react to a change is determined by a quantity known as a time constant. Circuits created with these time constants can be used to produce electronic timers.

In this activity, you will observe the time constant of a capacitor as it charges and discharges through a resistive path.

Equipment and Materials

- Multimeter
- Variable DC power supply
- Resistors – 100 K, 220 K Ohm
- Capacitor – 47 uF
- SPST switches (2) or one DPDT
- Connecting wires

Procedure

1. Construct the circuit in Figure 85.1.

DOI: 10.1201/9788743810124-91

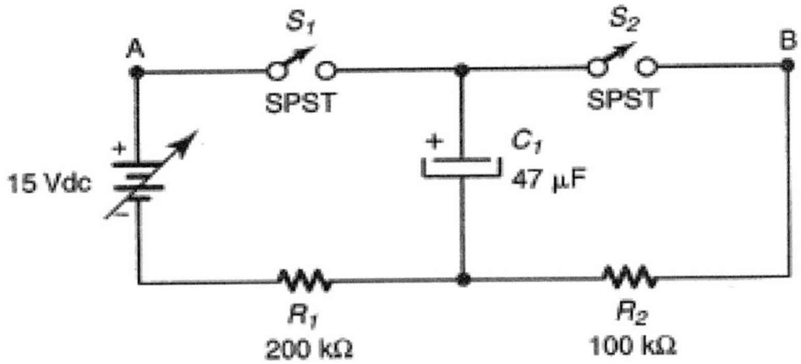

Figure 85.1 RC time constant test circuit.

2. When SPST switch is closed and SPST switch is open, the capacitor charges through. After the capacitor is charged, it is discharged through by opening switch and closing switch. Compute the time constant for charging and discharging the capacitor circuit.

   Charging = _______ s.

   Discharging = _______s.

3. Compute the total time required for the capacitor to charge to 25 V when S_1 is closed and S_2 is open.

   Total charge time ____s.

4. Compute the total time required for the capacitor to discharge 25 V when S_1 is open and S_2 is closed.

   Total discharge time ____s.

5. Compute the total charge and discharge currents for the circuit shown in Figure 85.1.

   Total charge current = _______ mA.

   Total discharge current = _______ mA.

6. Prepare your multimeter to measure direct current in the 1 mA range and connect it in series with R1 and C1 at point A in the circuit of Figure 85.1 (note: both S1 and S2 must be open).

7. Close and record the maximum charging current.

   Charging current = ________mA.

8. Describe and explain the action of the circuit current as the capacitor charges.

9. How did the computed charging current in Step 5 compare with the measured charging current in Step 7?

10. Disconnect the meter, open S1, and close S2.

11. With the meter prepared to measure the direct current described in Step 6, connect this meter in series with C1 and R2 at point B in the circuit (note: both S1 and S2 must be open).

12. Close S1 for five or six seconds. Open S1, close S2, and record the maximum discharge current.

    Discharge current = ________ mA.

13. Describe and explain the action of the current in the circuit as the capacitor discharged.

14. How did the computed discharge current in Step 5 compare with the measured discharge current in Step 12?

15. How did the charge and discharge currents of the capacitor compare? Why were they so different?

16. Disconnect the meter and open.

17. Prepare the meter to measure DC volts (25 V or higher) range and connect it across R1 in the circuit.

18. Close S1 and describe the voltage drop across R1 as the capacitor charges.

19. Disconnect the meter and place it across R2.

20. Open S1 and close S2. Describe the voltage drop across R2 as the capacitor discharges.

21. As the capacitor is charging, the current and the voltage across R1 decreases. When the charging current and voltage across R1 reach zero, the capacitor is fully discharged. Likewise, as the capacitor discharges, the discharge current and voltage drop across R2 decreases. When the discharge current and voltage across reach zero, the capacitor is fully discharged.

22. Disconnect the meter and open S1 and S2.

23. Connect the meter across R1, as described in Step 17.

24. Using a watch with a timer or a second hand, measure and record the time required for the capacitor to charge to 100 % of the source voltage when S1 is closed. This time corresponds to the time before the meter will measure zero voltage across R1.

    Charging time = ________s.

25. How did the measured time in Step 24 compare with the computed time in Step 3?

26. Connect the negative lead of the meter to the negative side of the capacitor. Do not connect the positive lead of the meter into the circuit at this time.

27. Open S1 and quickly connect the positive lead of the meter to the positive side of the capacitor. Record the maximum voltage across the capacitor at the instant the positive lead is connected to the capacitor.

    Voltage across ____________ V.

28. How did the voltage measured in Step 27 compare with the source voltage?

Analysis

1. What is the time constant of an RL circuit?

2. What is the time constant of an RC circuit?

3. How many time constants are required for the current through an RL circuit to reach its maximum value?

4. How many time constants are required for the voltage across a capacitor in an RC circuit to equal the source voltage?

5. How can the length of an RL time constant be increased? Decreased?

6. How can the length of an RC time constant be increased? Decreased?

Activity 86

Digital Timers

Objectives

Digital timers are a rather recent addition to the timer field. This type of timer employs a time base generator or oscillator, a counter, and a load driver. The counterpart of the system frequently has several output terminals that permit some selection of time ranges. Control ranges of 1T, 2T, 4T, 8T, 16T, and 32 T are typical. Each output is rated in terms of the time base (T) of the generator.

In this activity, you will employ the SE/NE555 timer IC as the time base generator of a simple digital timing system. The output of the generator will be counted and divided to some extent by the range of the system. Through this circuit investigation, you will see and actually work with the basic parts of a simple digital timing system. Commercially designed timers of this type are usually housed in an enclosure that does not permit access to component parts. These timers generally do not employ moving parts and are all solid state. The timing period is very accurate and usually ranges from microseconds to hours.

Equipment and Materials

- Power supply – 5 Vdc
- SE/NE555 IC
- SN7490 IC
- Resistors – 390 Ohm, 220 K Ohm (2)
- Capacitors – 0.01 uF, 4 uF, 50 uF
- Light-emitting diode

DOI: 10.1201/9788743810124-92

- SPST toggle switches (or push button switches)
- Circuit mounting board and connecting wires

Procedure

1. Construct the experimental digital timer circuit of Figure 86.1 by first connecting the time base generator. Do not energize the SN7490 at this time.

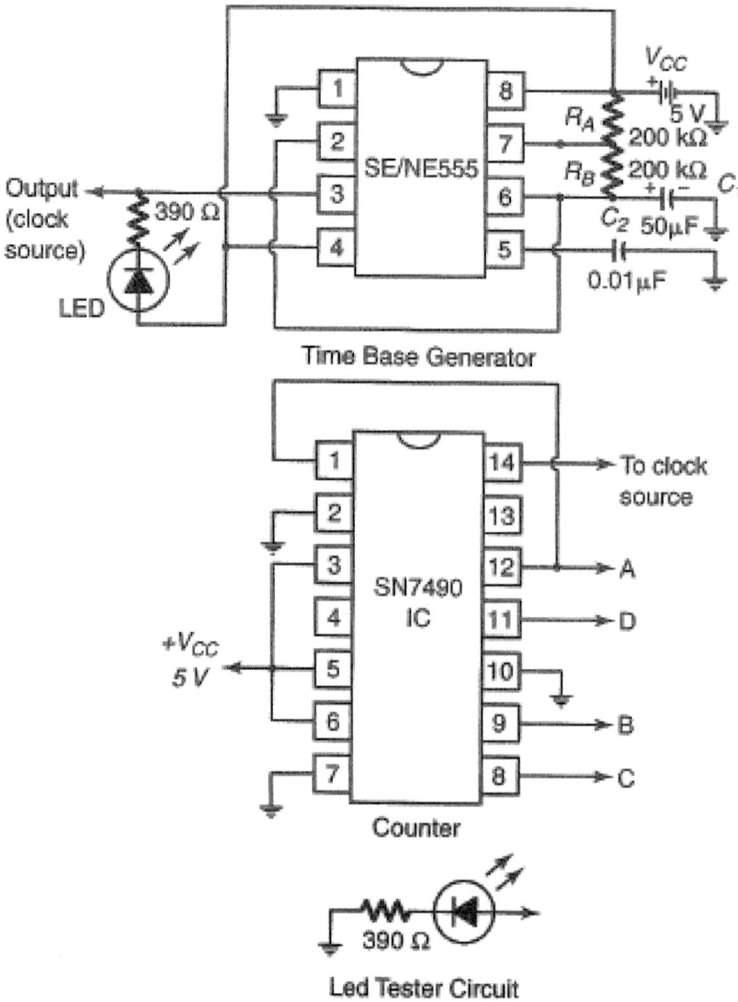

Figure 86.1 Digital timer circuit.

2. Calculate the charging time of the circuit. Test the accuracy of the circuit. Record your findings

3. Calculate the discharge time of the circuit. Test the accuracy of the circuit. Record your findings.

4. Turn off the power supply and complete the SN7490 counter IC. Connect it to the same power source used by the 555. Connect the second LED indicator to the D output of the 7490 IC. Connect the IC clock source to the output (pin 3) of the 555.

5. Turn on the power source and count the number of timing cycles needed to energize the 7490 LED readout. What mathematical function does this represent?

6. Turn off the power source and move the LED readout of the 7490 to output A. Turn on the power again and determine the cycles needed to energize the LED. _______ timing cycles.

7. What mathematical function does this represent? _______________

8. Turn off the power supply and alter the SN7490 circuitry as indicated in Figure 86.2. Change C_1 of the time base generator to 4uF.

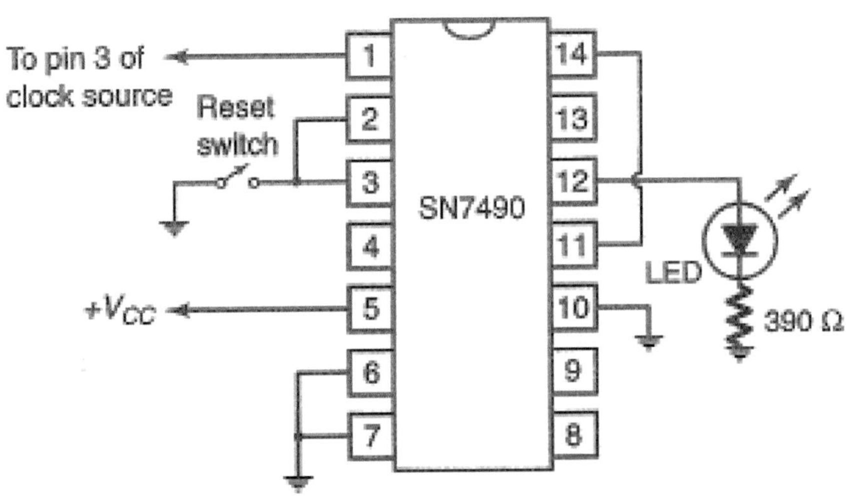

Figure 86.2 Modified counter circuit.

9. Turn on the power supply and open the reset switch. Close the switch during a time when the time base generator LED is off. Then count the number of timing cycles needed to energize the SN7490 LED.

   ________ timing cycles.

10. Count the number of timing cycles needed to turn off the LED. What does this represent as an output of the IC?

Analysis

1. How could the digital timer of this experiment be extended? Name two ways.

2. How can the digital timer of this experiment be made variable?

3. Would it be advantageous to have a reset switch for an actual digital timer? Why?

Activity 87

555 Astable Multivibrators

Objectives

An astable multivibrator repeats itself at the end of each operational cycle. A circuit of this type responds as a clock. The circuit triggers itself into a stage change automatically. Capacitor (C) charges through R_A and R_B and discharges through R_B. The operational cycle is dependent on these components. The charge time, discharge time, and total operational time can be determined by calculation as described in the text.

In this activity, you will:

1. Construct a clock with a 555 precision timer.
2. Analyze the operation of a 555 clock.
3. Determine the period of operation and frequency of a 555 clock.

Equipment and Materials

- Multimeter
- Power supply – 5 Vdc
- SE/NE555 IC
- Resistors – 100 Ohm (2), 1K Ohm (2)
- Capacitors – 100 uF, 1,000 uF
- Silicon diode – Texas Instruments 1N4001 or equivalent
- LEDs – 20 mA, 2 V (2)
- Circuit construction board and connecting wires

DOI: 10.1201/9788743810124-93

Procedure

1. Construct the astable multivibrator clock circuit of Figure 87.1.

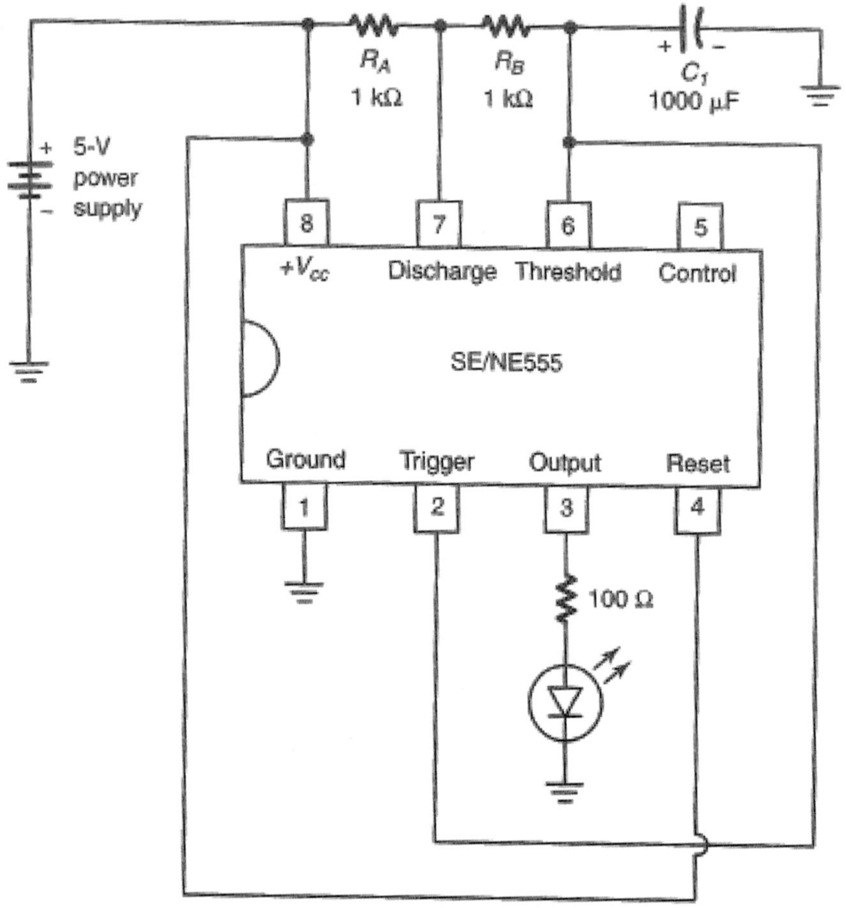

Figure 87.1 Multivibrator clock circuit.

2. Turn on the power supply and test the operation of the circuit. If the circuit is functioning properly, the LED will flash on and off. If this occurs, proceed with Step 3. If it does not, check each circuit connection for possible errors. Then measure the voltage at pins 4 and 8. It should be 5 V. To evaluate the operation of the RC circuit, measure the

voltage at pin 6 with respect to ground. This voltage should rise in value to 2/3 of V_{CC} and drop to 1/3 of V_{CC}. This voltage change is necessary to make the circuit change states. If it does not work, turn off the power and measure the resistance and continuity of each component.

3. Determine the charge time of the capacitor.

   Charge time = ________s.

4. Determine the discharge time of the capacitor.

   Discharge time = ___ ____s.

5. Determine the frequency of the circuit.

   Frequency = ________ Hz

6. The charge time of the capacitor depends on the voltage change that occurs at the threshold terminal (pin 6). This voltage should rise in value to 2/3 of V_{CC} and then fall in value to 1/3 of V_{CC}. With a multimeter, measure the threshold voltage (V_T).

   V_T = _______to ________V.

7. Calculate the change of the threshold terminal.

   Calculated threshold voltage change is ______ to ______ V.

   How does this compare with the actual measured voltage change of the circuit?

8. Turn off the power supply.

9. Modify the circuit of Figure 87.1 to conform to Figure 87.2. This change is rather minor. The value of C_1 is changed from 1000 uF o 100 uF. In addition to this, a diode is connected across, and a reverse-connected LED is attached to the output. The circuit will show the on and off times of the clock to be approximately the same. C1 will now charge through only R_A and discharge through R_B. The diode is forward-biased during the charging time and reverse-biased during the discharge time. As a result, the output will be a symmetrical square wave.

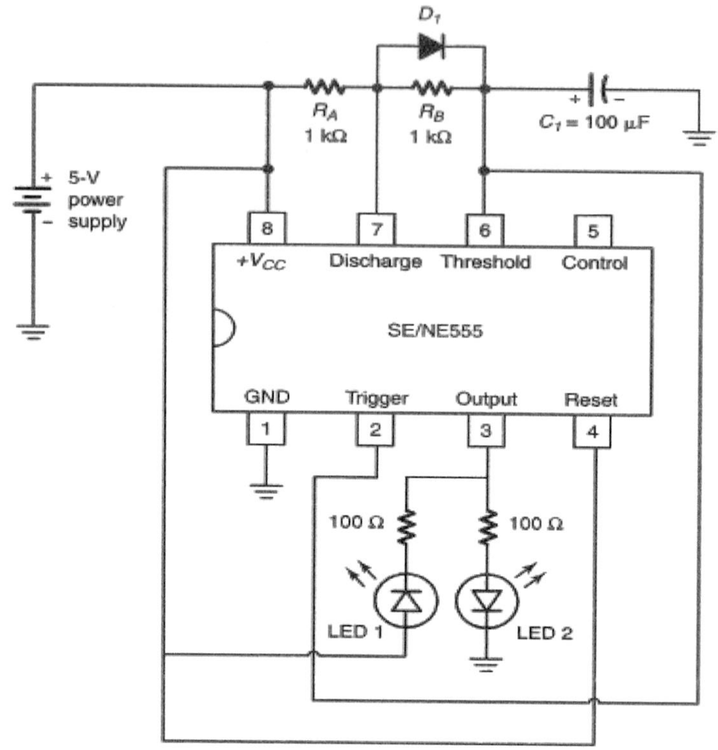

Figure 87.2 Modified multivibrator clock circuit.

10. Turn off the power supply. If the circuit is operating satisfactorily, the LEDs will alternately flash on and off. The on and off times should be equal. How does this circuit compare with Figure 87.1?

11. With a multimeter, measure the voltage across C1.

    Range of voltage change is ________ to ________ V.

12. Calculate the charge time and discharge time for this circuit.

    T_C = ______ s T_D = _______ s.

13. Determine the operational frequency of the circuit in Figure 87.2.

    f = ________ Hz.

14. Turn off the power supply and disconnect the circuit. Return all materials.

Analysis

1. If the clock circuit used in this activity has C changed to 1 uF, how would it alter the frequency of operation?

2. If resistors R_A and R_B were changed to 10 K Ohms, how would it alter the operation of the circuit?

3. Why does putting a diode across change the output to a symmetrical square wave?

Activity 88

IC Interval Timers

Objectives

The term interval timer is used to describe a function where the load is turned on at the start of an operational cycle and kept on for a specific period. At the end of the interval, the load is turned off automatically. Mixing operations, photographic timers, and automated machinery operate with this type of timer.

The 555 timer IC will be used in this experiment to demonstrate the interval timing principle. Essentially, this circuit is the same as the delay timer, with only the polarity of the load control voltage being changed. You will construct a fixed interval timer and a variable device and calculate the time constant. Through this experiment, you will see how the 555 can be used to achieve another unique timing operation.

Equipment and Materials

- Multimeter
- Power supply – 5 Vdc
- SE/NE555 IC
- Resistors – 390 Ohm, 220 K Ohm
- Potentiometer-1 M Ohm
- Capacitor – 0.01 uF, 50 uF
- SPST toggle switches (2) or push buttons
- Light-emitting diode
- Circuit construction board and connecting wires.

DOI: 10.1201/9788743810124-94

Procedure

1. Construct the interval timer of Figure 88.1.

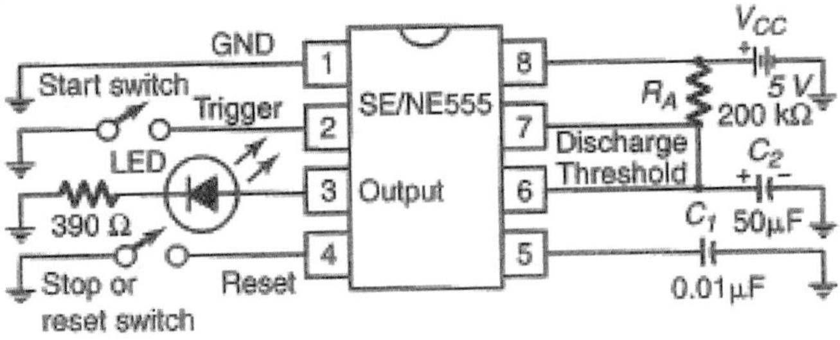

Figure 88.1 IC interval timer.

2. Turn on the 5-V power supply and momentarily turn on the start switch to start the **on** interval. Run several trial tests to verify the on interval.

   Average time is _____ seconds.

3. Calculate the interval as T = 1.1 RC = ___________ s.

4. The timing sequence can be reset after it has started by momentarily turning on the reset switch. The cycle can be started again without going through the complete interval.

5. Start the interval, then reset the operation to verify this function.

6. Using the nomograph in Figure 88.2, select a resistor-capacitor combination from any of the spares. Estimate the interval, insert R_A and C2 into the circuit, and test the circuit operation.

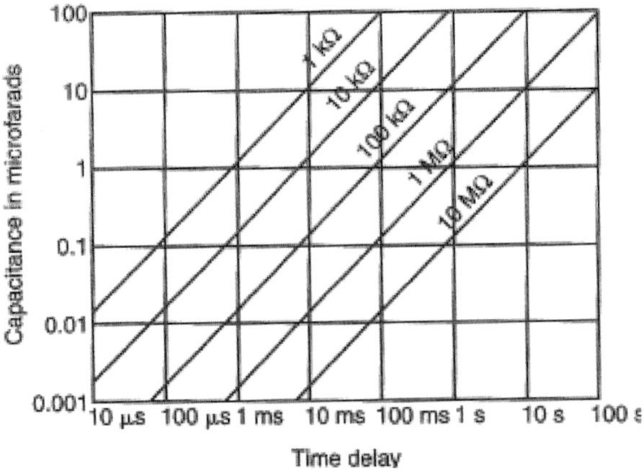

Figure 88.2 RC nomograph.

7. Prepare a multimeter to measure the DC voltage from pin 3 to ground. Start the interval and observe the time. How does it respond?

8. Place a 1 M potentiometer in place of R_A, making the interval variable. Test three positions of the potentiometer at the center, CCW, and CW for different time interval settings. If time permits, you may want to build a variable interval timer control that shows time on a calibrated dial.

9. Turn off the power supply and disconnect the circuit. Return all materials.

Analysis

1. What is the difference in an interval timer compared with a delay timer?

2. What are some of the practical limitations of an interval timer of this type?

3. What would be an application of this timer?

Activity 89

Delay Timer Loads

Objectives

The amount of load current controlled by an IC timer is a very important circuit consideration. The SE/NE555 IC timer can be used to control an output circuit of up to 200 mA of current. When larger current values are required, a relay or a driver transistor must be employed. The IC then serves as the driving source that actuates the load control device.

In this experiment, the 555 IC delay timer will first serve as a driver for a relay and then for a transistor load control device. The load can be energized by AC or DC in the relay circuit and only by DC in the transistor circuit. The transistor permits load control without moving parts or noise, which is a rather unique advantage over the relay control circuit. Control of AC and large currents is a unique advantage of the relay-type system. Through this activity, you will be able to see how the 555 timer IC can be used as a signal source for a load control system.

Equipment and Materials

- Multimeter
- DC power supply – 10 V
- AC power supply – 6.3 V
- SE/NE555 IC
- Resistors – 220 (2), 390, 200 K Ohm
- Potentiometer – 1 M Ohm
- Capacitors – 0.01 uF, 50 uF

DOI: 10.1201/9788743810124-95

- Transistor – 2N3397 or equivalent
- Diode – 1N4004 or equivalent
- Light-emitting diode
- Relay – 12 V, 125 Ohm Guardian 1335-2C-120D or equivalent
- SPST toggle switches (2) or push buttons
- Lamp – No. 47 or equivalent
- Circuit mounting board and connecting wires

Procedure

1. Construct the delay timer of Figure 89.1 with a relay.

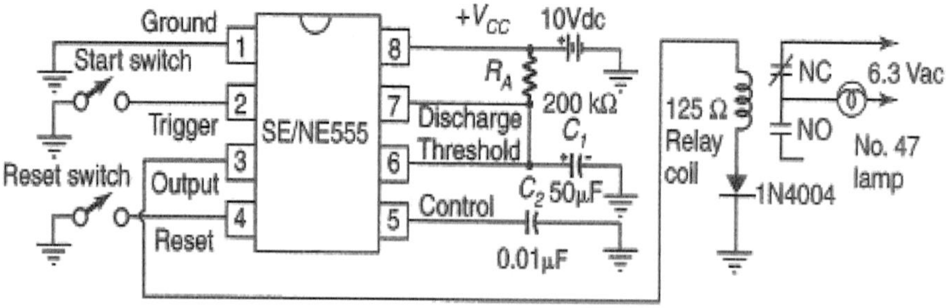

Figure 89.1 Delay timer controlling a relay.

2. The load to be controlled in this circuit is limited to approximately 2 A. This is based upon the contact point rating of the relay. The load lamp of this circuit is controlled by 6 Vac.
3. Initially, the load lamp should be on when the AC power is applied.
4. Turn on the DC power source. Test the circuit by momentarily closing the start switch. Then momentarily close the reset switch. Describe the switching action.
5. Calculate the delay time and verify its accuracy.

   Delay time (T) = __________ s.

6. Turn off the AC power source and DC power supply. Prepare the lamp for reverse control by reversing the NO and NC leads to the relay.

7. Apply AC power to the load and DC power to the circuit. Test the operation of the circuit. Describe the operating action of the circuit.

8. Turn off the power to the circuit and to the load lamp. Disconnect the relay from the circuit and replace it with a transistor lamp driver as indicated in Figure 89.2.

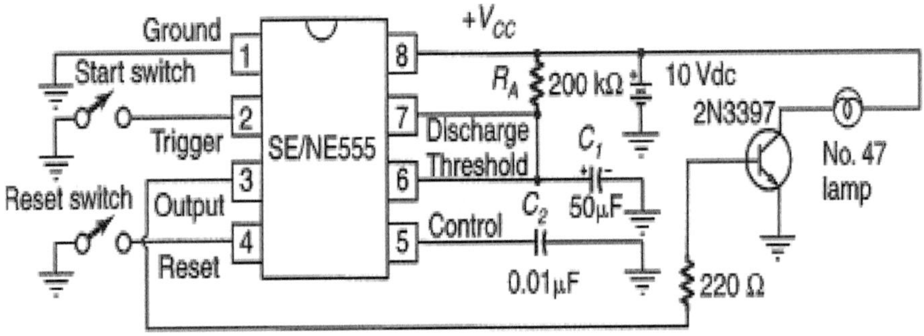

Figure 89.2 Delay timer controlling a lamp.

9. The amount of load current controlled by this circuit is primarily based upon the power rating of the driver transistor. In this circuit, the transistor will control only 1 A of current. The current handling capacity of the circuit could be increased with a higher current rating capacitor. Timers of this type are classified as solid-state electronic devices.

10. Turn off the power and disconnect the circuit. Return all materials.

Analysis

1. What are the advantages of an all solid-state timer over a relay-type delay timer?

2. Why is a diode placed in series with the relay coil?

3. Are there advantages of using solid-state control to drive a relay

4. Where could a delay-off timer be used in an industrial application?

5. What would be the problem in controlling the load of the transistor circuit with AC?

Activity 90

Mechanical Action Timers

Objectives

Mechanical action timers represent one classification of timing devices that achieve control by altering a mechanical change of some type. Fluid and air flows are typically forced to pass through small openings, called orifices, which restrict their flow. Hydraulic units, air or pneumatic devices, and mercury displacement are commonly used in this type of timer.

In this activity, you will construct a very simple pneumatic mechanical action timer that controls an incandescent lamp. The electrical circuit is switched on and off according to the delay action established by the timer. The device used in this activity is adjustable through a rather narrow range of time. You will test the switching action of the timer and adjust its delay action from one extreme to the other.

As a general rule, the pneumatic action device used in this activity is somewhat less expensive than comparable hydraulic and mercury displacement units. The resulting timing of the pneumatic unit is very similar in nearly all respects to the other units. After working with the pneumatic device, one should have a general idea of the capabilities of the other units without actually using the specific devices. All three mechanical action devices are very reliable and have a very long operational life expectancy. The accuracy is generally from 5 to 10 % of the set time position. Delay action usually does not exceed 60 minutes.

Equipment and Materials

- Multimeter
- AC power supply – 120 V

DOI: 10.1201/9788743810124-96

- Pneumatic timer – Eagle Signal 86-0-A0021 or equivalent
- Incandescent lamp and socket – 7.5 Watt
- Isolation transformer – 120 Vac

Safety

Safety is always a primary factor for laboratory activities. This lab should only be completed with supervision.

You may complete certain steps of the "Procedure" and the "Analysis" for better understanding without applying voltage.

Procedure

1. Prepare the multimeter to measure resistance in the R × 1 range. Measure the continuity between any two of the timer switch contacts. The recommended pneumatic timer for this activity should contain a SPDT switch. Determine the pivot terminal or hinge point of the electrical terminal.
2. Press the push button. If you have selected the correct hinge point of the switch, there should be a break in the continuity. The alternate switch terminal and hinge point should show continuity when the button is released.
3. After a short delay time, the switch will change back to its original condition. Make a sketch of the switch terminals in the space below. Point out the hinge or pivot terminal.
4. Connect the electrical part of the circuit shown in Figure 90.1.

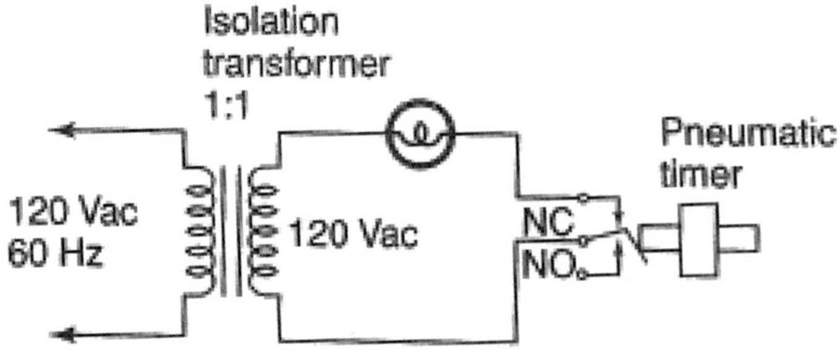

Figure 90.1 Pneumatic mechanical action timer.

5. Depress the push button. Explain the timing action immediately upon release of the push button. Test the circuit and record its initial delay time.

   Initial delay time = _______ seconds.

6. Turn the adjustment needle into the timer until it is firmly seated. Do not force this adjustment or use tools to turn it.

7. Depress the push button and determine the delay time.

   Delay time = _______ seconds.

8. Twist the adjustment needle out of the timer to its full limit of travel. Do not force this adjustment or use tools for this adjustment.

9. Depress the push button and determine the time delay.

   Delay time = _______ seconds.

10. The total range of adjustment is from _______ to _______ seconds.

11. Turn off the power supply and disconnect the circuit. Return all materials.

Analysis

1. Discuss the operation of a mechanical action timer.

Activity 91

Light-emitting Diodes

Objectives

In this activity, you will observe the characteristics of a light-emitting diode (LED). These devices are used for numerous applications in industry and elsewhere.

Equipment and Materials

- Multimeter
- Variable DC power source – 0–10 V.
- Light-emitting diode
- Resistor – 1 K Ohm

Procedure

1. Construct the circuit shown in Figure 91.1.

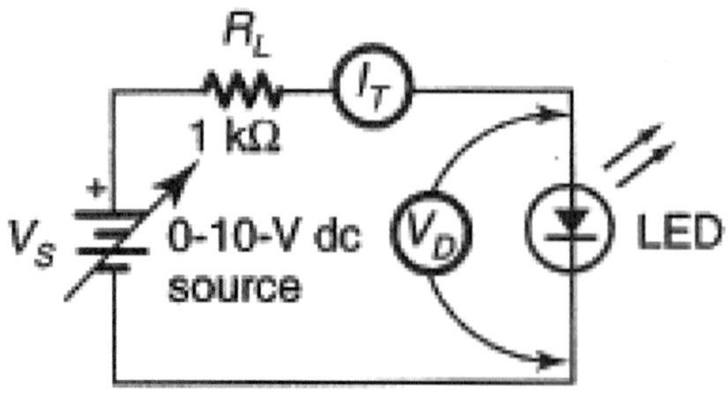

Figure 91.1 LED test circuit.

DOI: 10.1201/9788743810124-97

2. Before applying power, measure and record the forward and reverse resistance of the LED.

   Forward resistance = _______ Ohms.

   Reverse resistance = ________ Ohms.

3. Apply power and complete Figure 91.2.

4. Turn off the power supply and disconnect the circuit. Return all materials.

V_S	V_{R_L}	V_D	I_T	$R_D = V_D / I_T$	Light emitted*
1 V					
2 V					
3 V					
4 V					
5 V					
6 V					
7 V					
8 V					
9 V					
10 V					

*Light as detected by the human eye, i.e., dim, moderate, bright.

Figure 91.2 Table of LED characteristics. Fill in all values.

Analysis

1. Discuss the operation of an LED.

Activity 92

Photovoltaic Cells

Objectives

Many sources of electrical energy exist. These sources include chemical action (batteries, fuel cells), radioactive materials (atomic reactors), heat (thermocouples), pressure (piezoelectric crystals), conductive gases (magneto-hydrodynamic generators), electrical generators (moving coils and magnetic fields), and light.

Light is a form of energy that can be converted to electrical energy with very little effort. The most used device to convert light energy into electrical energy is the photovoltaic cell, sometimes called a sun or solar cell. Photovoltaic cells are used extensively in our space program to collect the rays of the sun and convert them into electrical energy. This converted energy is used to power the various circuits that control space satellites, lunar modules, and other spacecraft.

Photovoltaic cells are usually constructed of two layers of semiconductor material whose electrical characteristics have been altered by the addition of other elements or impurities. When the photovoltaic cell is exposed to light, the two layers interact to cause an excessive number of electrons to exist in one layer. The remaining layer, then, is deficient in electrons. This imbalance in the electrons causes a difference of potential (emf or voltage) to exist between the semiconductive layers. The difference in potential is directly proportional to the amount of light falling upon the cell. This voltage is used to cause current through a load. Thus, light can create electrical energy or power.

In this activity, you will examine the characteristics of the photovoltaic cell as it is used to convert light energy into electrical energy.

DOI: 10.1201/9788743810124-98

Equipment and Materials

- Multimeter
- Variable AC power supply
- Lamp with socket – 60 W
- Connecting wires
- Photovoltaic cell

Safety

Safety is always a primary factor for laboratory activities. This lab should only be completed with supervision.

You may complete certain steps of the "Procedure" and the "Analysis" for better understanding without applying voltage.

Procedure

1. Prepare your multimeter to measure DC volts. Complete the following connections as illustrated in Figure 92.1.

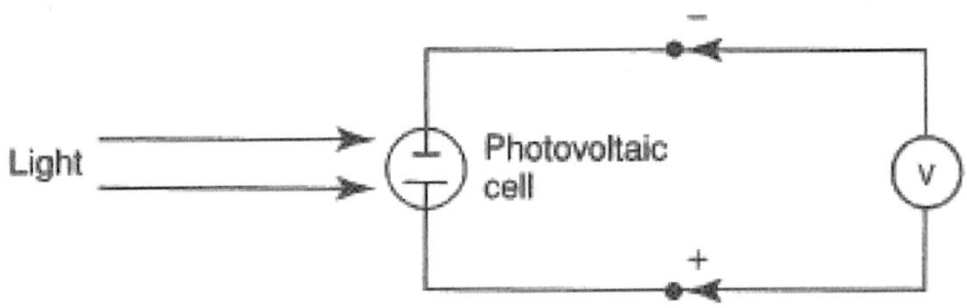

Figure 92.1 Setup for measuring the voltage produced by a photovoltaic cell.

2. Record the voltage output generated by the photovoltaic cell in total darkness and at

V = __________ V = __________

darkness room light

3. Connect the 60-watt lamp to the variable AC power supply as shown in Figure 92.2. The power supply must be adjusted to zero and must be turned off.

Figure 92.2 Light source for the photovoltaic cell.

4. Place the 60-watt lamp within ½ inch of the surface of the photovoltaic cell. Turn on the variable AC power supply and complete Figure 92.3 by adjusting the AC voltage as indicated. Chart your values in the graph of Figure 92.4.

Power supply voltage (Vac)	Generated cell voltage (mV)
20	
40	
60	
80	
85	
90	
95	
100	
105	
110	
115	
120	

Figure 92.3 Photovoltaic cell outputs.

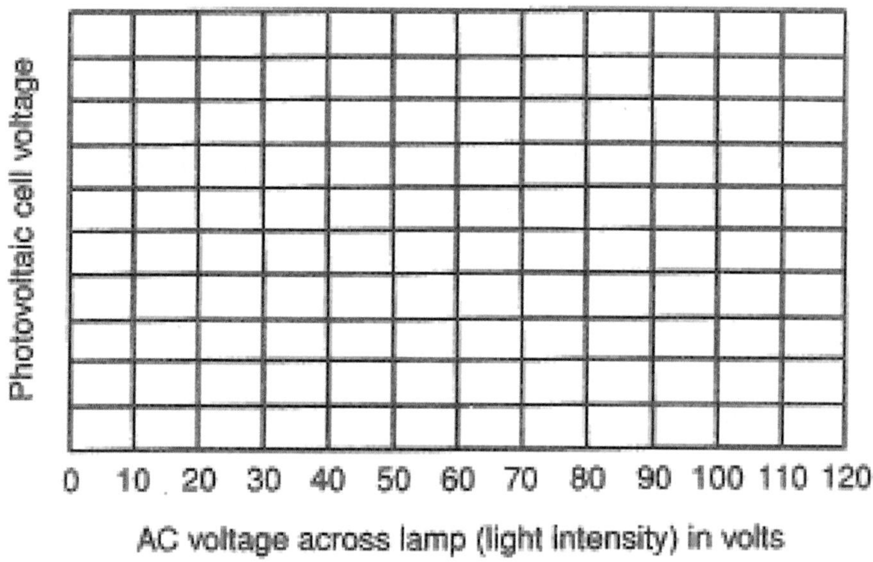

Figure 92.4 Graph of photovoltaic cell output voltage.

5. Using the graph, what is the relation between light intensity and generated photovoltaic cell voltage?
6. Turn off the power supply. Return all materials.

Analysis

1. List five sources of electrical energy.

2. Where might photovoltaic cells be used?

3. What are some advantages of converting light to electrical energy?

Activity 93

Light-activated Silicon Controlled Rectifier

Objectives

In this activity, you will observe the operational characteristics of a light-activated silicon controlled rectifier (LASCR). The LASCR is a four-layered solid-state device that can be triggered into conduction by focusing light onto its exposed surface. It operates as a conventional SCR since it requires a positive voltage on the gate to cause it to conduct. The major difference between the SCR and LASCR is that the LASCR will conduct with no positive gate voltage as long as a light source is focused onto its surface. The LASCR can be used to activate a larger SCR for control of larger loads.

Equipment and Materials

- Multimeter
- Oscilloscope
- DC power source – 6 V
- AC power source – 3 V
- LASCR
- Resistor – 22 Ohm
- Potentiometer – 1 M Ohm
- Lamp – 6 V

DOI: 10.1201/9788743810124-99

Procedure

1. Adjust the multimeter to the range. Connect the negative lead to the cathode and the positive lead to the anode of the LASCR. In Figure 93.1, record the anode–cathode resistance with and without light.

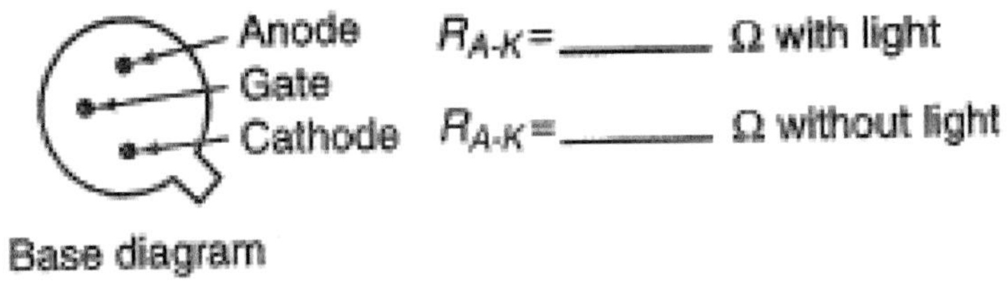

Figure 93.1 LASCR diagram and resistances.

2. Using the meter as the voltage source (connected as described in Step 1), trigger the LASCR with light, without using the gate. Now trigger the LASCR, without light, using the gate and jumper wire to the positive lead. It should be noted that the LASCR can be triggered with light and no gate connection, with gate potential and no light, or by using a combination of light and gate potential.

3. Construct the circuit shown in Figure 93.2.

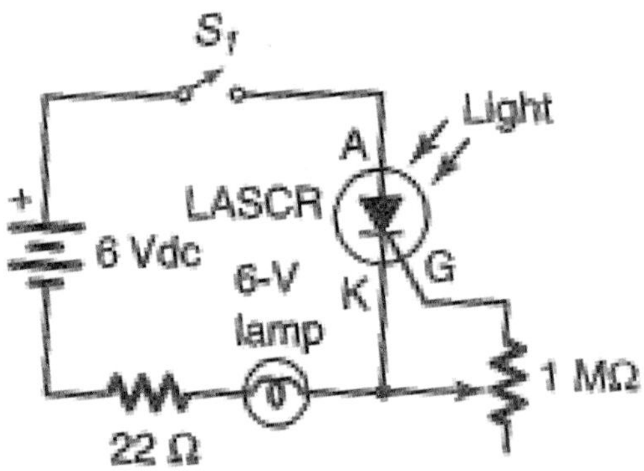

Figure 93.2 A DC LASCR circuit.

4. Place the LASCR in darkness and adjust the 1 M Ohm potentiometer to maximum resistance.

5. Close S1 and adjust the potentiometer until the LASCR activates. Disconnect the potentiometer and measure and record its resistance.

   R = _______ Ohms.

6. Open and adjust the potentiometer to maximum resistance.

7. Open and expose the LASCR to normal room light. Adjust the potentiometer until the LASCR activates. Disconnect the potentiometer and measure and record its resistance.

   R = ________ Ohms.

8. Construct the circuit shown in Figure 93.3.

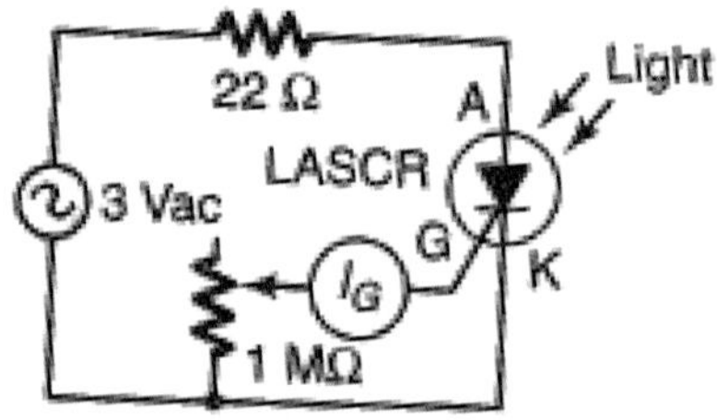

Figure 93.3 An AC LASCR circuit.

9. Adjust the potentiometer until the LASCR activates with normal room light.

10. Connect an oscilloscope across the LASCR (anode to cathode) and record in Figure 93.4 the waveform that appears.

11. Using an opaque object, partially shield the LASCR from light until the LASCR is delayed in firing.

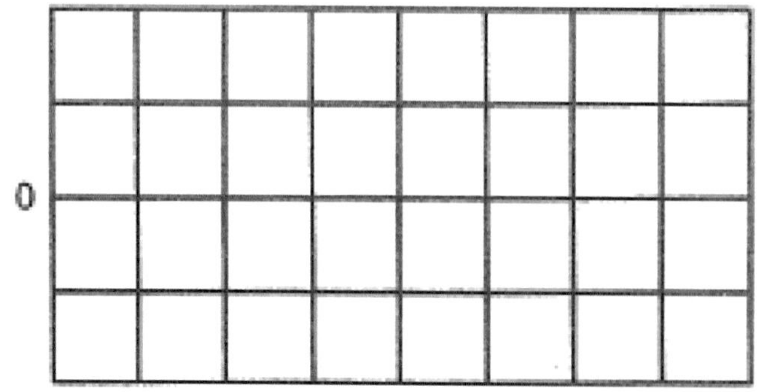

Figure 93.4 Observed LASCR waveform.

12. By altering the position of the opaque object, cause the waveform across the LASCR to appear as illustrated in Figure 93.5. Record the gate current that causes each waveform to be produced.

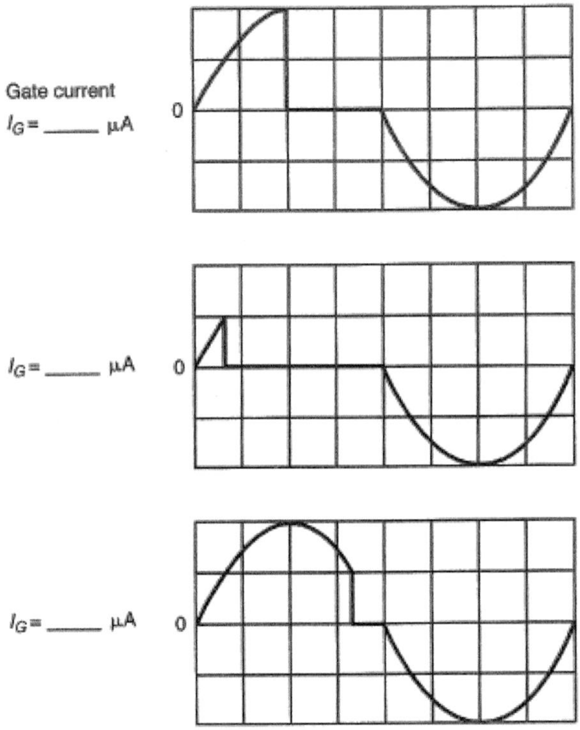

Figure 93.5 Observed gate current.

13. Turn off the power supply. Return all materials.

Analysis

1. Discuss the operation of the LASCR.

2. From the data of Steps 5 and 7, how do the resistances of the potentiometer compare? Why?

3. From the data obtained in the first circuit constructed, discuss the characteristics of the LASCR with and without light.

4. From the data of Step 12, as the conduction angle of the LASCR increases, what happens to the gate current?

5. Draw a circuit using a LASCR and an SCR that would control a high-current load.

Activity 94

Phototransistors

Objectives

In this activity, you will observe the characteristics of a phototransistor. These devices are ordinarily used to activate other semiconductor devices that are capable of handling greater amounts of current.

Equipment and Materials

- DC power source – 5 V
- Phototransistor – GE-X19 or equivalent
- Resistors – 1 K, 100 K Ohm
- Potentiometer – 50 K Ohm
- Lamp – 60 W

Safety

Safety is always a primary factor for laboratory activities. This lab should only be completed with supervision.

You may complete certain steps of the "Procedure" and the "Analysis" for better understanding without applying voltage.

Procedure

1. Using the lead diagram shown in Figure 94.1, measure and record the photo transistor emitter-base (E-B) resistance, collector-base (C-B)

DOI: 10.1201/9788743810124-100

resistance, and the emitter-collector (E-C) resistance when forward and reverse biased.

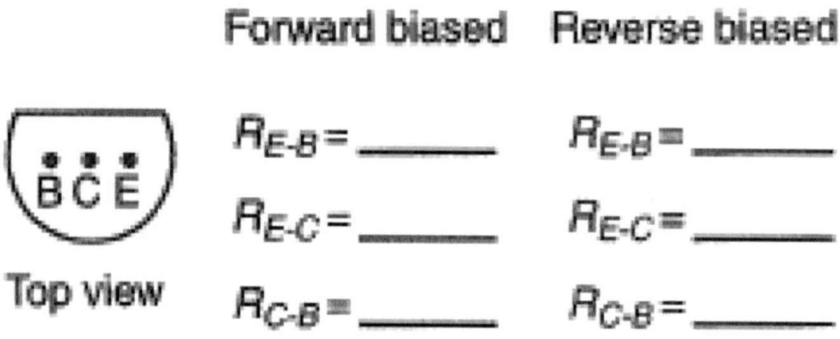

Figure 94.1 Phototransistor lead diagram and resistances.

2. Connect the positive lead of the meter to the collector and the negative lead to the emitter, and record this resistance with the transistor in darkness and in normal room light.

   Dark resistance = ________ Ohms.

   Light resistance = ________ Ohms.

3. Construct the circuit shown in Figure 94.2.

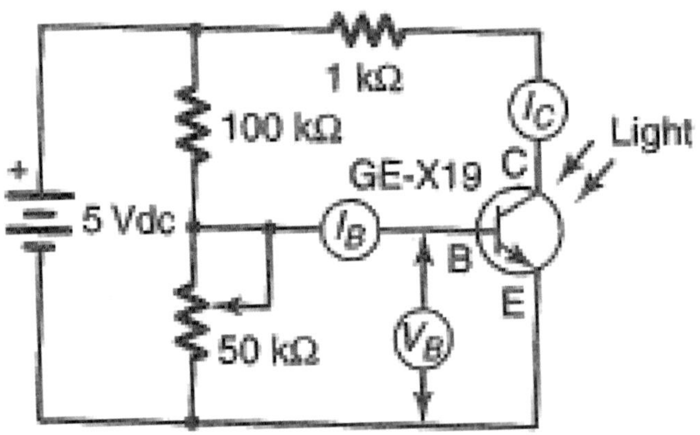

Figure 94.2 Photo transistor test circuit.

4. Adjust the potentiometer until the collector current is about 3 mA.

5. Place a lamp about 6 inches from the photo transistor (note: bulb should be adjusted to its maximum brightness).

6. Complete Figure 94.3 with and without the photo transistor exposed to light.

Light condition	V_B	$V_{E\text{-}C}$	I_B	I_C	$R_{E\text{-}C} = V_{E\text{-}C}/I_{E\text{-}C}$
With light					
Without light					

Figure 94.3 Photo transistor characteristics.

7. Turn off the power supply. Return all materials.

Analysis

1. As light intensity increases, according to the data of Figure 94.1, what happens to:

   a. Base voltage _____.

   b. Emitter-collector voltage _____.

   c. Base current _____.

   d. Collector current _____.

   e. Emitter-collector resistance _____.

2. Discuss the operation of a phototransistor.

Activity 95

Liquid Level Control

Objectives

Many industrial processes rely upon liquid level measurement. Sometimes variables such as fuel supply are monitored continuously. There are several techniques that can be used to measure the liquid level.

The measurement of liquid level is easy to accomplish by using transducers. Level changes result in the displacement of the top surface of the liquid. Many types of transducers are used to measure the liquid level. Resistive transducers can be used to measure the level of a conductive solution. A capacitive transducer can be used along with a movable plate whose position is determined by the level of the liquid. Photoelectric methods, radioactive methods, and ultrasonic methods can also be employed.

A simple type of level controller is the ball float system. This system uses a ball float to operate a lever. The lever is connected to a valve that regulates the liquid flow rate. Chemical industries commonly use differential pressure controllers to control the level of volatile liquids. The liquid pressure is proportional to its level in the enclosed container.

Level control can be accomplished by placing a light source at the same height above a conveyor line as the desired fill level of a container. In the illustration of Figure 95.1, containers on a conveyor line are positioned under a liquid dispenser. When the container is in position, the actuator causes the dispenser to allow liquid to pass into the container. When the liquid reaches the level of the light source, the light beam is interrupted. With no light striking its surface, the detector will cause a relay to activate. The activated relay will, in turn, cause the actuator on the dispenser to close. When another container is in position, the actuator will open once more. This liquid level control system ensures a uniform level of liquid in each container.

DOI: 10.1201/9788743810124-101

In this activity, you will construct and test a photoelectric circuit that can be used to measure the liquid level.

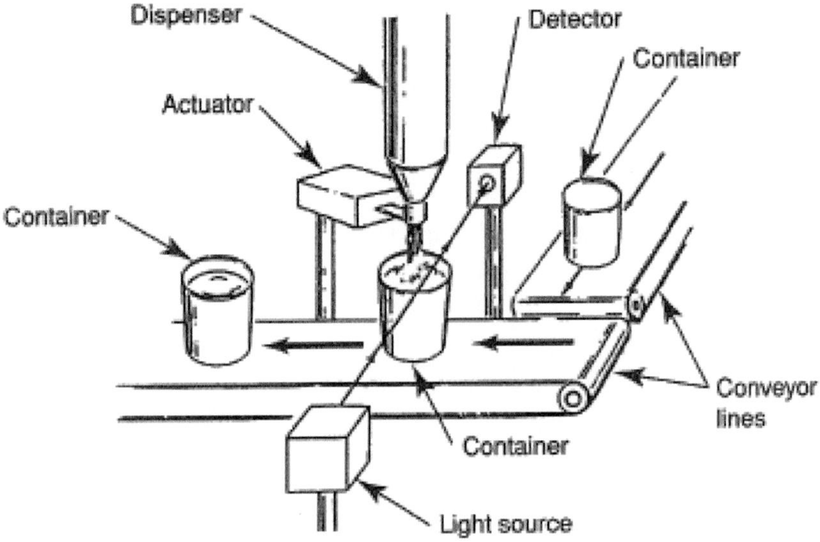

Figure 95.1 Method of liquid level control.

Equipment and Materials

- Multimeter
- Variable DC power supply
- AC power supply – 120 V
- Light-dependent resistor – GE-X6 or equivalent
- Container – glass or plastic
- Relay – 12 V – Guardian 1335-2C-120D or equivalent
- Lamp with holder – 7 W
- Lamp with holder – 60 W
- SPST switch

Safety

Safety is always a primary factor for laboratory activities. This lab should only be completed with supervision.

You may complete certain steps of the "Procedure" and the "Analysis" for better understanding without applying voltage.

Procedure

1. Construct the photoelectric liquid level control of Figure 95.2.

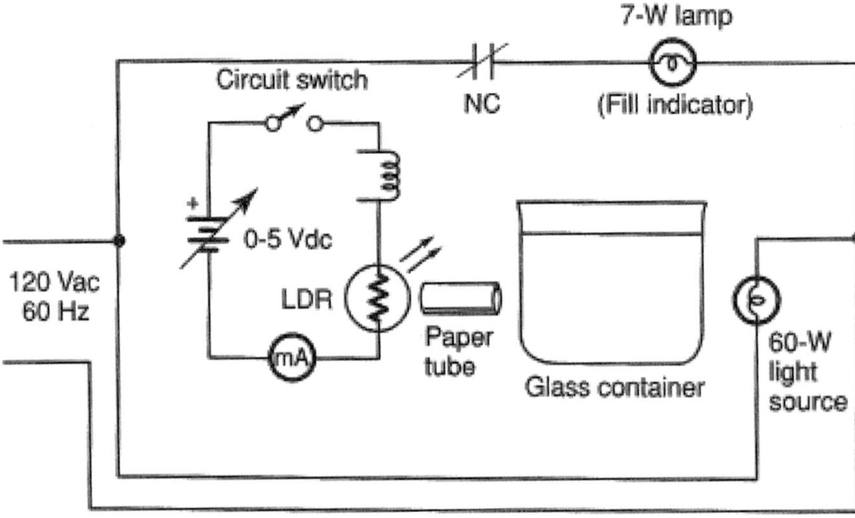

Figure 95.2 Photoelectric liquid level control.

2. Plug the light source into a 120 Vac power outlet and close the circuit switch. Position the light source near the bottom of the glass container before it is filled with water. Slide the paper tube around the light-dependent resistor (LDR) and place the open end against the container.

3. If the circuit is operating properly, the fill indicator lamp will light when the relay is actuated.

4. Adjust the DC source to alter the sensitivity of the circuit. Do not exceed 15 Vdc. The circuit should be able to detect a pencil passing in front of the paper tube.
5. Carefully fill the container with water until the indicator turns off. Avoid pouring water directly in front of the tube window area. You may need to try several trial runs to get the sensitivity to a level where the circuit will respond properly.
6. After the sensitivity has been adjusted, drain or siphon water from the container until the fill indicator is actuated again.
7. Test the liquid level control circuit two or three times to ensure that it operates properly.
8. Turn off the power. Return all materials.

Analysis

1. Discuss the operation of the circuit used in this activity.

2. What are some other types of circuits that could be used for liquid level control?

UNIT VII

Digital Control Basics and Applications

Activity 96

Binary Numbers

Objectives

Practically all of the electronic digital systems in operation employ binary numbers. This type of numbering system uses the number 2 as its base or radix. The two conditions or states of this system can be on or off, voltage or no voltage, or the number designations of 1 and 0. The process of changing from one state to the other can be easily achieved with switches, solid-state devices, or integrated circuits.

In this activity, a number of practice problems will be provided so that you can become more proficient in the use of the binary numbering system. The first problems deal with the conversion of base 10 numbers to binary equivalents. Then, binary numbers are converted to base 10 equivalents. The latter part of the activity is then devoted to binary coded decimal (BCD) conversions.

The practice problems provided in this activity are primarily designed to improve your ability to manipulate binary numbers. These numbers play an important role in the analysis of digital electronic system circuitry.

Procedure

1. Change the following decimal numbers (base 10) to equivalent binary numbers (base 2). Show all of the necessary steps needed to arrive at the final answer.

   12_{10} = ________.

   37_{10} = ________.

   62_{10} = ________.

DOI: 10.1201/9788743810124-103

59_{10} = ________.

60_{10} = ________.

15.5_{10} = ________.

22_{10} = ________.

41_{10} = ________.

47_{10} = ________.

83_{10} = ________.

102_{10} = ________.

9.75_{10} = ________.

2. Change the following binary numbers to their decimal equivalents.

1101_2 = ________.

101111_2 = ________.

10100010_2 = ________.

10.1_2 = ________.

11.11_2 = ________.

11100_2 = ________.

11010_2 = ________.

11000_2 = ________.

101.01_2 = ________.

110.111_2 = ________.

3. Express the following decimal numbers in binary coded decimal form.

$362._{10}$ = ________ BCD.

$683._{10}$ = ________ BCD.

$741._{10}$ = ________ BCD.

$405._{10}$ = ________ BCD.

$1248._{10}$ = ________ BCD.

$329._{10}$ = ________ BCD.

$197._{10}$ = ________ BCD.

$29._{10}$ = ________ BCD.

$758._{10}$ = ________ BCD.

4. Change the following BCD expressions into an equivalent decimal or base 10 value.

   $1001\text{-}0111\text{-}1001_{BCD}$ = ________.

   $0011\text{-}0010\text{-}0100_{BCD}$ = ________.

   $1001\text{-}0111\text{-}0110_{BCD}$ = ________.

   $1001\text{-}0010\text{-}0001_{BCD}$ = ________.

   $0101\text{-}0110\text{-}0011_{BCD}$ = ________.

   $1000\text{-}0111\text{-}0000_{BCD}$ = ________.

5. Perform the operations indicated.

   101_{10} = ________ Base 2.

   1101101_{2} = ________ Base 10.

   67_{10} = ________ Base 2.

   1110.11_{2} = ________ Base 10.

Analysis

1. In a positive logic system, +5 V would be assigned ______ value and 0 V would be assigned ______for a binary system.

2. What would be an advantage of expressing a decimal number in BCD form?

3. Why is the binary system particularly well-suited for electronic switching applications?

4. In a BCD expression, why is the largest number in any one position 9 when a number as large as 1111 or 15_{10} can be displayed by four binary numbers?

5. What is the radix of the decimal system? What is the radix of the binary system?

6. What is meant by the term "place value" in the description of a numbering system?

Activity 97

Octal and Hexadecimal Numbers

Objectives

Octal and hexadecimal numbers are commonly used in digital systems to process large numbers. These two systems are also very compatible with the binary system. Conversions from binary to octal or hexadecimal are accomplished by simply grouping values together in threes or fours. The octal system has a radix of 8 with place value assignments of 1s, 8s, 64s, 512s, 4096s 32,768s, etc. The hexadecimal system is similar but has a radix of 16. Place value assignments are 1s, 16s, 256s, 4096s. 65,536s, etc. Larger number values can obviously be represented by the hexadecimal system in fewer places.

In this activity, you will practice number system conversions. Binary, octal, hexadecimal, and decimal numbers all play an important role in digital systems. You will become more familiar with these conversion processes after working on the problems in this activity.

Procedure

1. Change the given octal numbers to decimal equivalents.

   $123._8$ = ____________________ 10.

   $4062._8$ = ____________________ 10.

   $712._8$ = ____________________ 10.

   $176._8$ = ____________________ 10.

   3120.8 = ____________________ 10.

   643.8 = ____________________ 10.

   541.8 = ____________________ 10.

DOI: 10.1201/9788743810124-104

674.8 =____________________10.

1026.8 =____________________10.

2. Change the given octal numbers to binary equivalents.

   125.8 =____________________2.

   73.8 =____________________2.

   36.8 =____________________2.

   136.8 =____________________2.

   46.8 =____________________2.

   54.8 =____________________2.

   412.8 =____________________2.

   67.8 =____________________2.

   107.8 =____________________2.

3. Change the given decimal numbers to octal (base 8) equivalents.

   4093_{10} = ________ Base 8.

   1017_{10} = ________ Base 8.

   3699_{10} = ________ Base 8.

   5012_{10} = ________ Base 8.

   4876_{10} = ________ Base 8.

   4162_{10} = ________ Base 8.

   3316_{10} = ________ Base 8.

   1491_{10} = ________ Base 8.

   1016_{10} = ________ Base 8.

4. Change the given binary numbers to octal equivalents.

   $100,110,100_2$ = ________.

   $101,111,111_2$ = ________.

   $110,110_2$ = ________.

$1{,}001{,}100_2$ = ________.

$1{,}001{,}000_2$ = ________.

$11{,}111{,}010_2$ = ________.

5. Convert the given hexadecimal (base 16) numbers to decimal equivalents.

   $12A3_{-16}$ = ________________10.

   $1BF_{-16}$ = ________________10.

   $217C_{-16}$ = ________________10.

   $1011B_{-16}$ = ________________10.

   $2BCD_{-16}$ = ________________10.

   $14BD_{-16}$ = ________________10.

   $1A_{-16}$ = ________________10.

   $15E_{-16}$ = ________________10.

   BCD_{-16} = ________________10.

6. Convert the given hexadecimal numbers to equivalent binary numbers.

   1479_{16} = ________.

   $10BA_{16}$ = ________.

   $144E_{16}$ = ________.

   $13AB_{16}$ = ________.

   $123A_{16}$ = ________.

   $129A_{16}$ = ________.

   $10AD_{16}$ = ________.

   BDE_{16} = ________.

   $157B_{16}$ = ________.

7. Convert the given decimal numbers to hexadecimal equivalents.

   1470_{10} = ________.

   3159_{10} = ________.

4153_{10} = ________.

5690_{10} = ________.

2137_{10} = ________.

1176_{10} = ________.

4910_{10} = ________.

1210_{10} = ________.

8. Convert the given binary numbers to hexadecimal equivalents.

$1\ 0110\ 1111\ 0100_2$ = ________.

$1\ 1011\ 1000\ 1110_2$ = ________.

$101\ 1011\ 0111_2$ = ________.

$1001\ 1111\ 1101_2$ = ________.

$1\ 0011\ 0110_2$ = ________.

$101\ 1111\ 0110_2$ = ________.

Analysis

1. Why are octal and hexadecimal number systems commonly used in digital systems?

2. What place values does an octal number with five digits have?

3. What place values does a hexadecimal number have with five digits?

Activity 98

Digital Logic Functions

Objectives

Control of a digital system is achieved by a variety of logic gates. Each gate has a particular output response to a combination of input signals. Binary data is represented by 1 s and 0 s. Multiple-input gates of the AND and OR variety and a single input/output NOT gate represent the three primary logic functions of a digital system. An understanding of these logic functions is essential when analyzing the operation of a digital system.

In this activity, each basic logic gate will be constructed with a single integrated circuit. Truth tables are then developed to show the relationship between the input alternatives and the corresponding output. Through this activity, you will gain experience in using the basic logic gates and see how the logic functions are achieved electronically.

Equipment and Materials

- Multimeter
- DC power supply – 5 V
- SN7408 IC or equivalent
- SN7411 IC or equivalent
- SN7432 IC or equivalent
- SN7404 IC or equivalent
- Resistors – 470 Ohm (4)
- Light-emitting diodes (4)

DOI: 10.1201/9788743810124-105

- SPST toggle switch
- SPDT toggle switches (3)
- IC Circuit construction board and connecting wires

Procedure

Part A: Gate testing with the SN7408 and SN7411

1. Connect the logic gate test circuit of Figure 98.1.

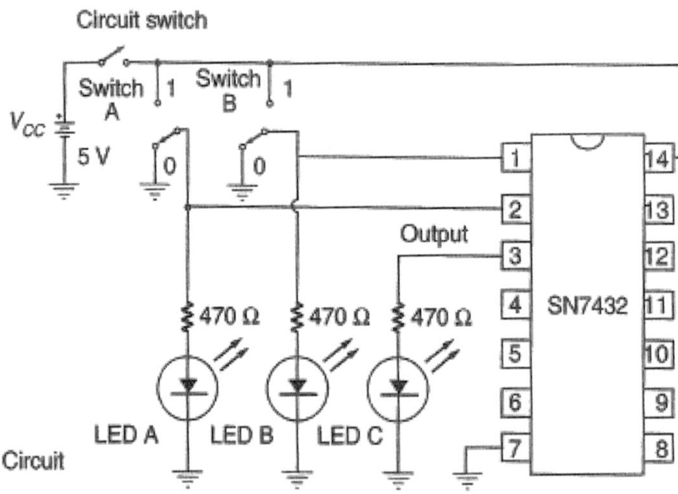

Truth table

LED		
A	B	C
0	0	
0	1	
1	0	
1	1	

Figure 98.1 SN7408 IC test circuit and truth table.

2. Before closing the circuit switch, turn on the power supply and adjust it to 5 V.

3. Turn on the circuit switch and test the IC gate according to the alternatives listed in the truth table of Figure 98.1.

4. Prepare a multimeter to measure voltage and determine the input and output voltages corresponding to 1 and 0.

   1 = ________ V, while a 0 = ________ V.

5. There are three other gates included in this IC chip. Pins 4 and 5 are the inputs, and pin 6 is the output of gate 2. Pins 9 and 10 are the inputs, and pin 8 is the output of gate 3. Pins 13 and 12 are the inputs, and pin 11 is the output of gate 4. Test these gates to see if they are working properly.

6. Make a gate symbol drawing on the blank IC dual in-line package layout of Figure 98.1 for the SN7408.

7. Open the circuit switch and remove the SN7408. In its place, connect an SN7411. Be certain that pin 14 is connected to the source and pin 7 is connected to ground. Connect the remainder of the circuit indicated in Figure 98.2.

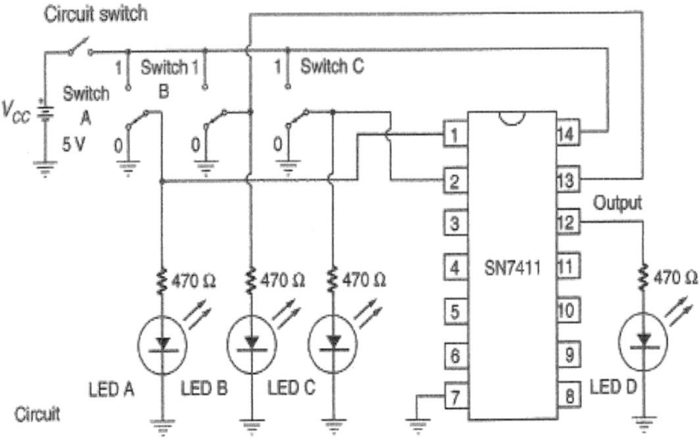

Truth table

LED			
A	B	C	D
0	0	0	
0	0	1	
0	1	0	
0	1	1	
1	0	0	
1	0	1	
1	1	0	
1	1	1	

Figure 98.2 SN7411 IC test circuit and truth table.

8. Close the circuit switch and test the SN7411. Complete the truth table for the listed input alternatives.

9. Open the circuit switch and disconnect the IC.

Part B: Gate testing with the SN7432

1. Construct the logic gate test circuit, which is shown in Figure 98.3.

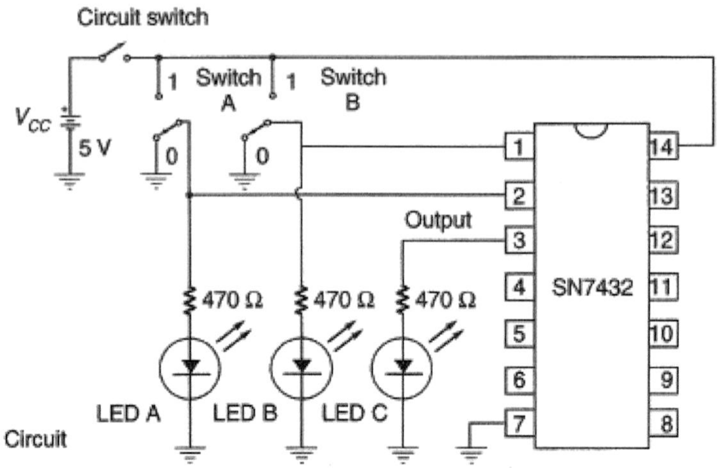

Truth table

LED		
A	B	C
0	0	
0	1	
1	0	
1	1	

Figure 98.3 SN7432 IC test circuit and truth table.

2. Before closing the circuit switch, turn on the power supply and adjust it to 5 V.

3. Turn on the circuit switch and test the IC gate according to those alternatives that are listed in the truth table of Figure 98.3.

4. What logic gate function is achieved by this gate?

5. There are three other gates included in this IC chip: pins 4 and 5 are inputs and pin 6 is an output; pins 9 and 10 are inputs and pin 8 is an output; and pins 12 and 13 are inputs and pin 11 is an output. Test these gates to see if they are working properly.

6. Make a gate symbol drawing on the blank IC dual in-line package layout of Figure 98.4, showing the input–output connections for the SN7432.

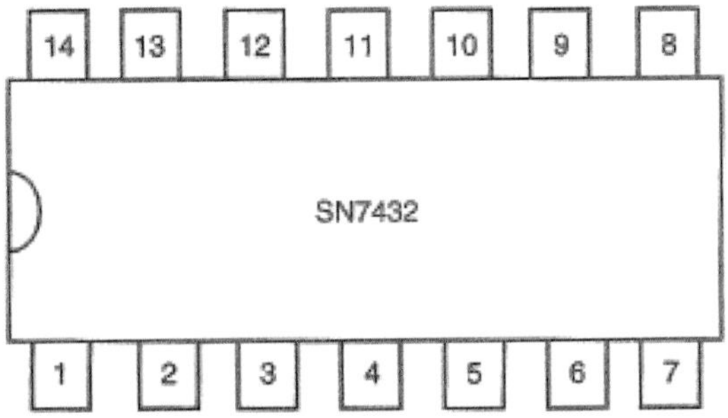

Figure 98.4 SN7432 IC pin connections.

7. Open the circuit switch and disconnect the circuit.

Part C: Gate testing with the SN7404

1. Connect the logic gate test circuit of Figure 98.5.

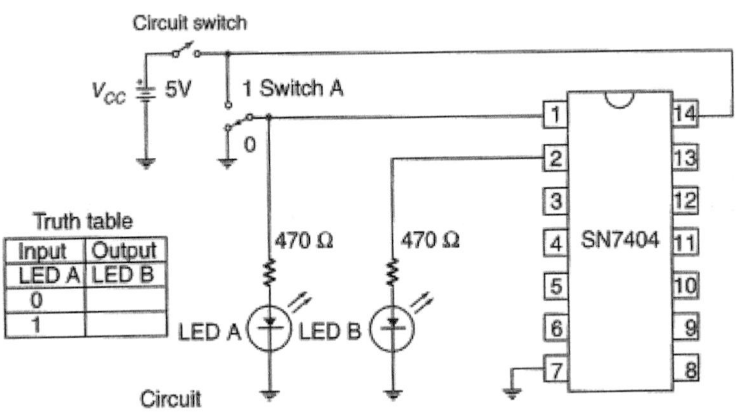

Figure 98.5 SN7404 IC test circuit and truth table.

2. Before closing the circuit switch, turn on the power supply and adjust it to 5 V.

3. Close the circuit switch and test the IC gate according to the alternate input conditions listed in the truth table of Figure 98.5. Record the respective output for each of the inputs.

4. What logic function is achieved by this gate?

5. Test the remaining five gates at pins 3-4, 5-6, 9-8, 11-10, and 13-12 (the pin number designation listed first is the input, with the last number indicating the output).

6. Make a gate symbol drawing on the blank dual inline package layout of Figure 98.6, showing the input–output connections for each gate of the SN7404.

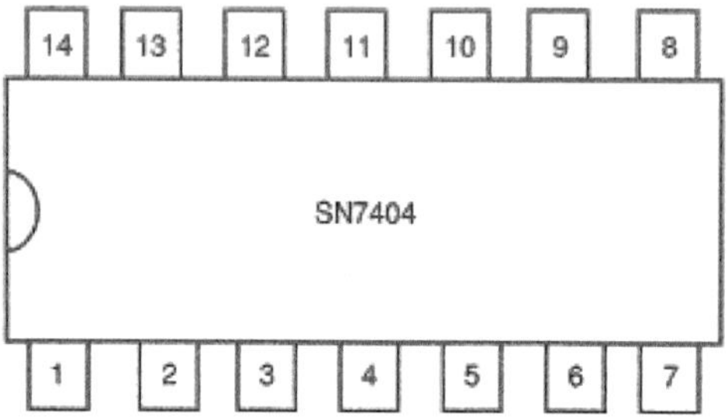

Figure 98.6 SN7404 IC pin connections.

7. Open the circuit switch and connect the output of gate 1 to the input of gate 2. Connect the output indicating LED to pin 4.

8. Close the circuit switch and test the logic circuit. What function does this indicate?

9. Open the circuit switch and disconnect the circuit. Return all parts.

Analysis

1. What logic functions are achieved by the gates studied in Part A?

2. What mathematical functions are achieved by the gates studied in Part A?

3. What are the symbolic representations of the gates studied in Part A?

4. Prepare a statement that describes the operation of the gates studied in Part A.

5. What mathematical function is achieved by the gate studied in Part B?

6. What are the two logic symbols that are used to represent the circuit constructed in Part B of this activity?

7. Prepare a statement that describes the operation of the gate studied in Part B.

8. What mathematical function is achieved by the logic gate studied in Part C?

9. What is meant by the terms negation and double negation?

10. What would be the resulting output of a 1 applied to three of the gates studied in Part C of this activity connected in series?

Activity 99

Combination Logic Rates

Objectives

When either an AND gate or an OR gate is connected to a NOT gate, two additional logic functions are achieved. A NOT-AND, or NAND, is one type of combination logic gate. The NOT-OR, or NOR, gate is representative of the second type of combinational logic gate achieved.

NOR, NAND, and NOT gates are often considered to be universal building blocks in digital systems. With these gates, it is possible to build four logic gates plus the original function. A person working with digital systems should be very familiar with the universal building block principle of combinational logic gates.

In this activity, you will investigate the NAND, NOR, and NOT gates in combinational logic gate construction operations. Only gates that have an inverting capability can be used in this building block technique.

Equipment and Materials

- Multimeter
- DC power supply
- SN7400 IC or equivalent
- SN7402 IC or equivalent
- SN7404 IC or equivalent
- SPST toggle switch
- SPDT toggle switches (2)

 DOI: 10.1201/9788743810124-106

- Resistors – 470 Ohm (3)
- Light-emitting diodes (3)
- IC circuit construction board and connecting wires

Procedure

Part A: NAND logic

1. Using the SN7400 quad NAND of Figure 99.1, build the IC test circuit and check gate 1.

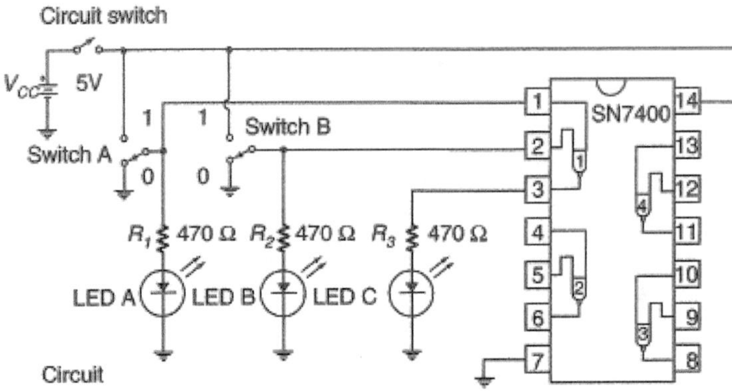

Truth table

Gate 1		
Inputs		Outputs
A	B	C
0	0	
0	1	
1	0	
1	1	

Figure 99.1 SN7400 IC test circuit and truth table.

2. Record the 1 s and 0 s of the output in the truth table for each of the input alternatives.
3. With a multimeter, measure and record voltage values represented by the 1 s and 0 s of this gate.

   1 = _______ V, while a 0 = _______ V.
4. Test the other three gates to make certain that they are functioning properly.
5. To build a NOT gate, remove the B circuit from pin 2 and connect pins 1 and 2 together. Use switch A as the input and test the output of the gate.
6. Connect NAND gate 1 as instructed in Step 1. Connect the output of gate 1 to pins 4 and 5 of NAND 2. Connect the output LED to pin 6. What function does this combination logic gate achieve?

7. Test the circuit to verify your theory. Did it work as you predicted?

8. Connect the combinational logic circuit of Figure 99.2. Complete the truth table showing the outputs at the indicated points.

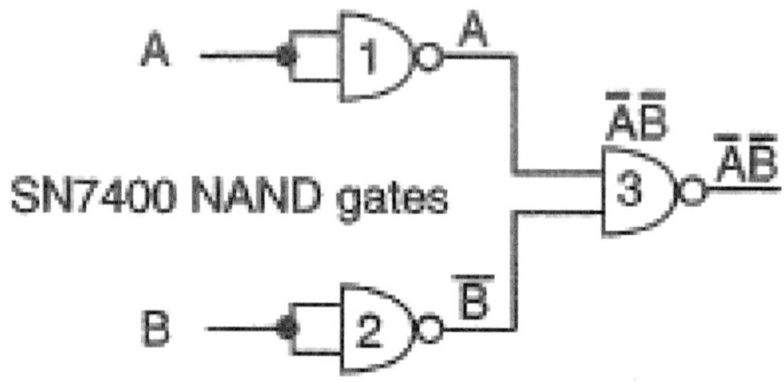

Input		Expression			Output
A	B	$\bar{A}$	$\bar{B}$	$\bar{A}\bar{B}$	$\overline{\bar{A}\bar{B}}$
0	0				
0	1				
1	0				
1	1				

Figure 99.2 Combination NAND gate circuit and truth table.

9. According to the truth table, what gate function is achieved by this combination logic gate? Test the gate to verify your prediction. Did it perform the function that you expected?

10. Open the circuit switch and disconnect the circuit.

Part B: NOR logic

1. Using an SN7402 quad NOR gate, Figure 99.3, build the IC test circuit and check gate 1.

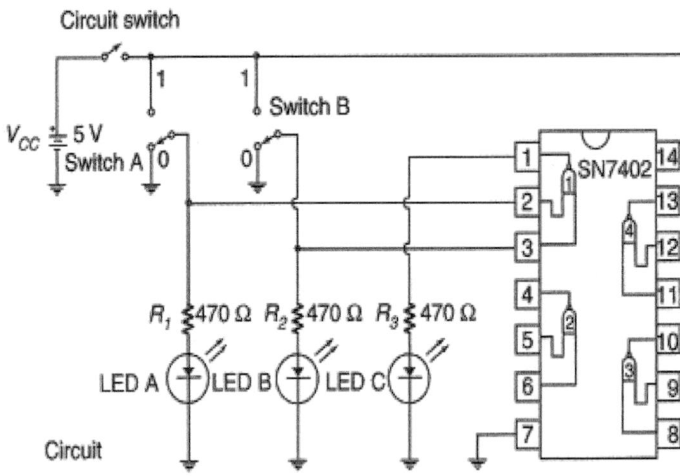

Truth table

Gate 1		
Input		Output
A	B	C
0	0	
0	1	
1	0	
1	1	

Figure 99.3 SN7402 IC test circuit and truth table.

2. Record the 1 and 0 outputs for the input alternatives listed in the truth table in Figure 99.3.

3. With a multimeter, measure the voltage values represented by the 1 s and 0 s of this gate.

   1 = ______ V, while a 0 = _______V.

4. Test the other three gates to make certain that they are functioning properly.

5. To build a NOT gate, disconnect the lead to pin 3 and then connect pin 2 to pin 3. Use switch as the input and test the output of the gate.

6. Connect NOR gate 1 as instructed in Step 1. Connect the output of NOR 1 to the input of NOR 2, with NOR 2 connected as a NOT gate. Connect the output LED to pin 4. What combination of logic is achieved by this configuration?

7. Test the circuit to verify your theory. Did it work as you predicted?

8. Connect the combinational logic circuit of Figure 99.4 using the SN7402.

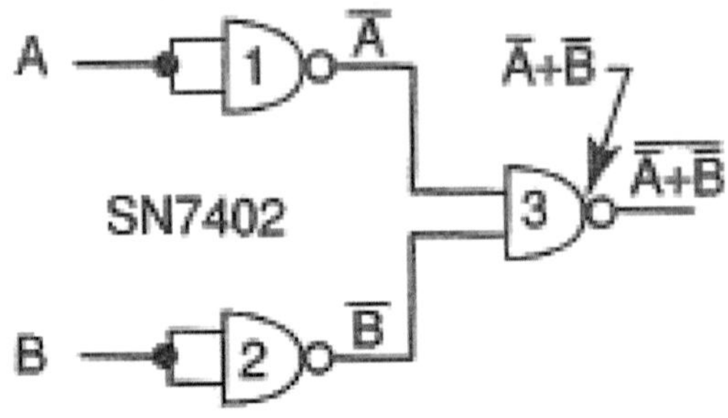

Inputs		Expression			Output
A	B	A	B	$\overline{A}+\overline{B}$	$\overline{\overline{A}+\overline{B}}$
0	0				
0	1				
1	0				
1	1				

Figure 99.4 Combination NOR gate circuit and truth table.

9. Complete the truth table in Figure 99.4 for outputs at the designated points. Then test the combinational logic circuit. Does it do what you predicted by the truth table?

10. Open the circuit power switch and connect NOR gate 4 as a NOT gate. Attach the output in Figure 99.4 to the input of NOR 4. Connect the LED to the output of NOR 4. Invert the last stage of the truth table of Figure 99.4 and predict the type of gate achieved by your modified circuit.

11. Close the circuit switch and test your prediction. Was it correct?

12. Open the circuit switch and disconnect the circuit.

Part C: NOT logic

1. Using the SN7404 hex-inverter IC of Figure 99.5 build the IC test circuit and check gate 1.

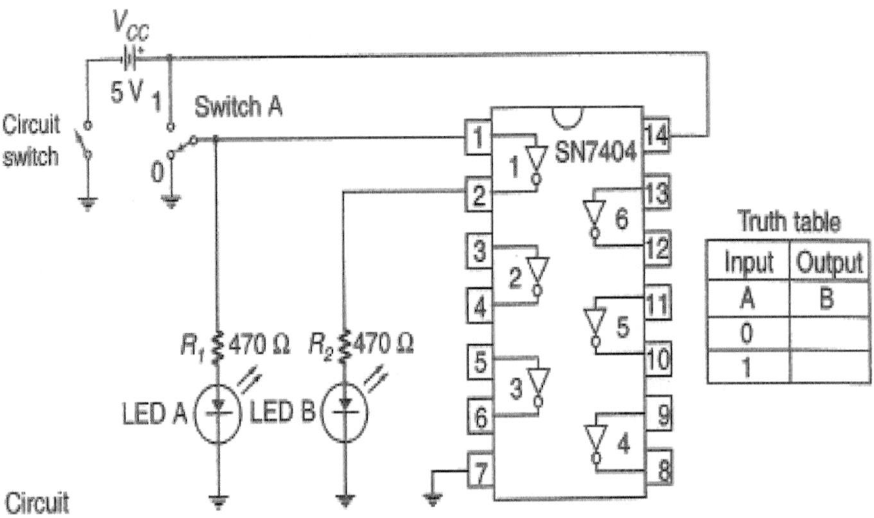

Figure 99.5 SN7404 IC test circuit and truth table.

2. Turn on the circuit switch and record the 1 s and 0 s of the output in the truth table of Figure 99.5 for the input alternatives.

3. With a multimeter, measure the voltage values of a representative 1 and 0 of this gate.

   1 = ________ V, while a 0 = ________ V.

4. Test the other five gates to verify that they are operating properly.

5. Open the circuit switch and combine gates 1 and 2 as indicated in Figure 99.6.

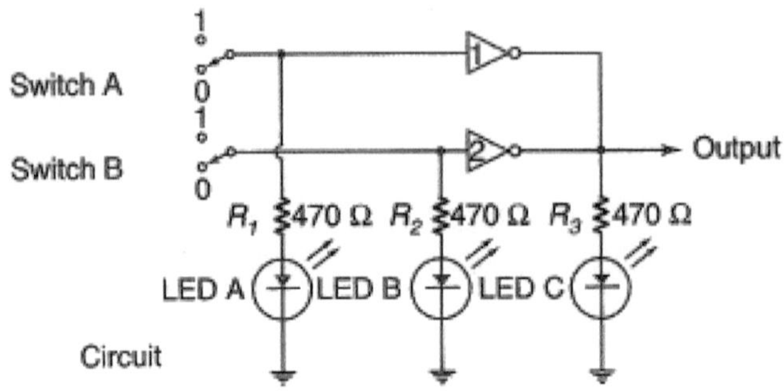

Truth table

Inputs		Output
A	B	C
0	0	
0	1	
1	0	
1	1	

Figure 99.6 Combination NOT gate circuit and truth table.

6. Close the circuit switch and test the combination gate. Record the 1 and 0 outputs for each of the input alternatives of the truth table in Figure 99.6. What logic function does this achieve?

7. Open the circuit switch and connect the output of the circuit in Figure 99.6 to the input of gate 3. Connect the LED-resistor to pin 6 and test the circuit. What logic function does it achieve?

8. Open the circuit switch and combine the gates of the inverter to form the circuit of Figure 99.7.

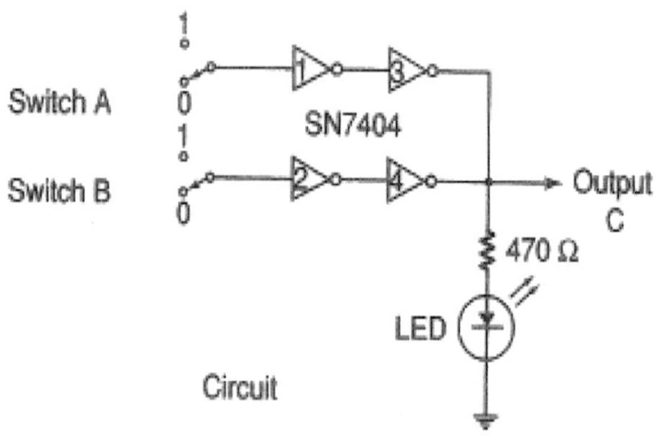

Truth table

Inputs		Outputs	
A	B	C	D
0	0		
0	1		
1	0		
1	1		

Figure 99.7 Cascaded double-inverter circuit.

9. Close the circuit switch and test the circuit. Record the 1 and 0 outputs of C in the truth table of Figure 99.7 for the given input alternatives. What gate function does this achieve?

10. Open the circuit switch and add the output of Figure 99.7 to the input of gate 5. Move the LED to the output of gate 5. What gate function do you predict the output will demonstrate?
11. Close the circuit switch and test the circuit. Record your findings at the D output of the truth table in Figure 99.7. Was your prediction correct?
12. Open the circuit switch and disconnect the components. Return all parts.

Analysis

1. Complete Figure 99.8 showing the outputs for each logic gate listed.

Inputs		Outputs			
A	B	NAND	AND	OR	NOR
0	0				
0	1				
1	0				
1	1				

2. Make a sketch of the five logic circuit combinations accomplished with the SN7402. Show the actual logic gate combinations used.
3. In any of the combined inverter gate combinations, why is the 0 considered a predominant factor?
4. Make a sketch of the five logic functions achieved by the SN7404.

Activity 100

R-S Flip-flops

Objectives

A bistable latch is often used to explain the operation of a flip-flop. The terms latch and flip-flop can be used interchangeably. The latch used here employs a pair of cross-connected two input logic gates. Either NOR or NAND gates can be used to accomplish this function. The logic gates of this configuration permit the circuit to remain in one state or the other, depending on whether it is set or reset. R-S flip-flops are used to debounce switches and to store data temporarily. R-S flip-flops are not available today on IC chips.

In this activity, you will:

1. Construct R-S flip-flops with logic gates.
2. Develop a truth table that shows the operation of an R-S flip-flop.
3. Evaluate the operation of an R-S flip-flop or latch.

Equipment

- Power supply – 0–5 Vdc
- SN7400 quad two-input NAND gate or equivalent
- SN7402 quad two-input NOR gate or equivalent
- Resistors – 100 Ohm (2), 1 K Ohm (2)
- LEDs – 20 mA, 2 V
- SPDT slide switches (2)
- Circuit construction board and connecting wires

DOI: 10.1201/9788743810124-107

Procedure

1. Connect the R-S flip-flop using two cross-connected NOR gates of an SN7402 in Figure 100.1. Any two of the four NOR gates can be used. The pin-out of gates 1 and 2 is shown. Pin 7 is connected to ground, and pin 14 to +5 V to energize the IC.

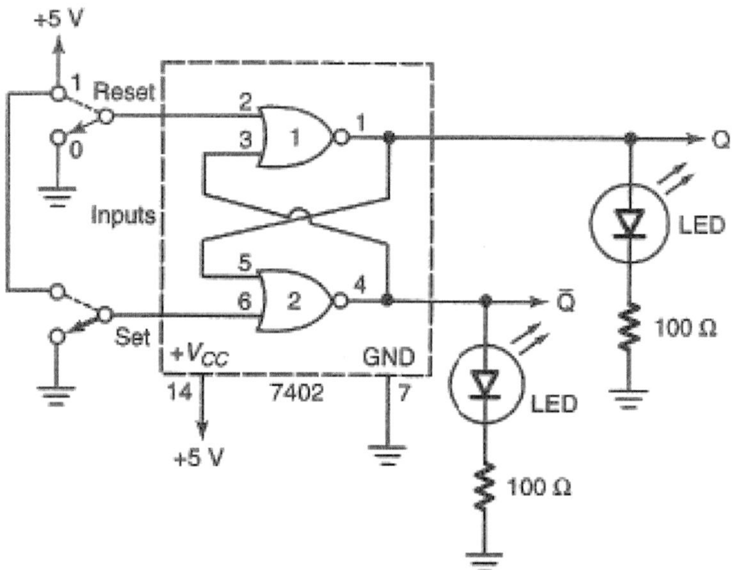

Truth table

S	R	Q	$\bar{Q}$
0	0		
0	1		
1	0		
1	1		

Figure 100.1 NOR gate R-S flip-flop. Complete the truth table.

2. Turn on the DC source and place the set and reset switches in the ground position. This represents the 0 V, or low-level, input. As a rule, the gate energized first causes its LED to light. This output is generally considered to be unpredictable. Note the behavior of the LEDs in the truth table for this condition of operation. An on LED is considered to be a 1, and an off LED indicates 0.

3. Change the operational states of the input to conform with the remaining three steps of the truth table. Record the status of the output in the truth table.

4. Alter the circuit of Figure 100.1 to conform to the circuit of Figure 100.2. In this case, only the input needs to be changed. With the switch in posi¬tion 1, record the status of Q and Q’ in the truth table. Change the switch to position 2. Record the status of Q and Q’.

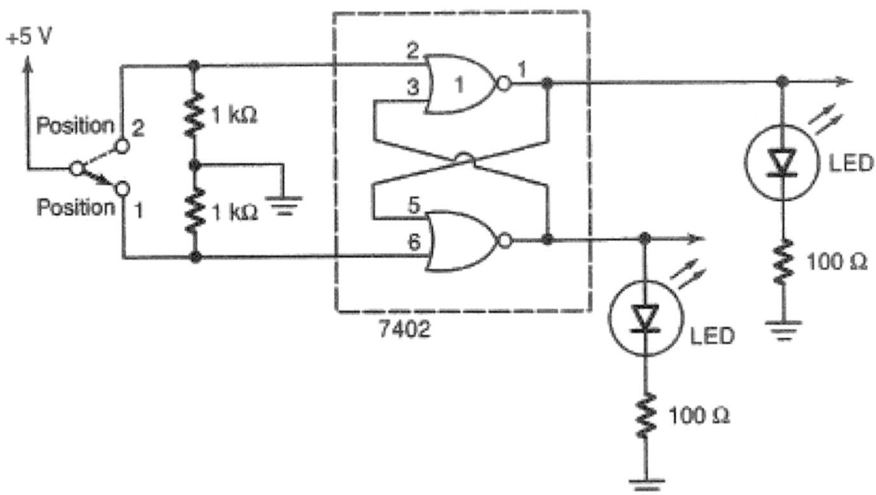

Truth table

Position	Outputs	
	Q	$\bar{Q}$
1		
2		

Figure 100.2 NOR gate circuit. Complete the truth table.

5. Turn off the power supply and remove the SN7402 from the circuit construction board. Replace it with an SN7400. The DC supply voltage is applied to the same pins of this IC. The other pins of the NAND gates are connected in a different manner. Connect the circuit according to Figure 100.3.

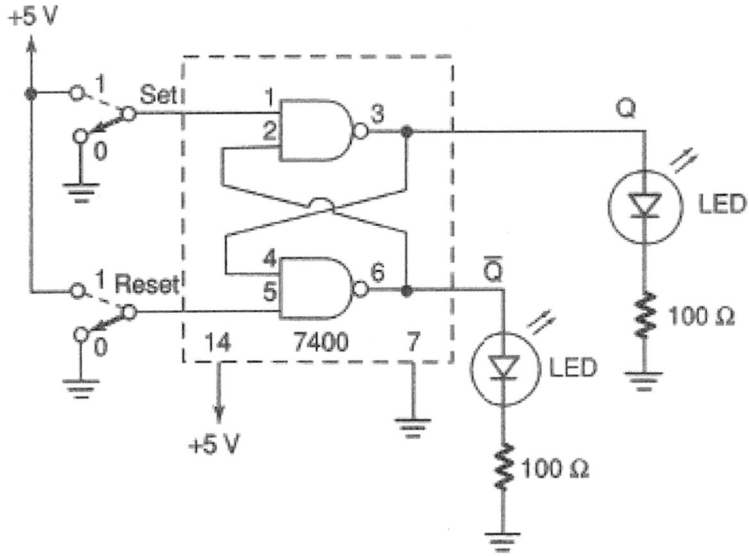

Truth table

S	R	Q	$\bar{Q}$
0	0		
0	1		
1	0		
1	1		

Figure 100.3 NAND gate R-S flip-flop. Complete the truth table.

6. Turn on the power supply and place the set and reset switches in the ground or 0 position. Note the behavior of Q and Q'. Record your findings in the truth table for this condition of operation.

7. Change the operational state of the inputs to conform to the other three steps of the truth table. Record the status of Q and Q' for each input step.

8. Alter the circuit of Figure 100.3 to conform to the circuit of Figure 100.4. Only the input of the circuit is changed. With the switch in position 1, record the status of Q and Q' in the truth table. Change the switch to position 2. Record the status of Q and Q' in the truth table.

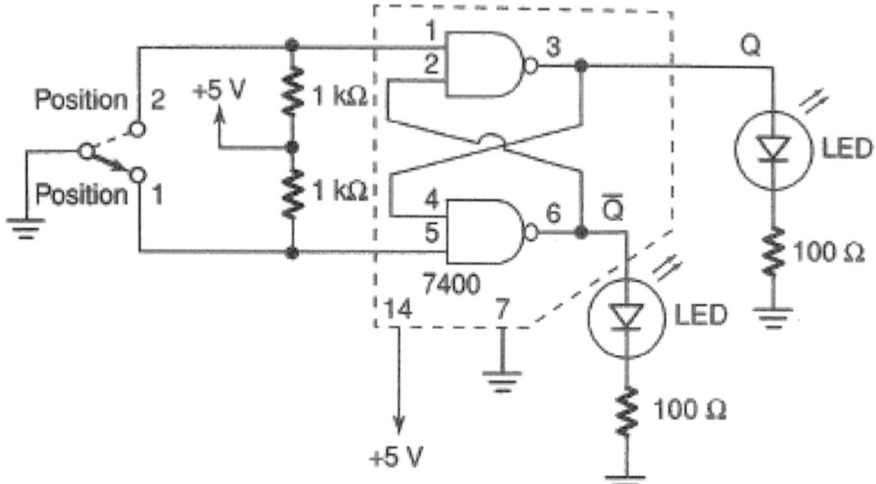

Truth table

Position	Output	
	Q	$\bar{Q}$
1	0	1
2	1	0

Figure 100.4 NAND gate circuit. Complete the truth table.

9. Disconnect the circuit. Return all materials.

Analysis

1. Explain what it takes to set an RS flip-flop constructed with NOR gates.

2. Explain what it takes to set an RS flip-flop constructed with NAND gates.

Activity 101

Binary Counter ICs

Objectives

The 74193 is a medium-scale integration (MSI) integrated circuit that is designed to achieve binary counting. This chip has four J-K flip-flops and is considered to be a synchronous up-down 4-bit binary counter. It can be connected as a ripple counter or as a parallel-loaded synchronous counter. MSI chips of this type permit a large number of components to be fabricated into functional devices. Counting operations are generally achieved by chips of this type.

In this activity, you will:

1. Connect and operate a binary counter with a single IC.
2. Connect a binary counter so that it will count up or down.
3. Demonstrate the operation of a binary counter.

Equipment and Materials

- Multimeter
- DC power supply – 5 V
- 74193 synchronous binary up-down counter or equivalent
- LEDs (4)
- Resistors – 100 Ohm (4)
- 1 N 4001 diode or equivalent
- Capacitor – 100 uF

DOI: 10.1201/9788743810124-108

- Circuit construction board and connecting wires
- Clock circuit – Any one of the following:
  - 555 timer
  - Square wave generator
  - Digital trainer with a built-in clock

Procedure

1. Connect the circuit in Figure 101.1. Refer to the pin connections of the 74193. Make connections as follows:

   Pins 15, 1, 10, and 9 (preset inputs) – open

   Pin 12 (carry out) – open

   Pin 13 (borrow out) – open

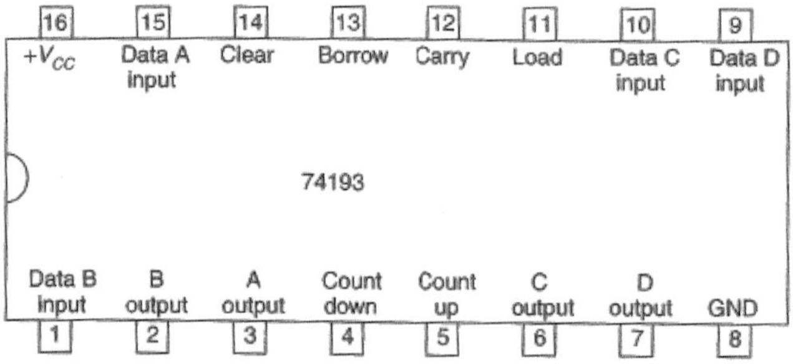

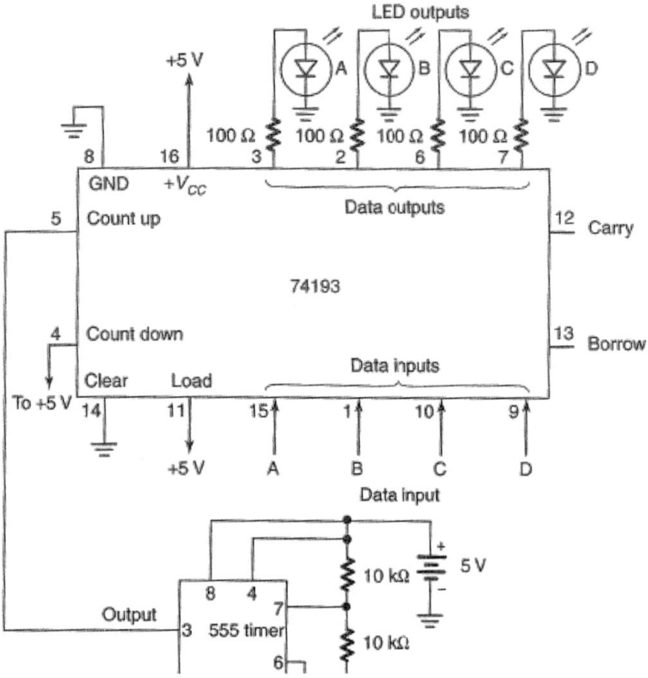

Figure 101.1 Pin layout and circuit setup for binary counter.

Pins 2, 3, 6, and 7 (outputs) – to LEDs

Pin 11 (load) – to +5 V

Pin 16 (+ V_{CC}) to +5 V

Pin 14 (Clear) – to ground

Pin 8 (ground) – to ground

Pin 4 (down) – to +5 V

2. Energize the circuit. If the output is not 0000_2, reset the circuit by momentarily disconnecting pin 14.

3. Connect the clock input to pin 5. This should start the circuit counting. If it does not count, check the output of the clock with a multimeter to see if it is changing states. Recheck the pin connections of the 74193 to see if they conform to the preceding connections.

4. The counting sequence is (up, down).

5. Momentarily disconnect the clock and clear the counter to 0000.

6. Connect the clock to pin 5. Describe the counting sequence in binary numbers.

7. Turn off the power supply and disconnect the clock from pin 5 and the +5 V supply from pin 4 of the counter. Connect pin 5 to the supply and pin 4 to the clock.

8. Energize the circuit and note the counting sequence. This represents an (up, down) counter.

9. Describe the counting sequence in binary numbers

10. Turn off the power supply. The counter circuit and clock should be retained for the next activity.

Analysis

1. What does it take to change the counting direction of a 74193?

2. Why is the circuit of this activity considered to be a ripple counter?

Activity 102

Binary-coded Decimal Counters

Objectives

One of the most significant operations of a digital system is counting. As a general rule, counting is achieved by binary numbers because of the ease with which two-state data can be controlled. Since most people are familiar with the decimal, or base 10, counting system, digital counters must make the transition from binary to decimal to be useful. Binary-coded decimal (BCD) counters are used to display decimal information in a binary code. Only the numbers 0 through 9 are displayed in binary form. Counts beyond 1001_2 or 9_{10} are invalid and are not used.

In this activity, a BCD counter is constructed with a single IC. The counter is manually triggered through its operation with a debounced switch. This permits circuit changes to be made while observing the response of the output. A clock signal is then applied to the counter to show how it responds automatically.

The 7490 BCD counter is used in this activity. It employs four flip-flops and a gate circuit to alter its count. The counter has four outputs, which are labeled A, B, C, and D. Output A is the LSB, and D is the MSB. This counter has a 0000_2 clear and a 10012 or 9_{10} reset. A decade output count can be removed from the output. When this output changes from its on state to zero, it indicates a count of $10_{2..}$ The 7490 is a modulo-10 counter.

In this activity, you will:

1. Become familiar with the operation of a BCD counter.
2. Construct a BCD counter and evaluate its operation.

DOI: 10.1201/9788743810124-109

Equipment and Materials

- Multimeter
- DC power supply – 5 V
- 7490 binary-coded decimal counter or equivalent
- SPDT switches (2)
- LEDs – 20 mA, 2 V
- Resistors – 100 Ohm (4), 1 K Ohm
- Pulser switch or a debounced R-S flip-flop – (7400 IC)
- Circuit construction board and connecting wires

Procedure

1. Connect the BCD counter of Figure 102.1.

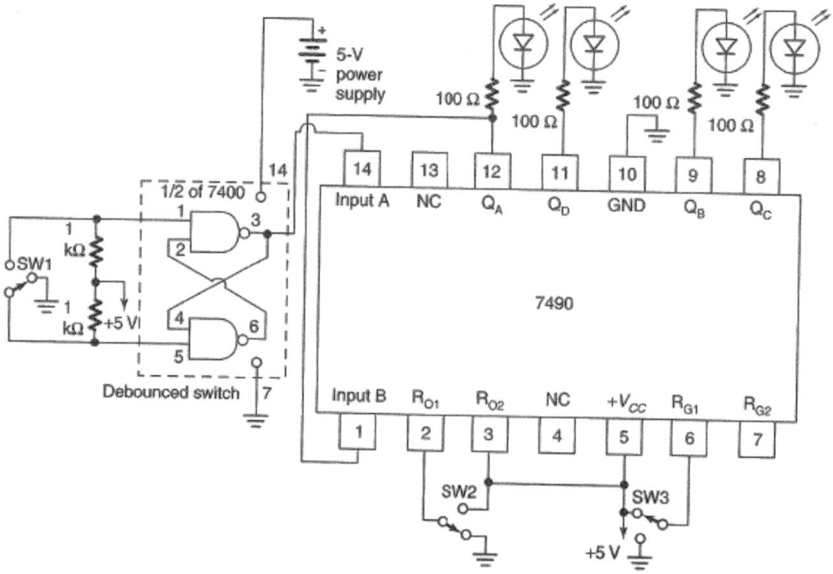

Pulse number	Display state			
	D	C	B	A
0	0	0	0	0
1				
2				
3				
4				
5				
6				
7				
8				
9				
10				
11				
12				

Figure 102.1 Circuit and truth table for a BCD counter.

2. The debounced switch, or pulser, is built with a 7400 quad NAND gate.
3. Turn on the power source and note the status of the LED display. If any one of the LEDs is on, it represents a random count. The count can be cleared by momentarily disconnecting pin 2, the input from ground. SW2 controls this operation.
4. With the LEDs all off, change the debounced switch of the 7400 on and then off. Note the response of the display. This is generally referred to as pulsing. The 7490 changes state on the negative-going state change. The pulsing switch operation has two states: high and low. The state of the counter changes on the negative transition of the debounced switch.
5. Pulse the input of the counter several times, noting the response of the display.
6. Reset the display to 0000. Pulse the switch one time. Record the status of the LEDs in the table of Figure 102.1. An on LED indicates 1, and an off LED is 0. Pulse the counts indicated by the table, recording the status of the LED display. The position of the LEDs is important in this display. LED QA is the LSB, and LED QD is the MSB. Complete the table for the indicated number of pulses. Note that the last count that appears on the display before it resets to 0000. This is referred to as? ______________

7. Pulse the counter so that the display shows 0011. Then clear the display with the switch. The clearing operation can be achieved on any count. Try some other count and clear the display. The letter designation refers to "reset to 0."

8. Pulse the number 0101 into the counter. Momentarily open the SW3. How does this operation alter the display?

9. Try several other numbers and reset the display with the switch. How does the display respond?

10. The designation RG1 refers to "reset to 1." The 7490 has two RG inputs. Either can be used independently, or they can be used in combination. Resetting to 1 permits some math operations to be achieved automatically.

11. Turn off the power and remove the connecting wire between pins 1 and 12. Change the debounced switch from input (pin 14) to input (pin 1). Disconnect the LED QA from pin 12.

12. Turn on the power. Reset the counter to 0000. Pulse the counter while observing the count on the display. LED QB is now the LSB, and LED QAD is the MSB. Indicate the counting sequence: ______________________________. This represents a modulo-__________________ counter.

13. Turn off the power. Remove the debounced switch from input B (pin 1) and connect it to input A (pin 14). Connect an LED to the output QA as in the original circuit.

14. Turn on the power. Reset the counter to 0000.

15. Pulse the counter one time while observing the display. How does the display respond?

16. Pulse the display several times while observing the display. Describe This represents what math function?

17. Turn off the power and disconnect the circuit. Return all materials.

Analysis

1. In order for a binary counter to be converted to a BCD counter, how many counts must it lose?

2. Why is BCD counting an important operation for a digital system?

Activity 103

Octal and Hexadecimal Counters

Objectives

Binary-coded octal (BCO and binary-coded hexadecimal counters are similar in many respects to the BCD counter. A BCO counter displays a maximum count of seven (or 111_2) before it clears and displays all Os.

When three separate counters are ganged together, the maximum count will be111'111'1112 or 7 × 64 + 7 × 8 + 7 × 1 or 511. Three BCH counters, by comparison, connected together can produce a maximum count of 1111, 1111, $1111_{2,}$ or 15 × 256 + 15 × 16 + 15 × 1 or 4095. BCO and BCH counters permit the manipulation of very large numbers that would be difficult to achieve with binary alone.

In this activity, you will construct a simple BCO counter and a BCH counter, with LEDs used to display the output. Two or more of these counters are then ganged together to achieve a greater counting capability. Through this activity, you will be able to observe the actual counting of octal and hexadecimal numbers, thus becoming more familiar with these counting methods.

Equipment and Materials

- DC power supply – 0–5 V, 1 A
- SN7493 counter IC or equivalent
- SE/NE555 timer/clock IC
- Capacitors – 0.01 uF, 100VDC, 4 uF, 25 VDC
- Resistors – 470 Ohm (4), 200 K Ohm (2)
- Light-emitting diodes (4)

DOI: 10.1201/9788743810124-110

- SPST switch
- IC circuit construction board and connecting wires

Procedure

1. Construct the binary-coded octal (BCO) counter of Figure 103.1.

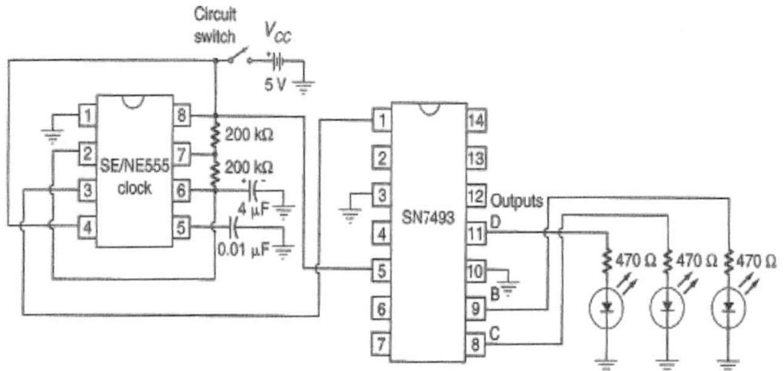

Figure 103.1 BCO counter and clock.

2. Before closing the circuit switch, turn on the power supply and adjust it to 5 Vdc.
3. Close the circuit switch, thus energizing the clock and the counter.
4. Explain the counting sequence displayed by the three LEDs. Label the LEDs A, B, and C from least significant digit to most significant digit, starting at the right and moving to the left.
5. With additional components, duplicate the counterpart of the circuit shown in Figure 103.2 and connect it to the original counter as indicated.

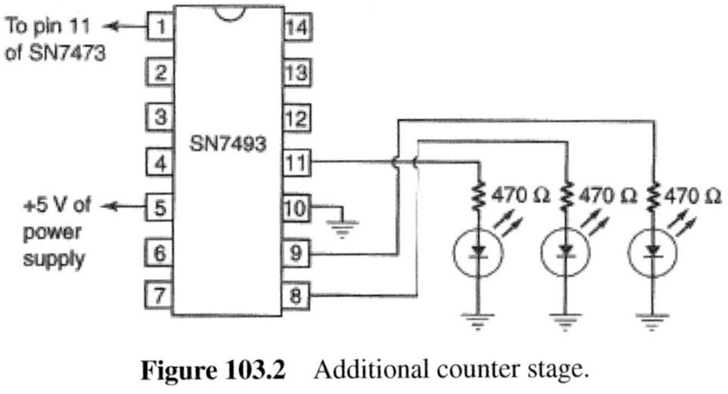

Figure 103.2 Additional counter stage.

6. Close the circuit power switch and describe the counting action.
7. The counter of Figure 103.1 will register units, while the one in Figure 103.2 will register 8 s. The maximum count that can be displayed by this two-stage counter is?
8. Open the circuit switch and alter the construction of the counter to conform with the circuit of Figure 103.3. Only the modifications are shown; the remainder of the circuit is unchanged.

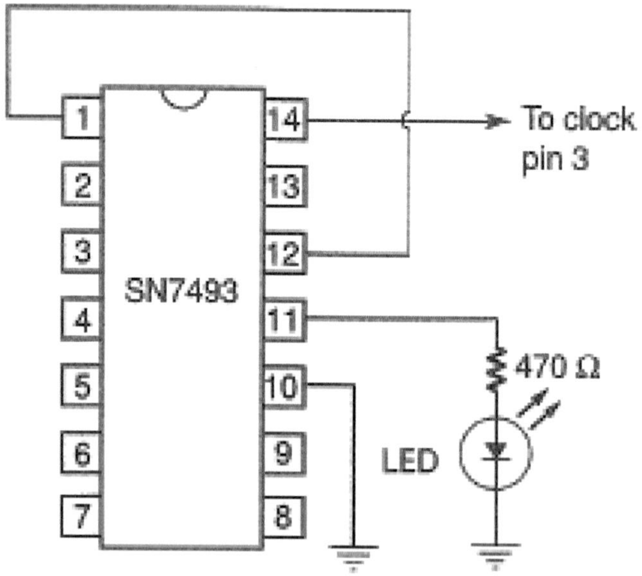

Figure 103.3 Additional counter stage.

9. Close the circuit switch and describe the counting action.
10. The added LED should appear as the least significant digit and be located to the right of the original three. What is the maximum count produced by the LED display?
11. With additional components, duplicate the counterpart of the circuit. Connect the clock input of the alternate counter to pin 11 of the counter of Figure 103.3. Use four LED readouts as in the modified circuit.
12. Close the circuit switch and describe the counting sequence.
13. Consider the original counter as the units part and the additional counter as the 16 s counter. The maximum count that can be achieved is?

14. If time permits, you may wish to connect a third counter to the two-stage counters. The maximum count that can be achieved by a three-stage counter is?
15. Open the circuit switch and disconnect the circuit. Return all parts.

Analysis

1. What is meant by the term binary-coded octal counter?

2. What is meant by the term binary-coded-hexadecimal counter?

3. Why does the second stage of a BCD or BCH counter change states when the first stage changes from all 1 s to 0 s?

Activity 104

Digital Counting Systems

Objectives

A digital counting system is the part of a digital system that is responsible for counting data and developing a readout display of the applied signal. Each new count applied to the system causes it to advance the previous count by one. Essentially, the counter is the last part of a digital system. Functionally, this part of the system is designed to change digital signals into information that will initiate a stage change of the readout. Binary signals are then decoded and used to produce a display of numbers.

There are four major sections of a counting system: signal source, counter, decoder-driver, and readout device. The counting signal for this system is generated by a clock circuit. In practice, this could come from a photoelectric detector, an electromechanical actuator, or some other physical transducer. The counter then responds to this signal by advancing once with each count. In this case, the counter has a binary input and BCD output. The BCD signal is then decoded and used to energize appropriate segments of the readout display.

The counting system used in this activity can only be used to count from 0 to 9. An output from the BCD circuit could, however, be used to initiate a 10's, 100's, 1000's, etc. count according to the demands of the system.

In this activity, you will build a one-digit counting system from discrete ICs. Each part of the system, starting with the readout, is built and tested. A similar counter built on a PC board is also tested and used to increase the count capability from 9 to 99, and possibly to 999. You will see how several counters are physically ganged together to increase counter size. The counting system principle is an extremely important concept today, with the growing emphasis on digital instruments.

 DOI: 10.1201/9788743810124-111

Equipment and Materials

- DC power supply – 5 V
- SN7490 IC or equivalent
- SN7447 IC or equivalent
- MAN-1 seven-segment readout or equivalent
- Resistors – 470 Ohm (7)
- SPDT logic switches (4)
- SPST toggle switch
- Single-phase counter circuit board
- IC circuit construction board and connecting wires
- Clock circuit (SN/NE555 IC)

Procedure

1. Connect the seven-segment LED readout of Figure 104.1.

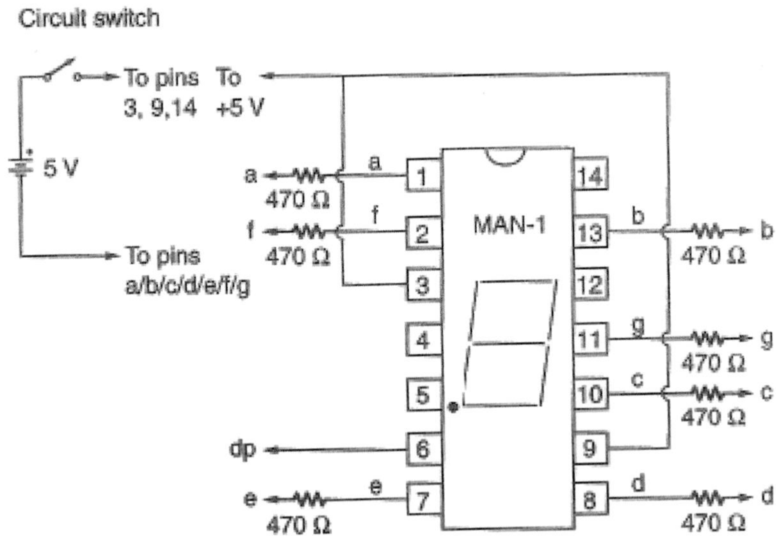

Figure 104.1 Seven-segment LED readout.

2. Close the circuit switch and test the negative side of the operation of each segment of the readout by momentarily touching each to the supply. If each segment is lit up properly, proceed to the next step. If not, test the circuit wiring for a possible error.

3. Connect the SN7447 seven-segment decoder/driver circuit of Figure 104.2.

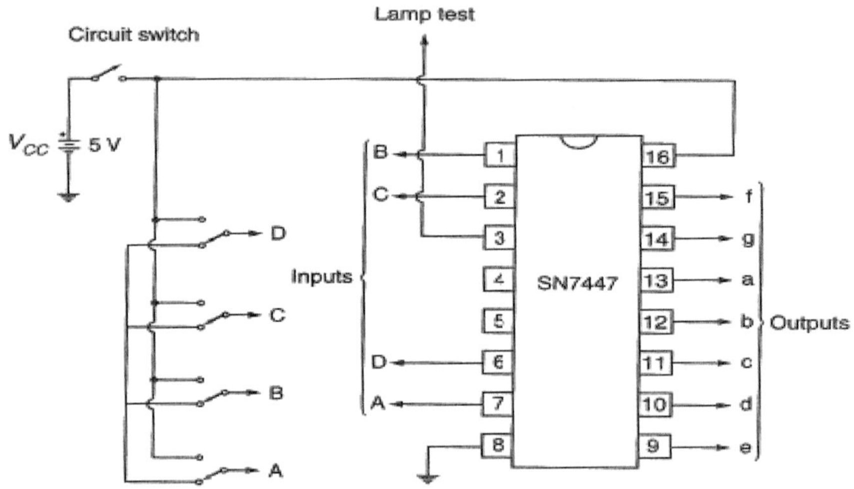

Figure 104.2 Seven-segment decoder/driver.

4. The lowercase letters to the right of the IC denote outputs that go to the readout accordingly. Connect the readout and SN7447 to the power supply.

5. Close the power supply circuit switch and momentarily connect the lamp test (pin 3) to ground. If the circuit is working properly, all seven segments will light up. Disconnect the lamp test wire from the circuit.

6. The A, B, C, and D logic switches are designed to control the numbers displayed on the readout. They are binary coded so that all 0 s will produce a 0 display.

7. Test the decoder-driver by placing the appropriate binary code into the switches that will produce the numbers 0 through 9 on the display. Counts 10 through 15 will produce coded indications for the numbers 10 to 15. Turn off the circuit switch and remove the four logic switches attached to A, B, C, and D.

8. Construct the SN7490 BCD counter of Figure 104.3.

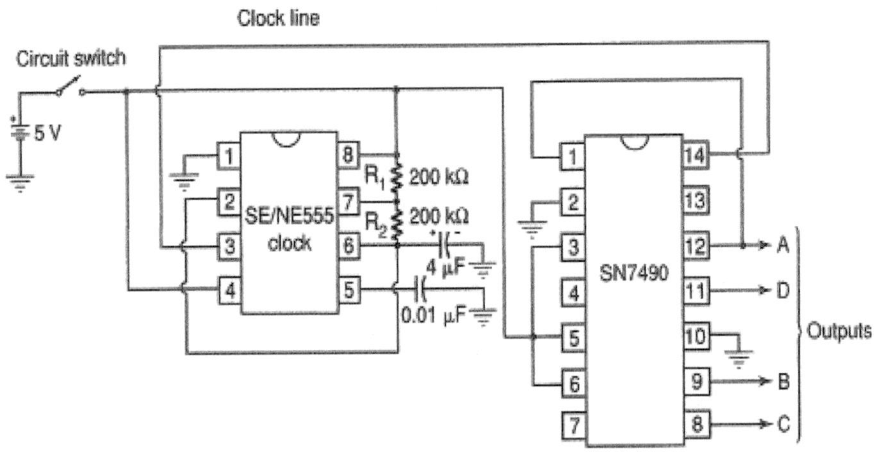

Figure 104.3 BCD counter circuit.

9. Connect the A-B-C-D outputs of the SN7490 to the A-B-C-D inputs of the SN7447, where the logic switches were connected. Close the circuit switch. If the counting system is operating properly, the numbers 0 through 9 will be repeatedly displayed. If it does not display this counting procedure, check the clock circuit and the wiring of the SN7490. Open the circuit switch.

10. If additional counting systems are available, you may want to connect a second unit to the circuit board. The clock input of the second board should be connected to the D output of the SN7490. When both boards are energized, a 99 count can be achieved. The count can be increased to 999 by adding the third circuit board. The clock input of the third board should be driven by the D output of the SN7490 of the second board. The process can be repeated by ganging additional boards together.

11. Disconnect the counting system from everything but the clock. Connect the counting system circuit board to the positive and negative sides of the power supply. Connect the clock to the circuit board and turn on the power supply. If the circuit functions properly, the counter should display a repeat number count of 0–9.

12. Turn off the power supply and disconnect the circuit board and clock circuit. Return all parts.

Analysis

1. Describe the function of the four basic parts of a digital counting system.

2. Explain why the D output of an SN7490 can be used to initiate a count for the next stage of a counter.

Activity 105

Oscilloscope Measurement

Objectives

The oscilloscope is one of the most important laboratory instruments in use today. It is essential for a technician to be able to use the oscilloscope competently. Oscilloscopes can be used to measure voltages, frequency, phase relationships, or to monitor waveform displays of various types.

In this activity, you will use the oscilloscope to measure the amplitude of an AC sine-wave voltage. The amplitude of an AC sine wave can be measured in three different ways: (1) peak-to-peak (p-p), (2) peak, and (3) root-mean-square (rms) or effective value. Peak-to-peak, as the term implies, is the measurement from the peak of the positive half of the waveform to the peak of the negative half, while peak is simply the measurement of either the positive or negative half. Root-mean-square voltage is the measurement of the amount of AC voltage that will do the same amount of work as a comparable DC voltage. The power-line voltage in the home, for example, is stated as an rms sine wave voltage, but the oscilloscope can measure voltages regardless of waveform type. This is a primary advantage of the oscilloscope.

Equipment and Materials

- Multimeter
- Oscilloscope
- AC power source
- Resistors – 270, 470, 1 K Ohm

DOI: 10.1201/9788743810124-112

Procedure

1. Turn on the power supply and adjust the voltage to 10 volts AC. Calculate the peak and p-p values of this voltage.

   Peak value ______ volts AC.

   Peak-to-peak value _____ volts AC.

2. Construct the circuit shown in Figure 105.1 and connect it to a 10-volt AC source. Measure the voltage across the following points with the meter. Calculate the peak and p-p values.

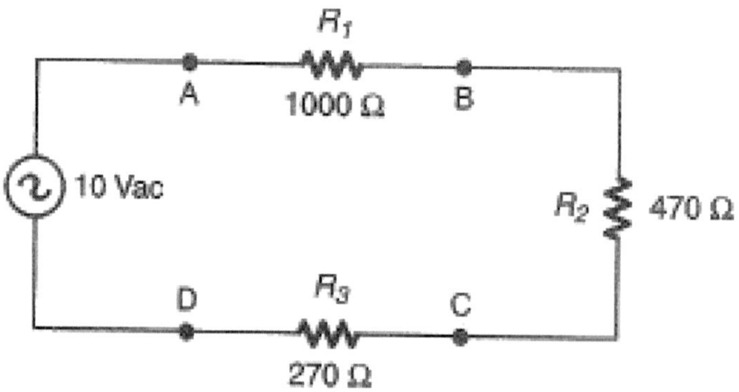

Figure 105.1 Test circuit for AC voltage measurements.

   Points A to B: _____ volts rms; _____ volts peak; _____ volts p-p.

   Points B to C: _____ volts rms; _____ volts peak; _____ volts p-p

   Points C to D : _____ volts rms; _____ volts peak; _____volts p-p

3. Prepare the oscilloscope for operation by adjusting the appropriate controls. Connect the ground lead to point and the vertical-input lead to point A. Properly center the waveform and adjust the horizontal frequency and sync controls to produce two sine waves. Adjust the vertical controls so that each square or horizontal line of the graticule on the scope screen equals a definite p-p voltage. As previously determined mathematically, the p-p voltage of the 10-volt source is approximately 28.2 volts. Calibrate the scope so the sine wave has a specific number of squares or lines of amplitude. Once the scope has been calibrated, do not adjust the vertical controls again (note: some scopes have internal calibration and the preceding is not required).

4. With the scope calibrated, measure the p-p voltage across the following points and record the values:

   Points to ____ volts p-p.

   Points B to ____volts p-p.

   Points C to ____ volts p-p.

Analysis

1. In order to measure p-p voltage, the scope must first be calibrated, and then it can be used to make any p-p measurements. How does changing the vertical attenuation or volts/division control affect the calibration?

2. Discuss the basic operation of the oscilloscope.

3. How can an oscilloscope be used to measure frequency?

4. Explain the procedure for calibrating an oscilloscope to measure p-p AC voltage.

Activity 106

Chart Recorders

Objectives

Chart recorders are used to monitor many types of electrical and physical quantities over a period of time. The advantage of a chart recorder is that a visible record of the monitored quantity can be kept for future use. There are many different types of chart recorders in use today.

In this activity, you will be able to study the construction and operation of a chart recorder. The recorder should be connected to some operational system so that you can observe its operation.

Equipment and Materials

- Chart recorder (any type available)

Procedure

1. Obtain a chart recording instrument and operating manual (if available). Record the manufacturer and model number.

   Manufacturer ______________________________.

   Model Number ______________________________.

2. What quantity or quantities does this instrument measure?

3. What is the range or scale of this instrument?

4. What method is used by this instrument to make a permanent record of the quantity measured?

DOI: 10.1201/9788743810124-113

5. What type of drive mechanism is used on this instrument?
6. What type of chart is used by this instrument?
7. Describe any special features this instrument has.
8. Prepare the chart recorder for operation. Connect the recorder to a circuit or system in which you can observe its operation over its range of measurements.
9. Have your supervisor/instructor check the instrument for operation.

   Approval: ________________________________.

Analysis

1. What are some specific applications where chart recorders might be used?

2. What are some types of drive mechanisms used for chart recorders?

3. What are some methods used by chart recorders for recording the measured quantity?

Activity 107

Single-phase Power Measurement

Objectives

In direct-current circuits, power (watts) is always equal to the product of voltage and current. This relationship is also true for resistive AC circuits. However, when reactive AC circuits that contain either inductance or capacitance are encountered, the power converted in the circuit is no longer equal to the product of voltage and current. It is necessary to use a watt meter to measure the true power (W) of these types of AC circuits. The product of voltage and current is referred to as the apparent power (VA) of a circuit. The power delivered from the source but not converted to another form of energy by the load is called reactive power (VAR). These three components form the power triangle relationship.

In this activity, you will use a single-phase watt meter to measure the true power of a motor. You will be able to measure the voltage, current, and power associated with the operation of a single-phase AC motor.

Equipment and Measurement

- Multimeter
- AC current meter
- Watt meter – single-phase (2000 W or higher)
- AC power source – 120 V, single-phase
- Single-phase AC induction motor

DOI: 10.1201/9788743810124-114

Safety

Safety is always a primary factor for laboratory activities. This lab should only be completed with supervision.

You may complete certain steps of the "Procedure" and the "Analysis" for better understanding without applying voltage.

Procedure

1. Obtain a single-phase AC motor of any type and a single-phase watt meter.
2. Check the nameplate data on the motor. Multiply the rated voltage of the motor by its rated current. Record the values.

   Rated voltage = _______ volts AC.

   Rated current = _______ amperes AC.

3. Multiply the voltage and current values to obtain a full-load apparent-power rating.

   Apparent power _______ VA.

4. Make sure that the watt meter will measure this amount of power.
5. Connect the watt meter to the motor as shown in Figure 107.1.

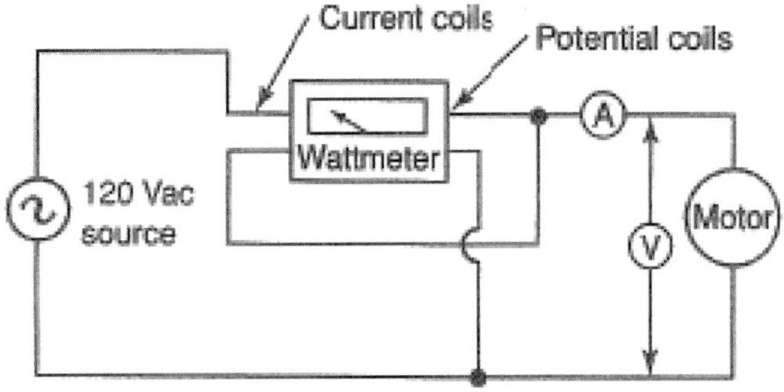

Figure 107.1 Test circuit for single-phase power measurement

6. Measure and record the AC voltage (V) applied to the motor.

   Voltage ______ volts AC.

7. Apply power to the motor. Observe the reading on the watt meter and current meter. Record the maximum power and current during the starting period.

   Starting power = ________ watts.

   Starting current = ________ amperes AC.

8. Measure and record the power and current while the motor is running.

   Power = ________ watts.

   Current = ________ amperes AC.

9. Apply a load to the shaft of the motor while observing the meters. Record the maximum power and current.

   Maximum power = ________ watts.

   Maximum current = ________ amperes AC.

Analysis

1. How is a single-phase watt meter constructed?

2. From the data of Step 7, calculate:

   a. Apparent power volts amperes = ________ VA.

   b. Power factor = ________.

   c. Reactive power = ________ VAR.

3. From the data of Step 8, calculate:

   a. Apparent power = ________ VA.

   b. Power factor ________

   c. Reactive power _____ VAR.

4. From the data of Step 9, calculate:
   a. Apparent power = _________ VA.
   b. Power factor = _______
   c. Reactive power = ____ _____ VAR.

5. As the load is increased on the single-phase AC motor, what happens to:
   a. Apparent power?
   b. True power?
   c. Power factor?
   d. Reactive power?

Activity 108

Three-phase Power Measurement

Objectives

Two single-phase watt meters can be used to measure the power converted by a three-phase load. When the meters are properly connected, the algebraic sum of the two readings will equal the total three-phase power converted by the load. However, if the power factor of the load device is less than 50 %, one watt meter will deflect in the wrong direction. In this case, the current coil connections on the meter should be reversed so that the meter will deflect properly. Then, the lower meter reading will be subtracted from the higher one.

In this activity, you will use two single-phase watt meters to measure the power converted by a three-phase motor.

Equipment and Materials

- Multimeter
- AC current meter
- AC power source – three-phase
- Watt-meters – single-phase (2)
- Three-phase AC induction motor

Safety

Safety is always a primary factor for laboratory activities. This lab should only be completed with supervision.

DOI: 10.1201/9788743810124-115

You may complete certain steps of the "Procedure" and the "Analysis" for better understanding without applying voltage.

Procedure

1. Obtain a 208-volt or a 240-volt three-phase AC motor and connect two single-phase watt meters to it as shown in Figure 108.1.

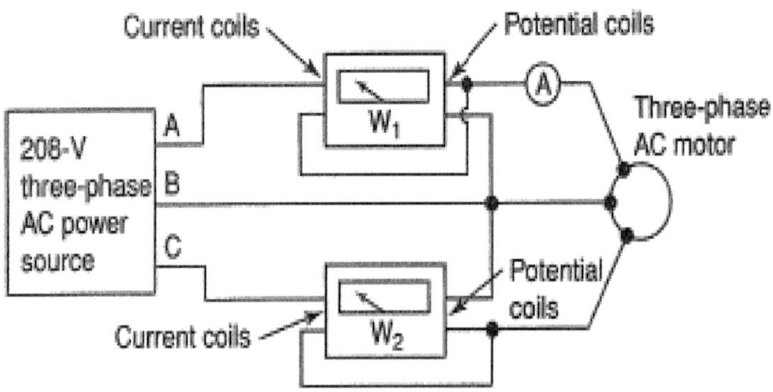

Figure 108.1 Test circuit for three-phase power measurement.

2. Measure and record the AC voltage applied to the motor.

   Voltage = ________ volts AC.

3. Apply power to the motor. Observe the watt meter and ammeter readings. Record the maximum value of each when the motor starts.

   Starting power watt meter 1 = _________ watts.

   Starting power (watt meter 2 = _________ watts.

   Current = _________ amperes AC.

4. Measure and record the same values while the motor is running.

   Power watt meter 1 = _________ watts.

   Power watt meter 2 = _______ watts.

   Current = _________ amperes AC.

5. Apply a load to the motor shaft while observing the meters. Record the maximum power and current.

   Maximum power watt meter 1 = ________ watts.

   Maximum power watt meter 2 = _________ watts.

   Maximum current = __________ amperes AC.

Analysis

1. What are some ways in which three-phase power can be measured?

2. From the data of Step 3, calculate:

   a. True power = _________ watts.

   b. Apparent power = _________ VA.

   c. Power factor = __________.

   d. Reactive power = __________ VAR.

3. As load is increased on a three-phase motor, what happens to:

   a. Apparent power

   b. True power

   c. Power factor

   d. Reactive power

4. From the data of Step 4, calculate:

   a. True power = ________ watts.

   b. Apparent power = __________ VA.

c. Power factor = __________.

d. Reactive power = _________ VAR.

5. From the data of Step 5, calculate:

    a. True power = ____ _____ watts.

    b. Apparent power = _________ VA.

    c. Power factor = ___________.

    d. Reactive power = ________ VAR.

Activity 109

Watt-hour Meter Familiarization

Objectives

Watt-hour meters are used to monitor the amount of electrical power used over a period of time. Any industrial plant or residence that uses electrical power has a watt-hour meter to measure the power used at that particular location.

In this activity, you will be able to observe the operation of a watt-hour meter. You can connect a load to the meter (such as a motor) to cause it to function. If a meter is not available in the lab, you can observe an operational watt-hour meter either near the laboratory or at home.

Equipment and Materials

- Watt-hour meter-single-phase
- Motor-single-phase AC (or other load device)

Procedure

1. Connect a watt-hour meter to the load as shown in Figure 109.1.

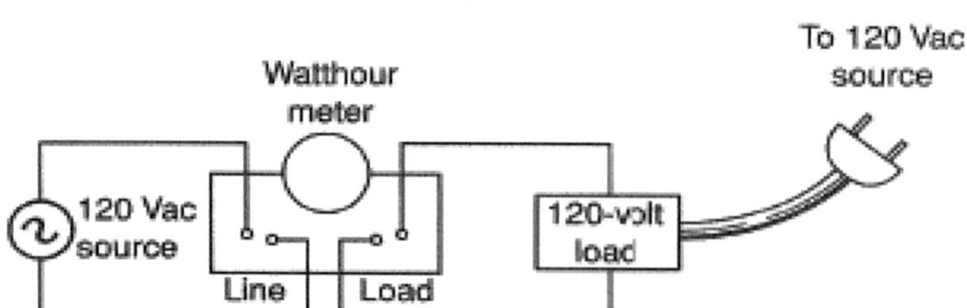

Figure 109.1 Circuit for observing the operation of a watt-hour meter.

DOI: 10.1201/9788743810124-116

2. Apply power to the load and observe the operation of the watt-hour meter. Note the speed of rotation of the aluminum disk.

   Speed = __________ (approximately).

3. Increase the load while observing the speed of the aluminum disk.

4. Have your supervisor/instructor check your meter for operation.

   Approval: ____________________.

Analysis

1. Discuss the operational principle of a watt-hour meter.

2. In Step 3, what happened to the speed of the aluminum disk? Why?

Activity 110

Dynamometer Analysis of Electric Motors

Objectives

A very thorough analysis of the operation of any motor under varying load conditions can be made by using a dynamometer. The dynamometer unit should be used to analyze and compare the characteristics of many types of fractional-horsepower motors. Most commonly small (¼ or ½ horsepower) motors are used on laboratory dynamometer units.

In this activity, you should analyze as many motors as possible. There is no particular order in which they should be completed. When you complete an analysis, you should submit to the instructor the following:

1. Motor nameplate data.
2. Motor analysis data.
3. Motor data graph with the following performance curves.
   a. Speed versus torque.
   b. Speed versus horsepower.
   c. Efficiency versus torque.
4. A written analysis in two parts:
   a. General information about the operational characteristics of this type of motor.
   b. Reference to the data obtained to discuss the characteristics of the motor (discuss each characteristic curve on the graph).

DOI: 10.1201/9788743810124-117

Equipment and Materials

- Multimeter
- Dynamometer unit
- Watt meter
- AC ammeter
- Electric motor(s)
- Graph paper (one sheet for each motor)

Procedure

1. Select a motor for analysis and record the nameplate data of the motor in Figure 110.1.

Motor type	
Manufacturing Co.	
Identification number	
Model number	
Frame type	
Number of phases (ac)	
Horsepower	
Cycles (ac)	
Speed (r/min)	
Voltage rating	
Current rating (amperes)	
Thermal protection	
Temperature rating–°C	
Time rating	
Other information	

Figure 110.1 Motor nameplate data.

2. Secure the motor onto the dynamometer unit. Make sure the shaft couplings are of the proper type for the motor you are analyzing.

3. Connect the motor to the current and power meters as shown in Figure 110.2. Be sure to connect the watt meter properly.

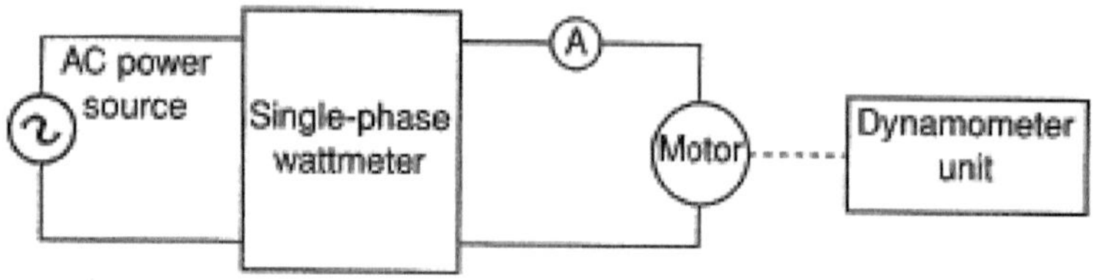

Figure 110.2 Connection diagram for dynamometer analysis of electric motors.

4. Adjust the load applied to the motor by the dynamometer through as many increments of torque as possible. Record the current and power for each value of torque in Figure 110.3.

5. Using the data recorded in Figure 110.3 and the rated voltage of the motor, calculate the horsepower, efficiency, and power factor for each value of torque. Record these calculated values in the appropriate columns of Figure 110.3.

Torque (ft • lb)	Speed (r/min)	True power (watts)	Line current (amperes)	Horsepower*	% Efficiency†	Power factor‡
1						
5						
10						
15						
20						
25						
etc.						

$$^{*}\text{ Horsepower} = \frac{\text{torque (ft}\cdot\text{lb) x speed (r/min)}}{5250}$$

$$\dagger\ \%\ \text{Efficiency} = \frac{\text{horsepower x 746}}{\text{true power (watts)}}$$

$$\ddagger\ \text{Power factor} = \frac{\text{true power (watts)}}{\text{apparent power (voltamperes)}}$$

Figure 110.3 Motor analysis data.

6. On a sheet of graph paper, plot the performance curves for the motor data graph by using the values of your recorded data. Refer to the sample graph of Figure 110.4 for the selection of the proper horizontal and vertical axes.

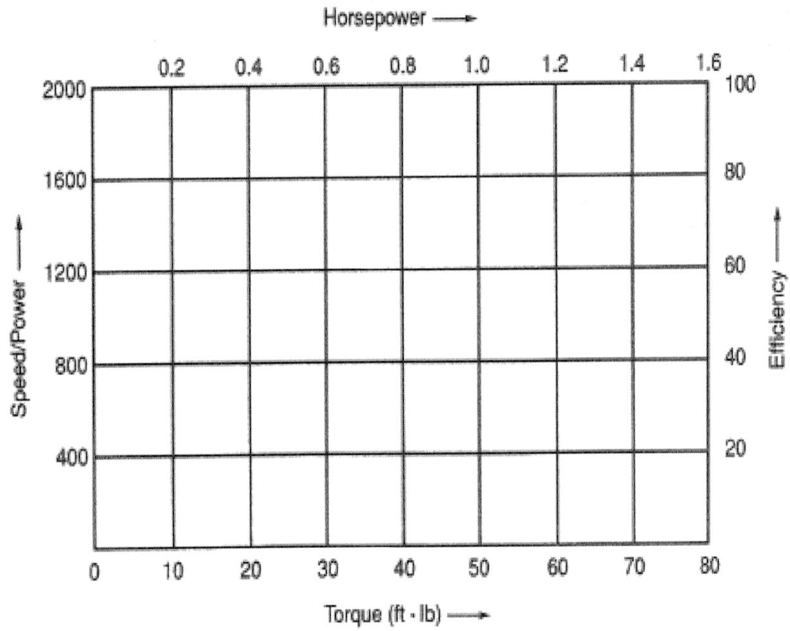

Figure 110.4 Motor data graph.

7. Prepare your data and analysis to be submitted for evaluation. Try to do a neat and accurate job on your data and graph, and a detailed job on the analysis.

8. Perform the same procedure on as many motors as possible. It may be interesting to compare the curves obtained from several different types of motors.

Analysis

1. Discuss the use of a dynamometer to analyze an electric motor.

2. Complete graphs and all of the information listed in the "Objectives" of this activity.

Appendix A

Arduino microcontroller Programs

Appendix A includes 25 **Arduino programs** that illustrate the tremendous control and communications capabilities this device offers. Taken together, these allow interaction with the Arduino development using various input switches and sensors, adding logic, timing, and memory functionality, controlling output devices such as LEDs, piezo speakers, and electric motors.

Lab -01	Original blink built-in LED	This is an initial sketch that runs when an Arduino Uno R3 is placed in the workspace and the "Start Simulation" button is pressed, with a few comments included explaining its behavior. It blinks the built-in LED connected to Pin 13 of the Arduino board, first switching it on, adding a 1s delay during which the LED stays on, then switching it off for 1s. The on-off sequence repeats.	 https://www.tinkercad.com/things/hA2qial1YJH
Lab-02	02a Change blink timing for built-in LED	Blink the built-in LED connected to Pin 13 of the Arduino board, in intervals of 0.5s	https://www.tinkercad.com/things/61IfvHSoalJ
	02b Change blink timing for built-in LED using constants	Blink the built-in LED connected to Pin 13 of the Arduino board, in intervals of 0.5s, using constants in the sketch	https://www.tinkercad.com/things/bZSBEejI0E1
Lab-03	03a Morse Code alphabet A with built-in LED	We will be sending a single character (A) in Morse Code using the built-in LED. Refer to Wikipedia – https://en.wikipedia.org/wiki/Morse_code – for more information about the International Morse Code.	 https://www.tinkercad.com/things/d7WEFBjfhxD

	03b Morse Code letter A with built-in LED using functions	We will be sending a single character (A) in Morse Code using the built-in LED and custom functions for signaling the dot-dash (dit-dah) durations.	 https://www.tinkercad.com/things/iZDOtRk8Bnf
Lab-04	Sending a name in Morse Code with a built-in LED	We will be sending a word (name) in Morse Code using the built-in LED and custom functions for signaling the dot-dash (dit-dah) durations.	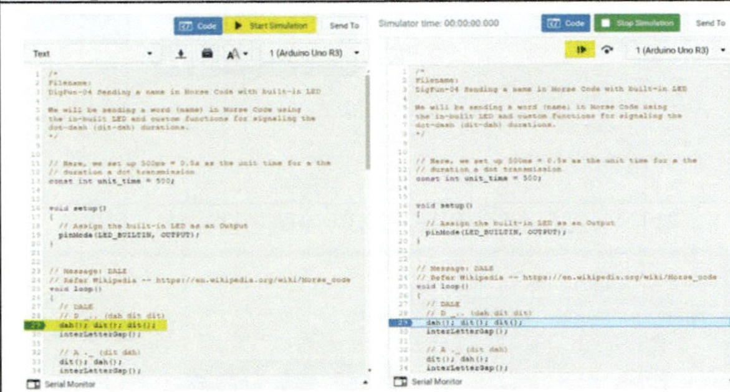 https://www.tinkercad.com/things/h6EwAvncYCT
Lab-05	Sending a Morse Code message with a built-in LED	We will be sending a message (Thanks Dale) in Morse Code using the built-in LED and custom functions for signaling the dot-dash (dit-dah) durations.	https://www.tinkercad.com/things/9HIW01F67nv

Lab-06	06a Blink external LED	Blink an external LED connected to Pin 13 of the Arduino board, in intervals of 0.5s	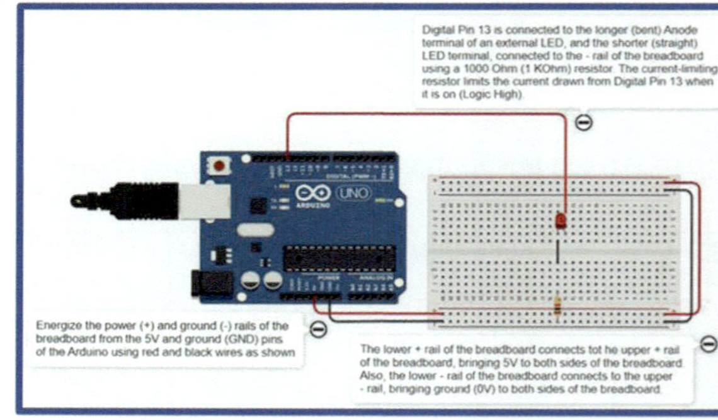 https://www.tinkercad.com/things/krJ2Psxct4e
	06b Blink external LED and segment A of a 7-segment	Blink an external LED and one segment (A) of a common-cathode 7-segment display, connected to Pin 13 of the Arduino board, in intervals of 0.5 s	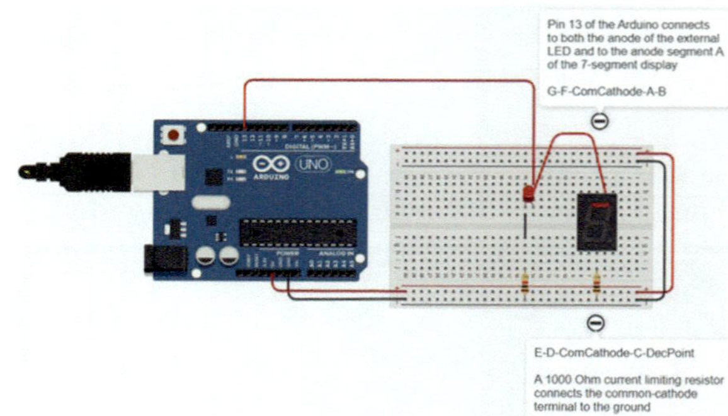 https://www.tinkercad.com/things/14xaQ54GSo1

<table>
<tr>
<td>Lab-07</td>
<td>Press the pushbutton to blink external LED</td>
<td>Read in 1 digital (Boolean-valued) input (A) wired using a pushbutton switch, and if pressed, turn an external LED on and off in intervals of 0.5 s</td>
<td>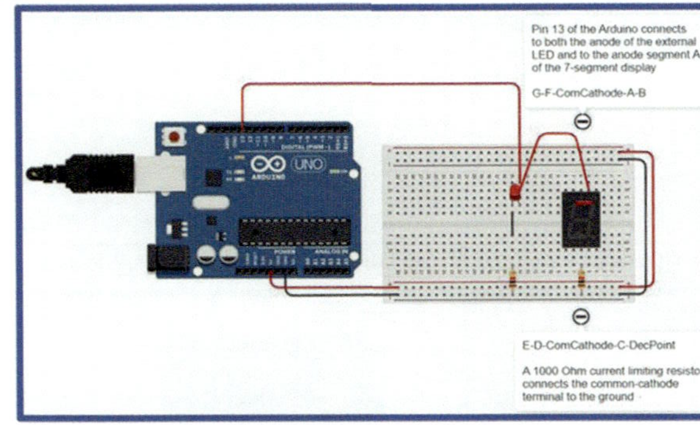

https://www.tinkercad.com/things/3E4QwUXQJr2</td>
</tr>
<tr>
<td></td>
<td></td>
<td></td>
<td>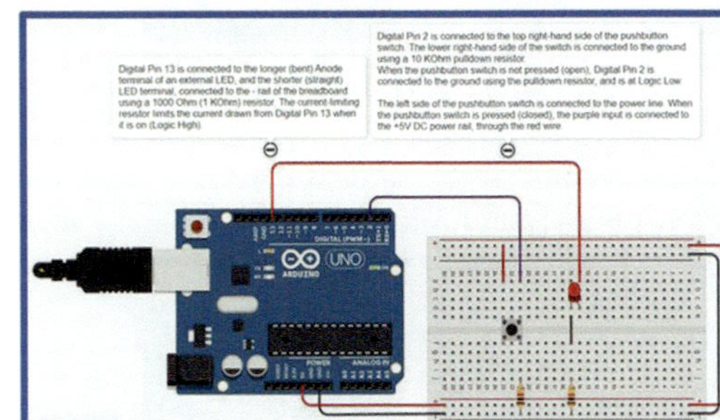
</td>
</tr>
</table>

Lab-08	Close the slide switch to blink external LED	Read in 1 digital (Boolean-valued) input (A) wired using a slide switch, and if closed, turn an external LED on and off in intervals of 0.5 s	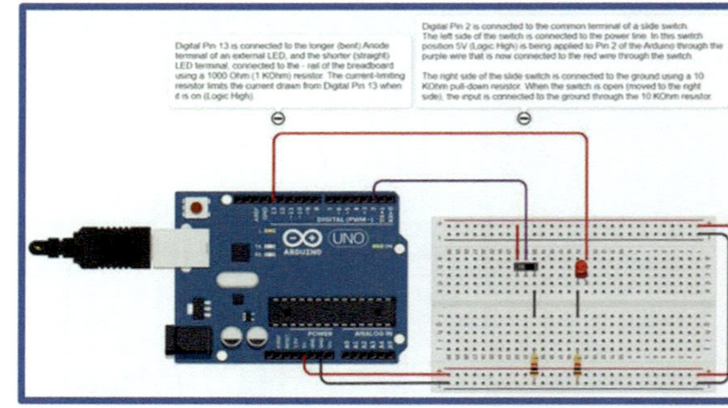 https://www.tinkercad.com/things/1Iu643ACiXJ
Lab-09	09a NOT logic gate	Read in 1 digital (Boolean-valued) input (A) wired using a slide switch, and develop a digital output, Z = A'	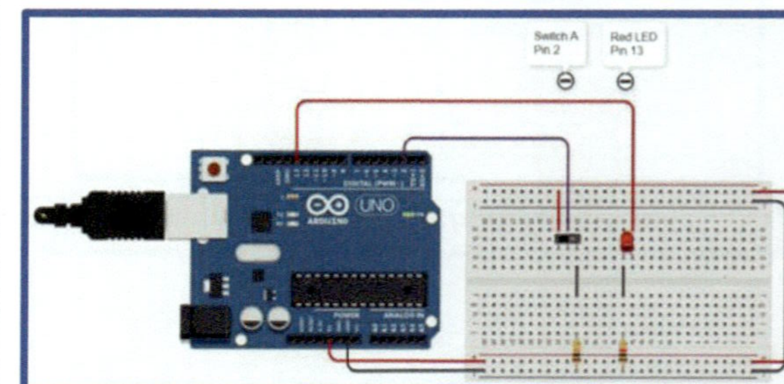 https://www.tinkercad.com/things/c33wnOpvfdb
	09b NOT logic gate with status display on Serial Monitor	Read in 1 digital (Boolean-valued) input (A) wired using a slide switch, and develop a digital output, Z = A'. The status of the input switch, A, and the output LED, Z, is shown on the Serial Monitor (located below the Code section)	 https://www.tinkercad.com/things/gKyjxl508Qk

Lab-10	AND logic gate	Read in 2 digital (Boolean-valued) inputs (A and B) wired using slide switches, and develop a digital output, Z = AB	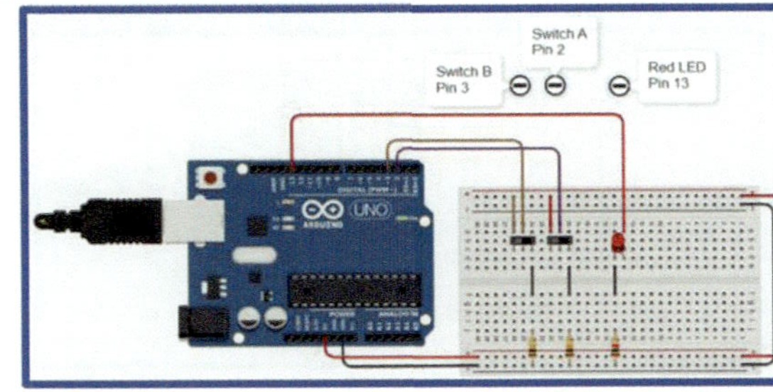 https://www.tinkercad.com/things/7r2aXXah4RV
Lab-11	OR logic gate	Read in 2 digital (Boolean-valued) inputs (A and B) wired using slide switches, and develop a digital output, Z = AB	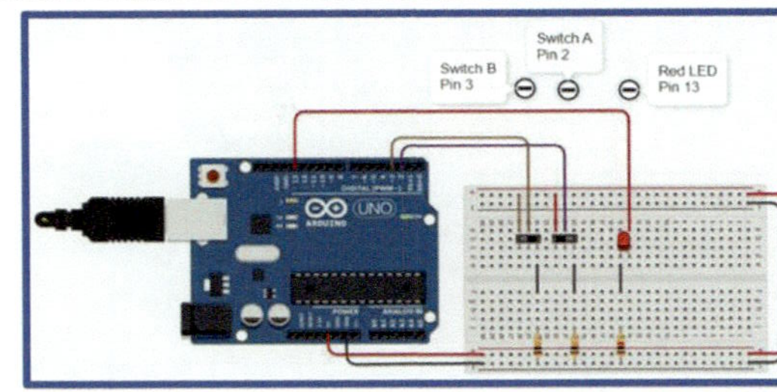 https://www.tinkercad.com/things/jW8FsA6vU8K
Lab-12	NAND logic gate	Read in 2 digital (Boolean-valued) inputs (A and B) wired using slide switches, and develop a digital output, Z = (AB)' = A' + B'	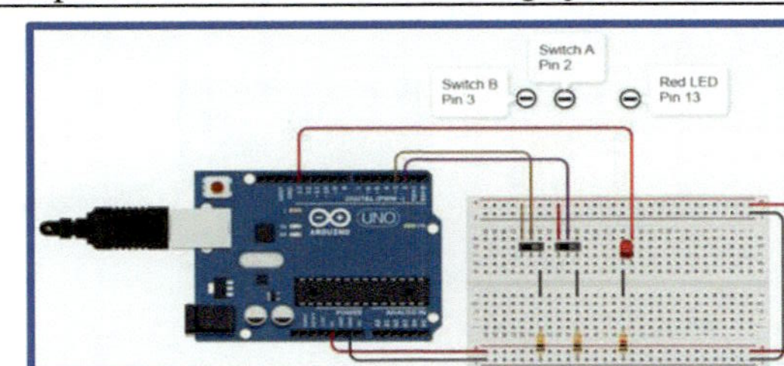 https://www.tinkercad.com/things/6qQxgZnOKQw

Lab-13	NOR logic gate	Read in 2 digital (Boolean-valued) inputs (A and B) wired using slide switches, and develop a digital output, Z = (A + B)' = A'B'	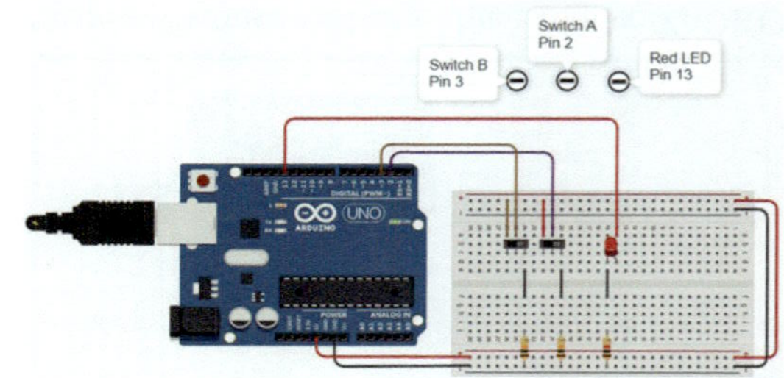 https://www.tinkercad.com/things/dE2Y59wcCIv
Lab-14	XOR logic gate	Read in 2 digital (Boolean-valued) inputs (A and B) wired using slide switches, and develop a digital output, Z = AB' + A'B	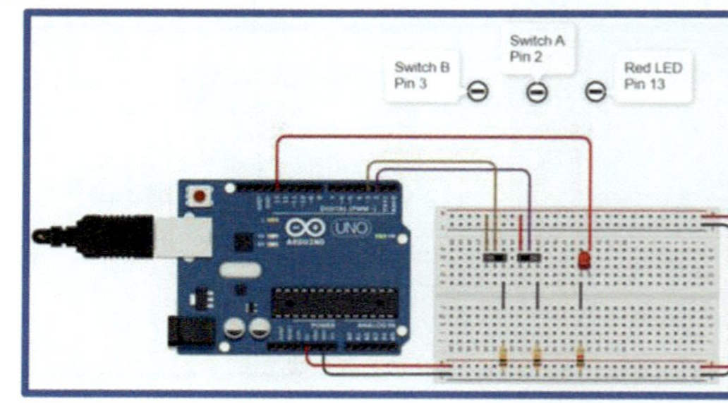 https://www.tinkercad.com/things/cZEwK8qgzMW
Lab-15	XNOR logic gate	Read in 2 digital (Boolean-valued) inputs (A and B) wired using slide switches, and develop a digital output, Z = A'B' + AB	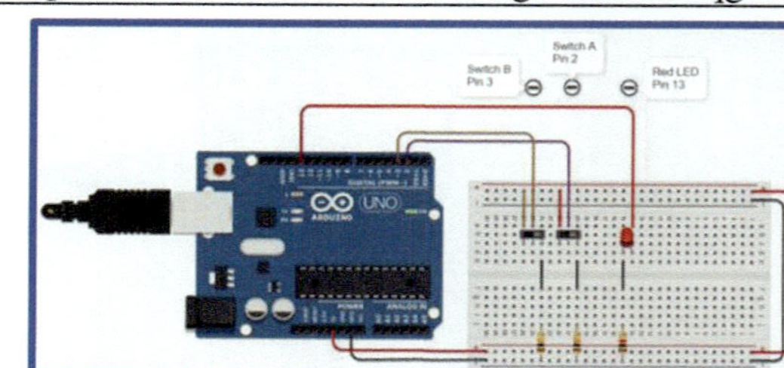 https://www.tinkercad.com/things/8Ov6MsY9wkj

Lab-16	16a Combinational logic gates	Read in 3 digital (Boolean-valued) inputs (A, B, C) wired using slide switches, and develop a digital output, Z = A + BC', simulating the operation of combinational logic gates, NOT, AND, and OR.	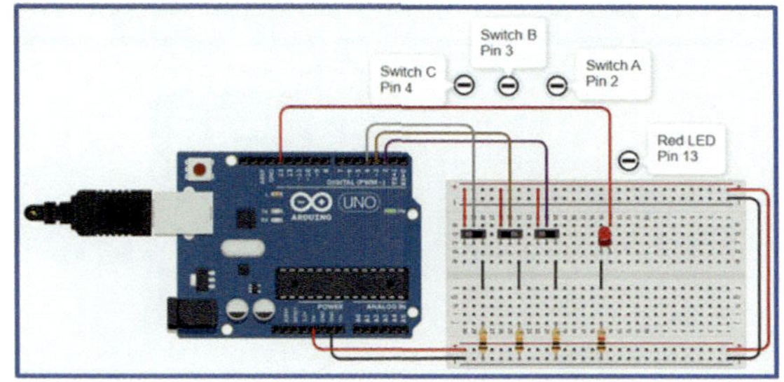 https://www.tinkercad.com/things/bjYIK0Xx5vb
	16b Combinational logic gates and status display	Read in 3 digital (Boolean-valued) inputs (A, B, C) wired using slide switches, and develop a digital output, Z = A + BC', simulating the operation of combinational logic gates, NOT, AND, OR. I/O status is shown on screen using Serial Monitor.	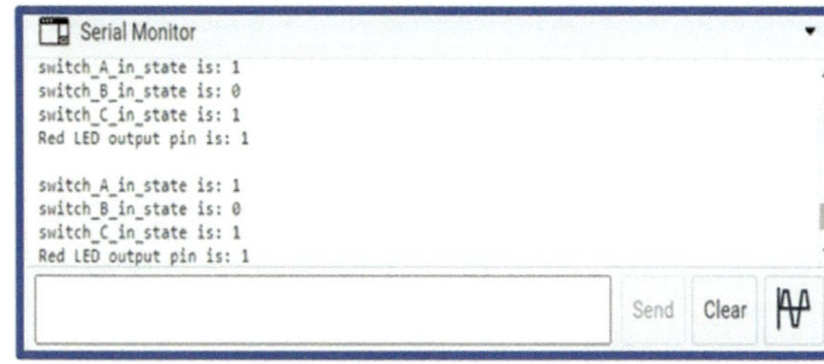 https://www.tinkercad.com/things/2fnivjQIXYR
Lab-17	17a Rotary pot control of LED and Serial Monitor display	Read in analog inputs (A0 of Arduino) using a potentiometer, and when the voltage crosses a set threshold, develop a digital output, Z	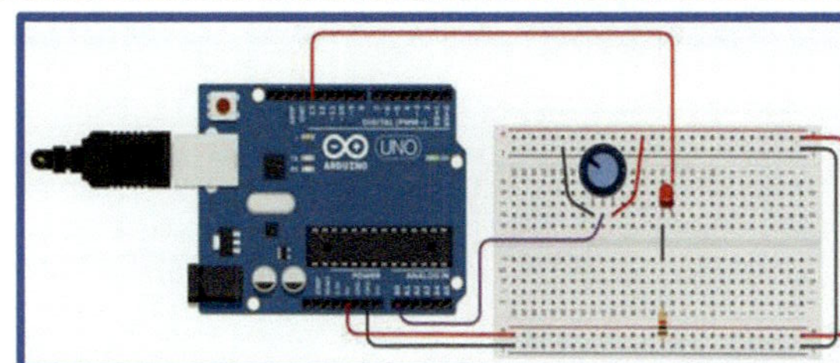 https://www.tinkercad.com/things/92NnFIArSgt

	17b Rotary pot control of LED on-off duration and Serial	Read in analog input (A0 of Arduino) using a potentiometer, and control the on-time for digital output, Z	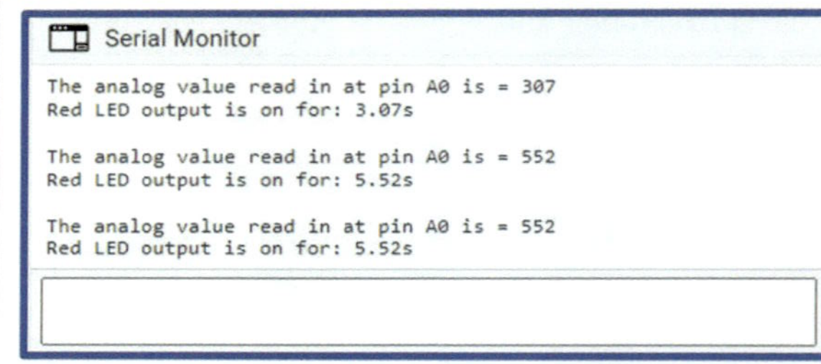 https://www.tinkercad.com/things/9MkpOmTK6Bu
Lab-18	18a Photoresistor control of LED and Serial display	Read in analog input (A0 of Arduino) using a photoresistor (light dependent resistor), and when the ambient value crosses a set threshold, develop a digital output, Z	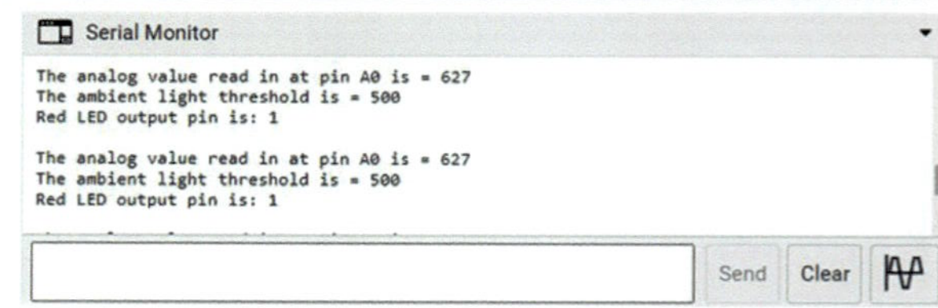 https://www.tinkercad.com/things/ksntUrmpLBw
	18b Photoresistor control of LED using PWM	Read in analog input (A0 of Arduino) using a photoresistor (light dependent resistor), and based on the ambient lighting, develop a pulse width modulated (PWM) output for brightening or dimming the output of an LED	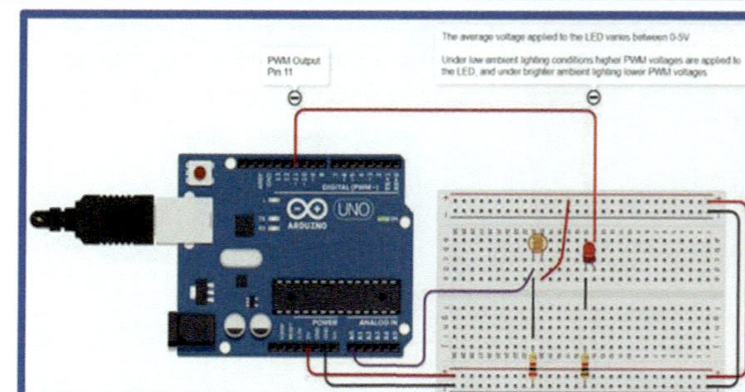 https://www.tinkercad.com/things/5NekpLSYYKJ

Lab-19	Switch control of 7-segment display	Read in 2 digital (Boolean-valued) inputs (A and B) wired using slide switches, and display the count on a 7-segment display	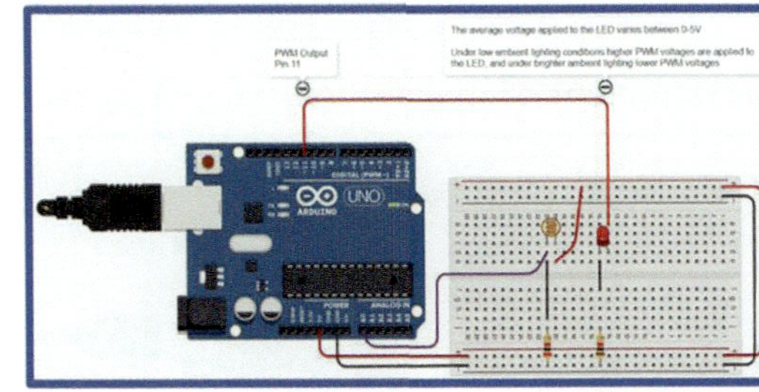 https://www.tinkercad.com/things/4lN2yNsMDuT
Lab-20	Memory remembering the status of inputs and Toggle mode	Read in 2 digital (Boolean-valued) inputs (Set/J and Reset/K) wired using slide switches, and develop a digital memory output that remembers the previous state of the output. When both inputs are active, the output enters Toggle mode, turning the output on and off at regular intervals.	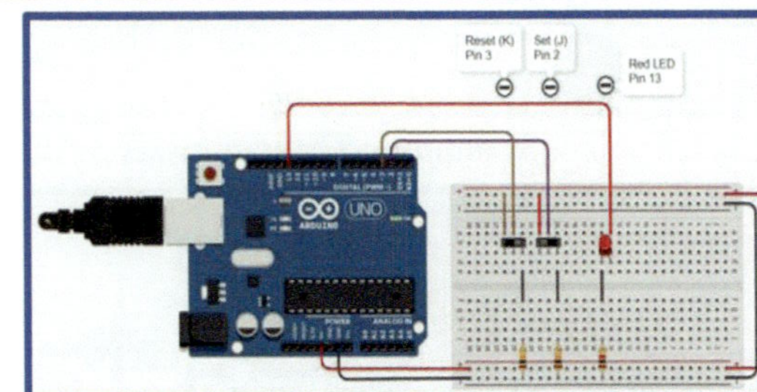 https://www.tinkercad.com/things/4kK0zbjROT2
Lab-21	21a Simple 2-bit up counter with reset	4 states (0 to 3) of a 2-bit up counter are implemented using 2 LEDs, with a time delay between different states	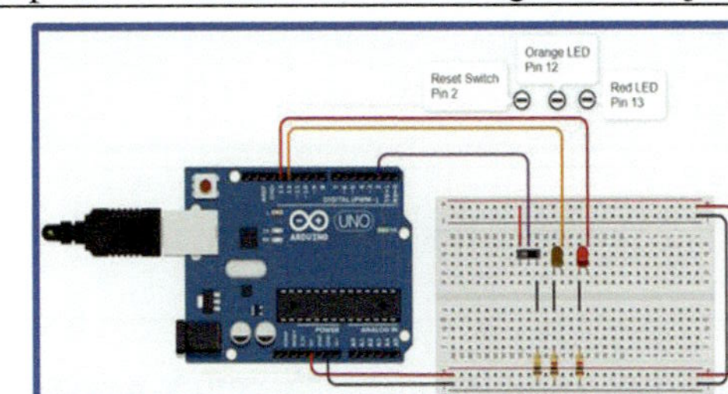 https://www.tinkercad.com/things/irhgsNGvyR9

	21b A 2-bit up counter with reset using for loop	4 states (0 to 3) of a 2-bit up counter are implemented using 2 LEDs, with a loop moving between different states	https://www.tinkercad.com/things/5JAneCPhn9v
	21c Alt 2-bit up counter with reset using for loop	4 states (0 to 3) of a 2-bit up counter are implemented using 2 LEDs, with a loop moving between different states	https://www.tinkercad.com/things/0msisOjfxpf
Lab-22	Combining sequential and combinational logic	Decode specific states of an up-counter for turning on an output LED	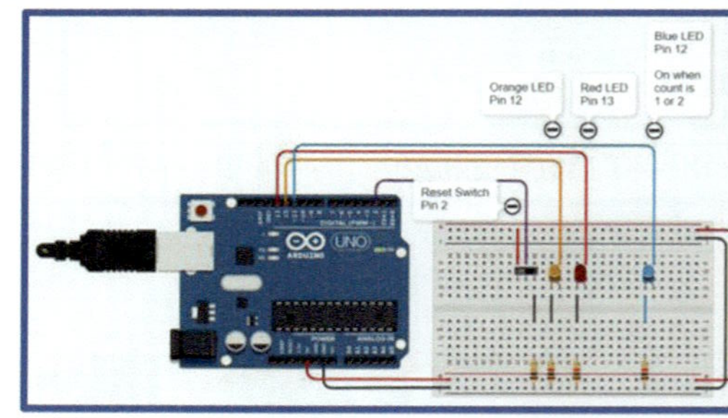 https://www.tinkercad.com/things/boYhHqzu3wB
Lab-23	23a Control Piezo buzzer frequency and duration	Read in 1 digital (Boolean-valued) input (A) wired using a slide switch, and based on the value, play either one tone or another for different durations	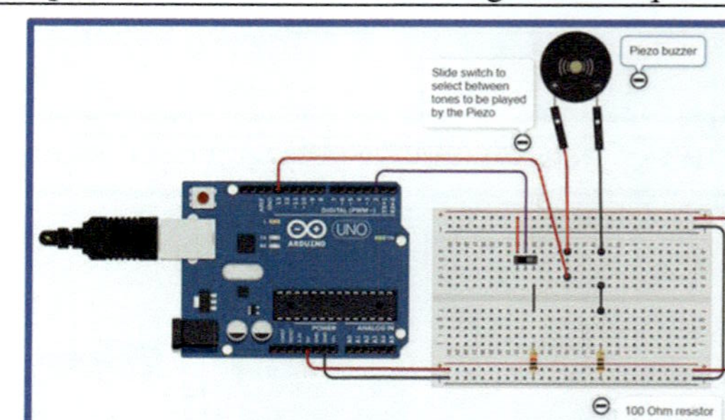 https://www.tinkercad.com/things/dLz1JM5EN82

	23b Play musical notes or other sound on Piezo buzzer	Read in 1 digital (Boolean-valued) input (A) wired using a slide switch, and based on the value, play either do-re-mi or a low-frequency tone	https://www.tinkercad.com/things/5XfuH8ZIurl
Lab-24	24a Control a higher voltage DC motor	Read in 1 digital (Boolean-valued) input (A) wired using a slide switch, and if closed, turn a higher voltage DC motor on using a transistor	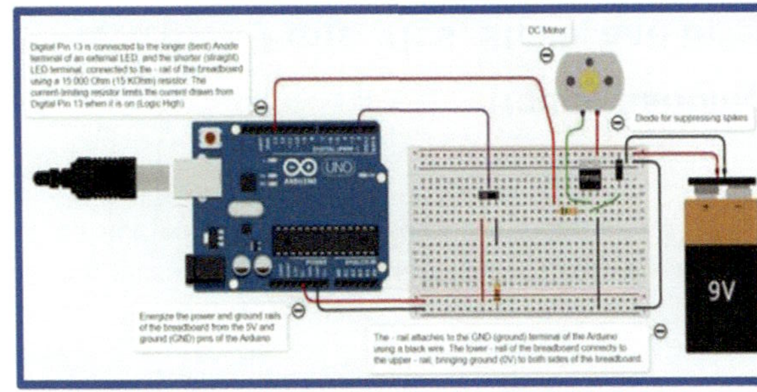 https://www.tinkercad.com/things/jX9X0S5m58G
	24b Control the position of a servo motor	Read in 1 digital (Boolean-valued) input (A) wired using a slide switch, and if closed, rotate a servo motor first one way by a specified number of degrees, and then the other way; otherwise return to home position	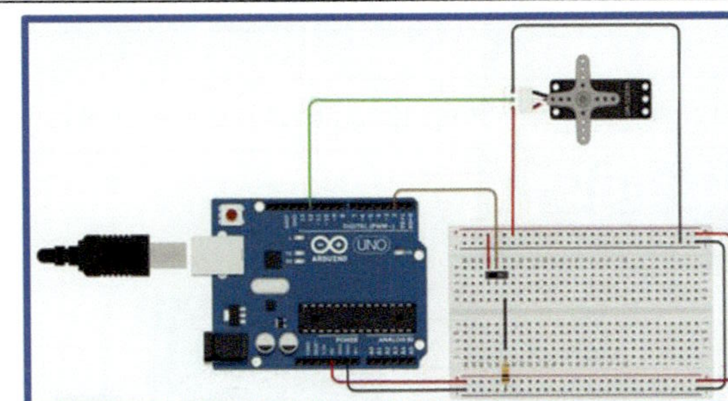 https://www.tinkercad.com/things/ikhYhNPHsKD

Lab-25	Message on LCD	This sketch initially printed "Hello World!" to the LCD and showed the time in seconds since the program started. It is now modified to print: "Lab it to Learn!" "Time Flies: ###s; where # represents the time elapsed in seconds since the program was first run or reset.	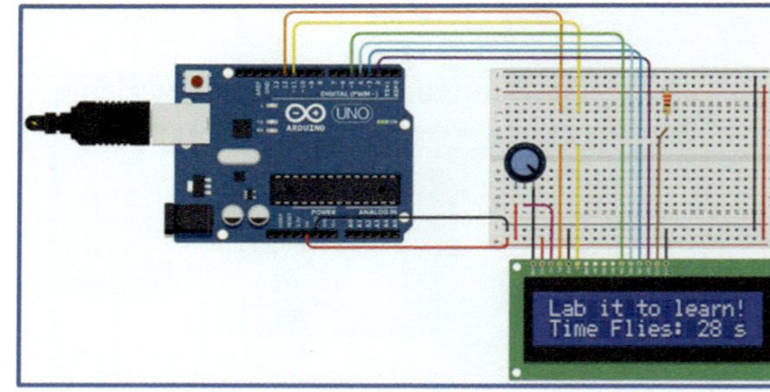 https://www.tinkercad.com/things/0c74U7GJc71

Dear Lab'venturers!
We wish you the best as you continue your tinkering exploration! Tinkercad offers numerous Arduino Uno (and other Microcontroller electronics) activities to lab in a safe, simulated environment.

Share the joy of the lab fun you are experiencing with a friend, family member, class, group, or acquaintance. Learning in a community of fellow enthusiasts on increasingly realistic projects of your choice will create a sense of competency, of contentment.

Happy Labbing!

Profs. Vigs, Steve, and Brian

February, 2023

Lab-01 Original blink built-in LED

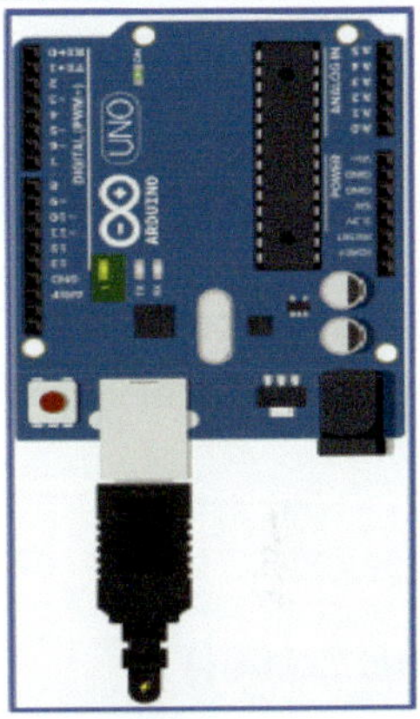

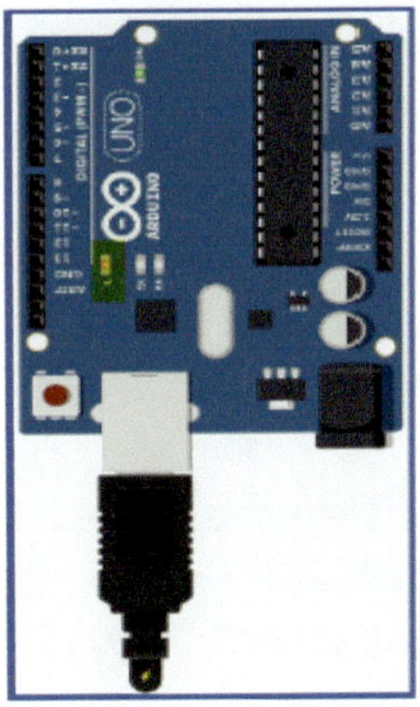

```
/*
Filename:
DigFun-01 Original blink built-in LED

This is initial sketch that runs when an
Arduino Uno R3 is placed in the workspace
and the "Start Simulation" button is
pressed with a few comments included
explaining its behavior. It blinks the
built-in LED connected to Pin 13 of the
Arduino board, first switching it on,
adding a 1s delay during which the LED
stays on, then switching it off for 1s.
The on-off
sequence repeats.
*/


// The setup() function runs once when
// the program is first
// started or the Arduino board is reset.
// In the setup() we
// typically assign user understandable
// names to Arduino I/O
// pins, declare variables for monitoring
// the status/values
// of quantities that when the program
// runs, and assign user-
// understandable pin names as Inputs
// Outputs, or as shown below
// assign a predefined component name for
// use as an output)
void setup()
{
  // Assign the built-in LED as an Output
used in this sketch
  pinMode(LED_BUILTIN, OUTPUT);
}
// The loop() function runs continuously
void loop()
{
  // Use the digitalWrite(Specify-Arduino-Digital-Pin, value)
  // function to specify which Digital output should be set
  // at Logic High or Logic Low
  digitalWrite(LED_BUILTIN, HIGH);
  delay(1000); // Wait for 1000 milliseconds = 1s
  digitalWrite(LED_BUILTIN, LOW);
  delay(1000); // Wait for 1000 milliseconds = 1s
}
```

DigFun-02a Change blink timing for built-in LED

```
/*
Filename:
DigFun-02a Change blink timing for built-in LED

Blink the built-in LED connected to Pin 13 of the Arduino
board, in intervals of 0.5s
*/


void setup()
{
// Assign the built-in LED as an Output
pinMode(LED_BUILTIN, OUTPUT);
}


void loop()
{
// Turn the built-in LED on
digitalWrite(LED_BUILTIN, HIGH);
delay(500); // Wait for 500 milliseconds = 0.5s
// Turn the built-in LED off
digitalWrite(LED_BUILTIN, LOW);
delay(500); // Wait for 0.5s
}
```

Lab-02b Change blink timing for built-in LED using constants

```
/*
Filename:
DigFun-02b Change blink timing for built-in LED using
constants

Blink the built-in LED connected to Pin 13 of the Arduino
board, in intervals of 0.5s using constants in the sketch
*/


// Recommend using "constant" for values that do not change
// while the program runs as it does not use RAM
// Declare and assign these constants preferably before
// the functions setup() and loop()
// 500ms = 0.5s values for the LED on time and off time
const int on_time = 500;
const int off_time = 500;


void setup()
{
 // Assign the built-in LED as an Output
 pinMode(LED_BUILTIN, OUTPUT);
}


void loop()
{
 // Turn the built-in LED on
 digitalWrite(LED_BUILTIN, HIGH);
 // Wait time with LED on is determined by "on_time" value
 delay(on_time);
 // Turn the built-in LED off
 digitalWrite(LED_BUILTIN, LOW);
 // Wait time with LED off is determined by "off_time" value
 delay(off_time);
}
```

Lab-03a: Morse Code letter A with built-in LED

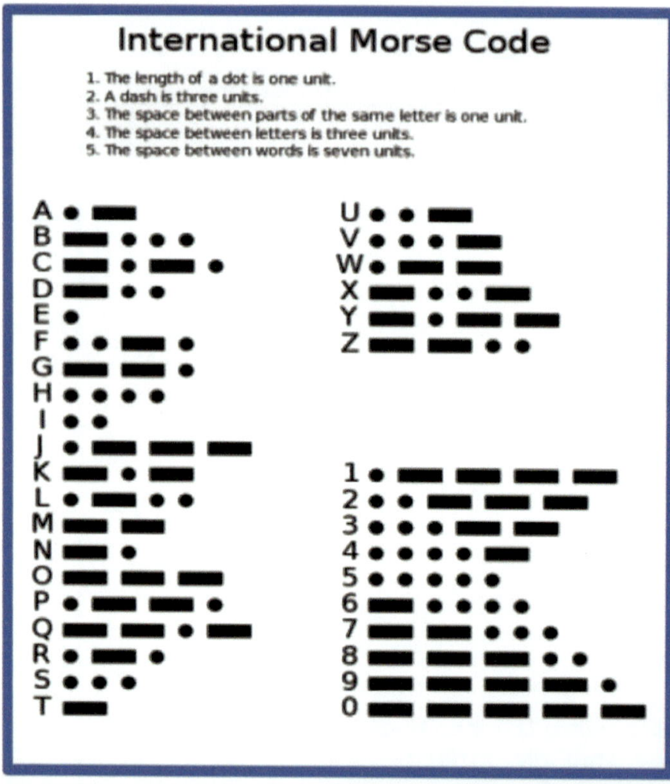

Illustration source: Public domain image from Wikimedia Commons:
https://commons.wikimedia.org/wiki/File:International_Morse_Code.svg

```
/*
Filename:
DigFun-03a Morse Code letter A with built-in LED

We will be sending a single character (A) in Morse Code using
the in-built LED. Refer to Wikipedia --
https://en.wikipedia.org/wiki/Morse_code – for more
information
about the International Morse Code.
*/

// A series of short pulses (dots/dit) and long pulses
// (dashes/dah) that are 3 times the duration of the short
// pulses are used in Morse Code.
// There is also a gap between the letters of a word, equal
// to the duration of a short pulse.
// There is a gap of 7 times the duration of dots between
words
// Here, we set up 500ms = 0.5s as the unit time for a the
// duration a dot transmission
const int unit_time = 500;

void setup()
{
 // Assign the built-in LED as an Output
 pinMode(LED_BUILTIN, OUTPUT);
}

void loop()
{
 // Sending the first letter of the English Alphabet - A
 // using Morse code -- encoded as a "dot dash"

 // LED on for dot duration
 digitalWrite(LED_BUILTIN, HIGH);
 delay(unit_time); // dit
 // Following the dot, there is a period of signal absence
 // equal in duration to the that of the dot
 digitalWrite(LED_BUILTIN, LOW);
 delay(unit_time); // signal absence
 // LED on for dash duration
 digitalWrite(LED_BUILTIN, HIGH);
 delay(3*unit_time); // dah
 digitalWrite(LED_BUILTIN, LOW);
```

```
  // Since the loop repeats, sending successive "A" there
should
  // be a 3500 ms period of signal absence between successive
  // transmissions of the letter/word A
  delay(7*unit_time); // signal absence

  /*
  International Morse code, one of the first digital codes
  invented, primarily encodes A-Z and 0-9, using a series of
  dots (typically called dits) and dashes (called dahs)
  The on-duration of the dit -- whether through an electrical
  signal, light, sound -- is used to set the timing for
  the Morse code communication system.

  The duration of the dah is 3 times that of the dit
  So, if the dit sound lasts for 500 ms, the dah is 1500 ms

  Letters and number often include a combination of dots and
  dashes. There is a period of signal absence between parts
  of the same letter or digit, equal to that of a dit,
  in our case 500 ms. For example, in Morse Code, "A or a" is
  encoded as "dit dah" or "500ms on, 500 ms off, 1500 ms on"

  The duration between letters in a word, or digits in a
  number is 3 dot (dit) units, in our case 1500 ms

  The duration between words is 7 dot (dit) units,
  in our case 7 * 500ms = 3500 ms = 3.5 s

  The letter "A" in Morse Code is "._" (dit dah)
  The period of silence gap between the dit and the dah for "A"
  is the duration of a dit: Se we have 500ms on (dit),
  500ms off (silence gap), and 1500ms on (dah)

  Refer Wikipedia:
  https://en.wikipedia.org/wiki/Morse_code
  */
}
```

Lab-03b Morse Code letter A with built-in LED using functions

```
/*
Filename:
DigFun-03b Morse Code letter A with built-in LED using
functions

We will be sending a single character (A) in Morse Code using
the in-built LED and custom functions for signaling the
dot-dash (dit-dah) durations.
*/


// Here, we set up 500ms = 0.5s as the unit time for a the
// duration a dot transmission
const int unit_time = 500;


void setup()
{
 // Assign the built-in LED as an Output
 pinMode(LED_BUILTIN, OUTPUT);
}


void loop()
{
 dit(); // The dit now also includes unit_time for signal
      // absence following the dit duration
 dah(); // The dah now also includes unit_time for signal
      // absence following the dah duration

 // Since the loop repeats, sending successive "A" there
should
 // be a 3500 ms period of signal absence between the words
 inter_word_gap();
}
```

```
void dit()
{
  digitalWrite(LED_BUILTIN, HIGH);
  delay(unit_time);

  // including a period of signal absence along with dit
  digitalWrite(LED_BUILTIN, LOW);
  delay(unit_time); // signal absence
}

void dah()
{
  digitalWrite(LED_BUILTIN, HIGH);
  delay(3*unit_time); // dah

  // including a period of signal absence along with dah
  digitalWrite(LED_BUILTIN, LOW);
  delay(unit_time); // signal absence
}

void inter_word_gap()
{
  // The total gap (signal silence) between successive words
  // is 7*unit_time. However, since each dit or dah also
  // includes a single unit_time of signal silence, we just
  // need 6*unit_time following either a dit or dah at the end
  // of a word
  digitalWrite(LED_BUILTIN, LOW);
  delay(6*unit_time); // signal absence of 6 * dit duration
}
```

DigFun-04 Sending a name in Morse Code with built-in LED (example: Dale, honoring our colleague, Prof. Dale Patrick)

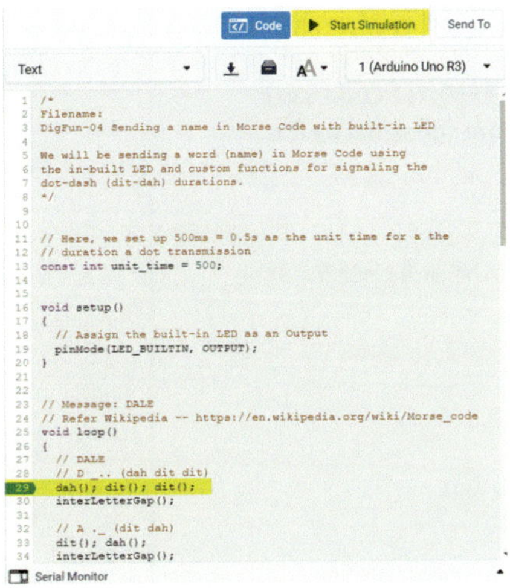

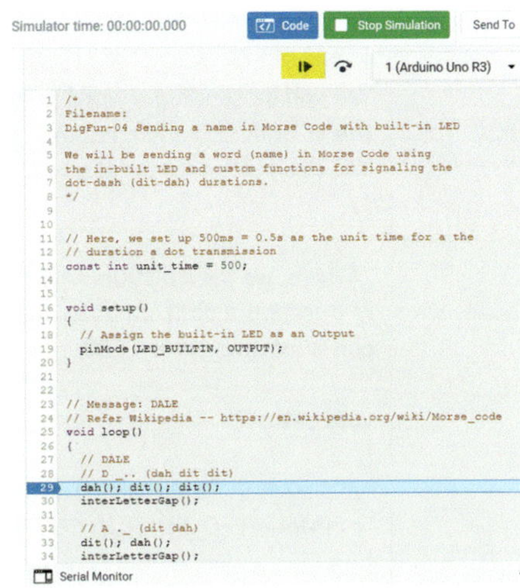

Clicking on the line number allows us to set breakpoint where the program will pause when simulation is started by pressing the "Start Simulation" button.

Now, when the simulation starts, it will stop at the first breakpoint, and at successive ones, if these are set. The simulation waits for the user to press the "Resume Execution" button (black bar with forward facing arrow). This allows us to view the "dah dit dit" pattern using the inbuilt LED clearly, before moving on to the Morse Code for the next letter in the sequence.

```
/*
Filename:
DigFun-04 Sending a name in Morse Code with built-in LED

We will be sending a word (name) in Morse Code using
the in-built LED and custom functions for signaling the
dot-dash (dit-dah) durations.
*/

// Here, we set up 500ms = 0.5s as the unit time for a the
// duration a dot transmission
const int unit_time = 500;

void setup()
{
 // Assign the built-in LED as an Output
 pinMode(LED_BUILTIN, OUTPUT);
}

// Message: DALE
// Refer Wikipedia --
https://en.wikipedia.org/wiki/Morse_code
void loop()
{
 // DALE
 // D _.. (dah dit dit)
 dah(); dit(); dit();
 interLetterGap();

 // A ._ (dit dah)
 dit(); dah();
 interLetterGap();

 // L ._.. (dit dah dit dit)
 dit(); dah(); dit(); dit();
 interLetterGap();

 // E (dit)
 dit();
 interLetterGap();

 // Interword gap of 3500 including 500 ms following
 // the last dit of the "E" in Dale
 interWordGap();
}
```

```
//Function for the dit sound
void dit()
{
 digitalWrite(LED_BUILTIN, HIGH);
 delay(unit_time);

 // including the period of signal absence along with dit
 digitalWrite(LED_BUILTIN, LOW);
 delay(unit_time); // signal absence
}

//Function for the dah sound
void dah()
{
 digitalWrite(LED_BUILTIN, HIGH);
 delay(3*unit_time); // dah

 // including the period of signal absence along with dah
 digitalWrite(LED_BUILTIN, LOW);
 delay(unit_time); // signal absence
}

// Function for the period of signal absence between
// letters of a word including period of signal absence
// following the dit and dah functions
// So, inter-letter gap totals 3*dit
void interLetterGap()
{
 digitalWrite(LED_BUILTIN, LOW);
 delay(2*unit_time); // signal absence
}

// Function for the period of signal absence of
// 6 * dit duration as each dit or dah function also includes
// signal absence at end
// So, total inter-word gap totals 7*dit)
void interWordGap()
{
 digitalWrite(LED_BUILTIN, LOW);
 delay(6*unit_time);
}
```

Lab-05: Sending a Morse Code message with built-in LED (example: Thanks Dale)

```
/*
Filename:
DigFun-05 Sending a Morse Code Message

We will be sending a message (Thanks Dale) in Morse Code
using
the in-built LED and custom functions for signaling the
dot-dash (dit-dah) durations.
*/

// Here, we set up 500ms = 0.5s as the unit time for a the
// duration a dot transmission
const int unit_time = 500;

void setup()
{
 // Assign the built-in LED as an Output
 pinMode(LED_BUILTIN, OUTPUT);
}

// Message: Thanks Dale
// Lowercase and uppercase letters are treated the same in
Morse
// Refer Wikipedia --
https://en.wikipedia.org/wiki/Morse_code
void loop()
{
 // T _ (dah)
 dah(); // The dah also includes signal absence

 // h .... (dit dit dit dit)
 dit(); dit(); dit(); dit();
 inter_letter_gap();

 // a ._ (dit dah)
 dit(); dah();
 inter_letter_gap();

 // n _. (dah dit)
 dah(); dit();
 inter_letter_gap();
```

```
void dah()
{
 digitalWrite(LED_BUILTIN, HIGH);
 delay(3*unit_time); // dah

 // including the period of signal absence along with dah
 digitalWrite(LED_BUILTIN, LOW);
 delay(unit_time); // signal absence
}


void inter_letter_gap()
{
 // Period of signal absence between letters of a word
 // including time of signal absence following dit and dah
 digitalWrite(LED_BUILTIN, LOW);
 delay(2*unit_time); // signal absence
}


void inter_word_gap()
{
 // Period of signal absence between letters of a word
 // including time of signal absence following dit and dah
 digitalWrite(LED_BUILTIN, LOW);
 delay(6*unit_time); // signal absence
}
```

Lab-06a Blink external LED

Adding an external LED to Digital Pin 13 for viewing the blinking LED clearly

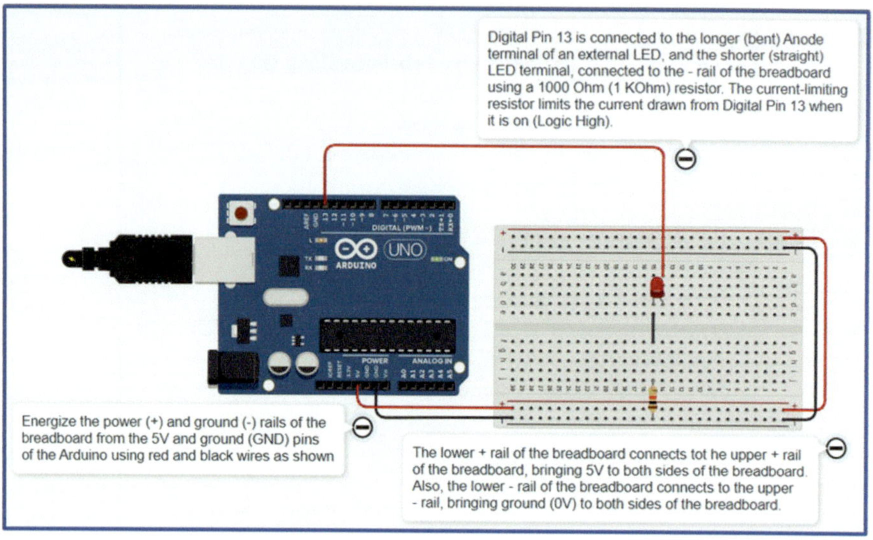


```
/*
Filename:
DigFun-06a Blink external LED

Blink an external LED connected to Pin 13 of the Arduino
board,
in intervals of 0.5s
*/

// Assign user-understandable names to Arduino Output pin
const int red_led_out_pin = 13;

// 500ms = 0.5s values for the LED on time and off time
const int on_time = 500;
const int off_time = 500;

void setup()
{
  // Assign user-understandable pin names as Inputs or
Outputs
  pinMode(red_led_out_pin, OUTPUT);
}
```

```
void loop()
{
 // Turn the external LED on
 digitalWrite(red_led_out_pin, HIGH);
 // Wait time with LED on is determined by "on_time" value
 delay(on_time);

 // Turn the external LED off
 digitalWrite(red_led_out_pin, LOW);
 // Wait time with LED off is determined by "off_time" value
 delay(off_time);
}
```

Lab-06b Blink external LED and segment A of a 7-segment

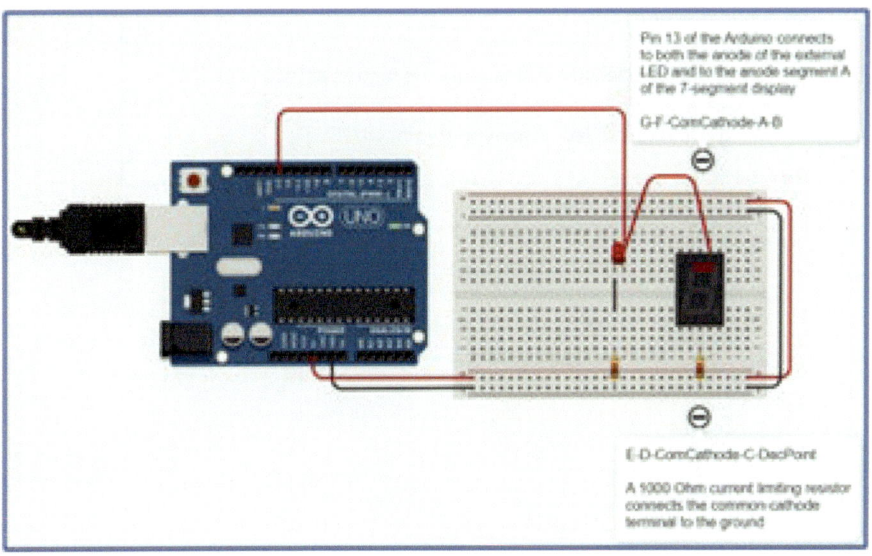

```
/*
Filename:
DigFun-06b Blink external LED and segment A of a 7-segment

Blink an external LED and one segment (A) of a common-
cathode
7-segment display, connected to Pin 13 of the Arduino board,
in intervals of 0.5s
*/

// Assign user-understandable names to Arduino Output pin
const int red_led_display_out_pin = 13;

// 500ms = 0.5s values for the LED on time and off time
const int on_time = 500;
const int off_time = 500;


void setup()
{
 // Assign user-understandable pin names as Inputs or
Outputs
 pinMode(red_led_display_out_pin, OUTPUT);
}
```

```
void loop()
{
 // Turn the external LED on
 digitalWrite(red_led_display_out_pin, HIGH);
 // Wait time with LED on is determined by "on_time" value
 delay(on_time);

 // Turn the external LED off
 digitalWrite(red_led_display_out_pin, LOW);
 // Wait time with LED off is determined by "off_time" value
 delay(off_time);

 /* A typical common-cathode 7-segment display, consists of
 7 individual rectangular-shaped LEDs and 1 round-shaped
LED,
 the cathodes (negative terminals) of all are connected
 together forming a common cathode. In order to light up a
 specific segment a suitable voltage must be connected to the
 segment. The top segment is "A." Moving clockwise, the
segment
 on the top-right is "B," bottom-right "C," bottom "D,"
 bottom-left "E,", top-left "F," and center "G." The decimal
 point is located on the bottom right.

 As shown in the circuit diagram, the display we are using has
 two common cathode terminals, one located at the top of
the
 display, the other at the bottom. We are limiting the current
 drawn by the 7-segment display by connecting its
 common-cathode to a 1000 Ohm current-limiting resistor
 connected to the ground.

 Refer to the datasheet of the specific 7-segment display to
 determine which pins are connected to which of the anodes
of
 the specific LED segments.
 */
}
```

Lab-07 Press the pushbutton to blink external LED

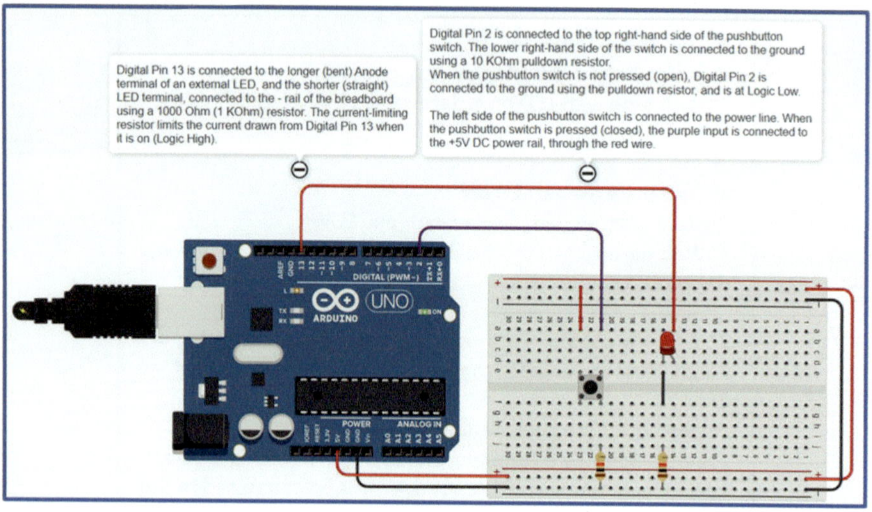


```
/*
Filename:
DigFun-07 Press pushbutton to blink external LED

Read in 1 digital (Boolean-valued) input (A) wired using a
pushbutton switch, and if pressed turn an external LED on and
off in intervals of 0.5s
*/

// Assign user-understandable names to Arduino I/O pins
const int switch_A_in_pin = 2;
const int red_led_out_pin = 13;

// 500ms = 0.5s values for the LED on time and off time
const int on_time = 500;
const int off_time = 500;

// Status of Input pin is monitored by a variable
boolean switch_A_in_pin_state;
```

```
void setup()
{
 // Assign user-understandable pin names as Inputs or
Outputs
 pinMode(switch_A_in_pin, INPUT);
 pinMode(red_led_out_pin, OUTPUT);
}


void loop()
{
 // Read in values of input switch, storing current state
 switch_A_in_pin_state = digitalRead(switch_A_in_pin);

 // Evaluate status of switch A
 if ( switch_A_in_pin_state == HIGH ){ // pushbutton pressed
   // If switch A is closed, turn output LED on and off
   digitalWrite(red_led_out_pin, HIGH);
   delay(on_time); // Wait for on_time duration
   digitalWrite(red_led_out_pin, LOW);
   delay(off_time); // Wait for off_time duration
 }

 else { // switch A off
   // If switch A is open, turn output LED on
   digitalWrite(red_led_out_pin, LOW);
 }

}
```

Lab-08 Close the slide switch to blink external LED

Adding an external slide switch for blinking the LED connected to Digital Pin 13 only when the switch is in the ON position

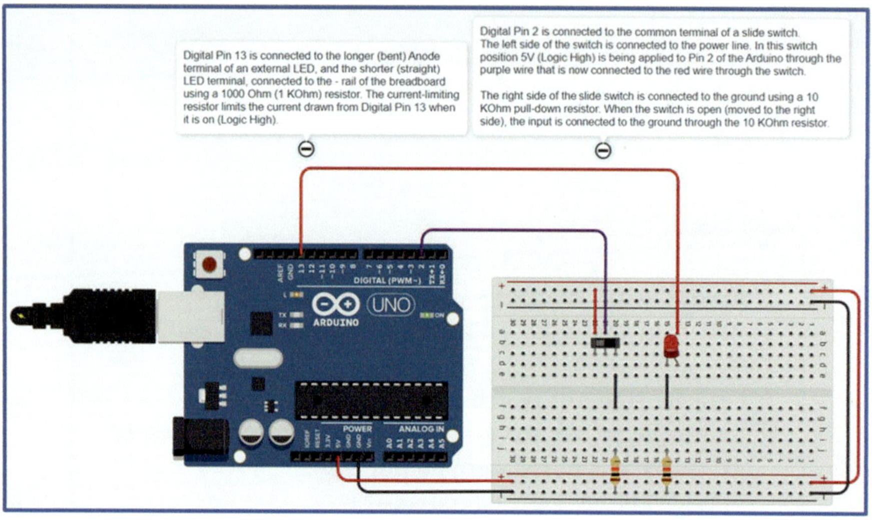


```
/*
Filename:
DigFun-08 Close slideswitch to blink external LED

Read in 1 digital (Boolean-valued) input (A) wired using a
slide switch, and if closed turn an external LED on and off
in intervals of 0.5s
*/


// Assign user-understandable names to Arduino I/O pins
const int switch_A_in_pin = 2;
const int red_led_out_pin = 13;

// 500ms = 0.5s values for the LED on time and off time
const int on_time = 500;
const int off_time = 500;

// Status of Input pin is monitored by a variable
boolean switch_A_in_pin_state;
```

```
void setup()
{
 // Assign user-understandable pin names as Inputs or
Outputs
 pinMode(switch_A_in_pin, INPUT);
 pinMode(red_led_out_pin, OUTPUT);
 }


void loop()
{
 // Read in values of input switch, storing current state
 switch_A_in_pin_state = digitalRead(switch_A_in_pin);

 // Evaluate status of switch A
 if ( switch_A_in_pin_state == HIGH ){ // switch closed
   // If switch A is closed, turn output LED on and off
   digitalWrite(red_led_out_pin, HIGH);
   delay(on_time); // Wait for on_time duration
   digitalWrite(red_led_out_pin, LOW);
   delay(off_time); // Wait for off_time duration
 }

 else { // switch A off
   // If switch A is open, turn output LED on
   digitalWrite(red_led_out_pin, LOW);
 }

}
```

Lab-09a NOT logic gate

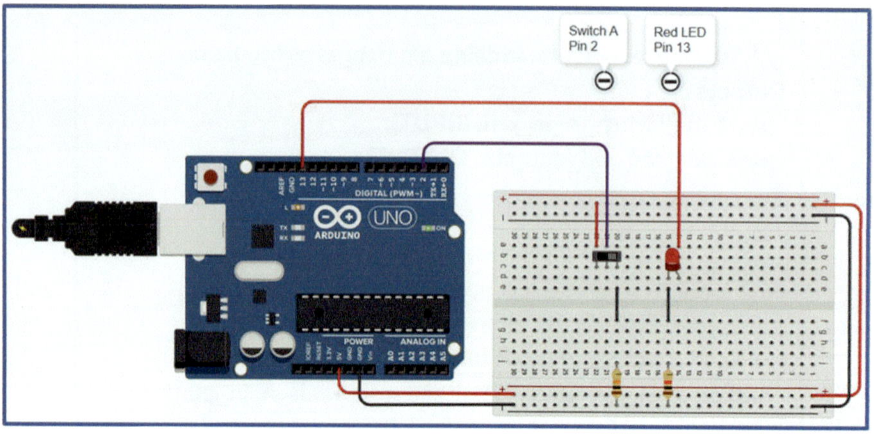


```
/*
Filename:
DigFun-09a NOT logic gate

Read in 1 digital (Boolean-valued) input (A) wired using a
slide switch, and develop a digital output, Z=A'
*/

// Assign user-understandable names to Arduino I/O pins
const int switch_A_in_pin = 2;
const int red_led_out_pin = 13;

// Status of Input pin is monitored by a variable
boolean switch_A_in_state;

void setup()
{
  // Assign user-understandable pin names as Inputs or
Outputs
  pinMode(switch_A_in_pin, INPUT);
  pinMode(red_led_out_pin, OUTPUT);
}
```

```
void loop()
{
 // Read in values of input switch, storing current state
 switch_A_in_state = digitalRead(switch_A_in_pin);

 //  A Z=A' = (Not A = Invert A)
 //  0 1
 //  1 0

 // Evaluate status of switch A
 if ( switch_A_in_state == HIGH ){
  // If switch A is closed, turn output LED off
  digitalWrite(red_led_out_pin, LOW);
 }
 else {
  // If switch A is open, turn output LED on
  digitalWrite(red_led_out_pin, HIGH);
 }

}
```

Lab-09b NOT logic gate with status display on Serial Monitor

```
/*
Filename:
DigFun-09b NOT logic gate with status display on Serial Monitor

Read in 1 digital (Boolean-valued) input (A) wired using a
slide switch, and develop a digital output, Z=A'.
Status of the input switch, A, and the output LED, Z, is
shown on the Serial Monitor (located below the Code section)
*/

// Assign user-understandable names to Arduino I/O pins
const int switch_A_in_pin = 2;
const int red_led_out_pin = 13;

// Status of Input pin is monitored by a variable
boolean switch_A_in_state;
boolean red_led_out_state;

// 0.5s delay between successive runs of the loop() function
// for easier viewing of the Serial Monitor display
const int loop_delay_time = 500;

void setup()
{ // Assign user-understandable pin names as Inputs or Outputs
  pinMode(switch_A_in_pin, INPUT);
  pinMode(red_led_out_pin, OUTPUT);

  // initialize serial communication at 9600 bits per sec
  // The Serial Monitor tool is typically used for displaying
  // information from the Arduino (pin status or variables) and
  // for providing text/numeric user input to the program
  Serial.begin(9600);
}
```

```
void loop()
{
 // Read in values of input switch, storing current state
 switch_A_in_state = digitalRead(switch_A_in_pin);

 // Serial.print() prints values and stays on same line
 Serial.print("switch_A_in_state is: ");
 // Anything within double quotes is printed as is

 // Serial.println() prints values and
 // moves the cursor to the next line for printing;
 Serial.println(switch_A_in_state); // print variable value

 //  A Z=A' = (Not A = Invert A)
 //  0 1
 //  1 0

 // Evaluate status of switch A
 if ( switch_A_in_state == HIGH ){
  // If switch A is closed, turn output LED off
  digitalWrite(red_led_out_pin, LOW);
 }
 else {
  // If switch A is open, turn output LED on
  digitalWrite(red_led_out_pin, HIGH);
 }

 // Read value of output Z, storing current state
 red_led_out_state = digitalRead(red_led_out_pin);

 // print out variable Z value
 Serial.print("Red LED output pin is: ");
 Serial.println(red_led_out_state);
 Serial.println(); // blank lines between output for clarity
 delay(loop_delay_time);
}
```

Lab-10 AND logic gate

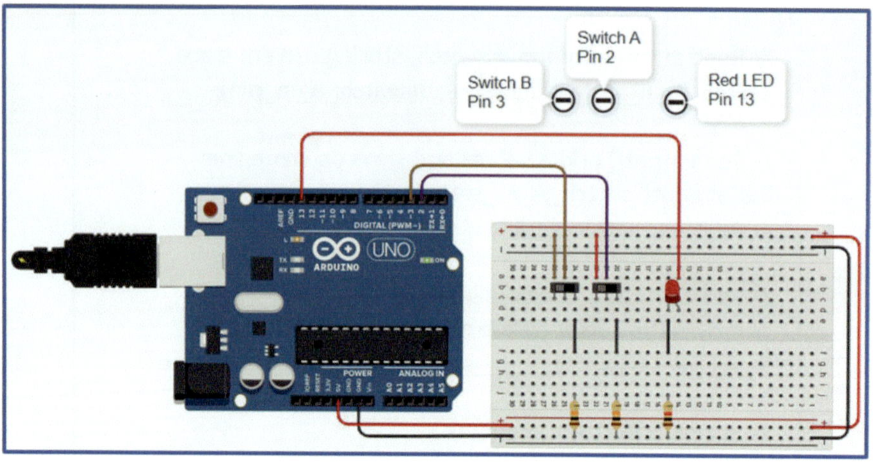


```
/*
Filename:
DigFun-10 AND logic gate

Read in 2 digital (Boolean-valued) inputs (A and B) wired using
slide switches, and develop a digital output, Z=AB
*/

// Assign user-understandable names to Arduino I/O pins
const int switch_A_in_pin = 2;
const int switch_B_in_pin = 3;
const int red_led_out_pin = 13;

// Status of Input pins is monitored by variables
boolean switch_A_in_state;
boolean switch_B_in_state;

void setup()
{
 // Assign user-understandable pin names as Inputs or
Outputs
 pinMode(switch_A_in_pin, INPUT);
 pinMode(switch_B_in_pin, INPUT);
 pinMode(red_led_out_pin, OUTPUT);
}
```

```
void loop()
{
 // Read in values of input switches, storing current state
 switch_A_in_state = digitalRead(switch_A_in_pin);
 switch_B_in_state = digitalRead(switch_B_in_pin);

 // B A Z=AB (AND)
 // 0 0 0
 // 0 1 0
 // 1 0 0
 // 1 1 1

 // Evaluate expression: Z=AB
 if ( switch_A_in_state == HIGH &&
    switch_B_in_state == HIGH){
  // When both switches are closed, turn output LED on
  digitalWrite(red_led_out_pin, HIGH);
 }
 else {
  // When either switch A is open or B is open or
  // both are open, turn output LED off
  digitalWrite(red_led_out_pin, LOW);
 }

}
```

```
/*
Filename:
DigFun-11 OR logic gate

Read in 2 digital (Boolean-valued) inputs (A and B) wired using
slide switches, and develop a digital output, Z=AB
*/

// Assign user-understandable names to Arduino I/O pins
const int switch_A_in_pin = 2;
const int switch_B_in_pin = 3;
const int red_led_out_pin = 13;

// Status of Input pins is monitored by variables
boolean switch_A_in_state;
boolean switch_B_in_state;

void setup()
{
 // Assign user-understandable pin names as Inputs or
Outputs
 pinMode(switch_A_in_pin, INPUT);
 pinMode(switch_B_in_pin, INPUT);
 pinMode(red_led_out_pin, OUTPUT);
}
```

Lab-11 OR logic gate

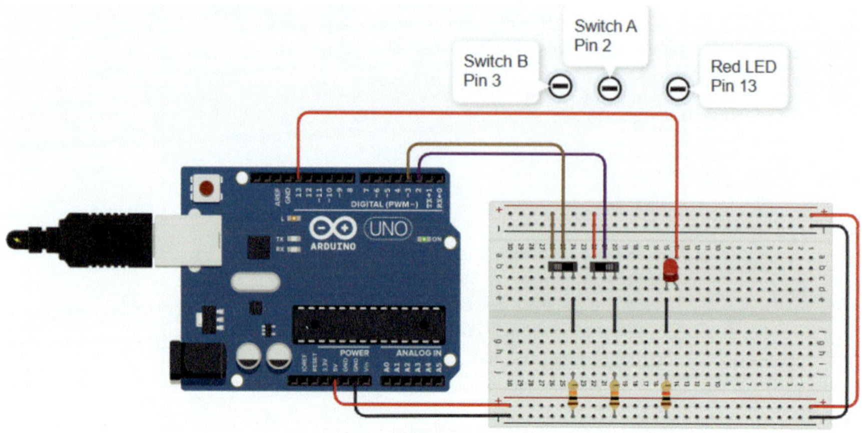


```
void loop()
{
  // Read in values of input switches, storing current state
  switch_A_in_state = digitalRead(switch_A_in_pin);
  switch_B_in_state = digitalRead(switch_B_in_pin);

  // B A Z=A+B (OR)
  // 0 0 0
  // 0 1 1
  // 1 0 1
  // 1 1 1

    // Evaluate expression: Z=A+B
  if ( (switch_A_in_state == HIGH) ||
      (switch_B_in_state == HIGH) ) {
    // When either or both switches are closed,
    // turn output LED on
    digitalWrite(red_led_out_pin, HIGH);
  }
  else {
    // When both switches A and B are open, turn output LED
off
    digitalWrite(red_led_out_pin, LOW);
  }

}
```

Lab-12 NAND logic gate

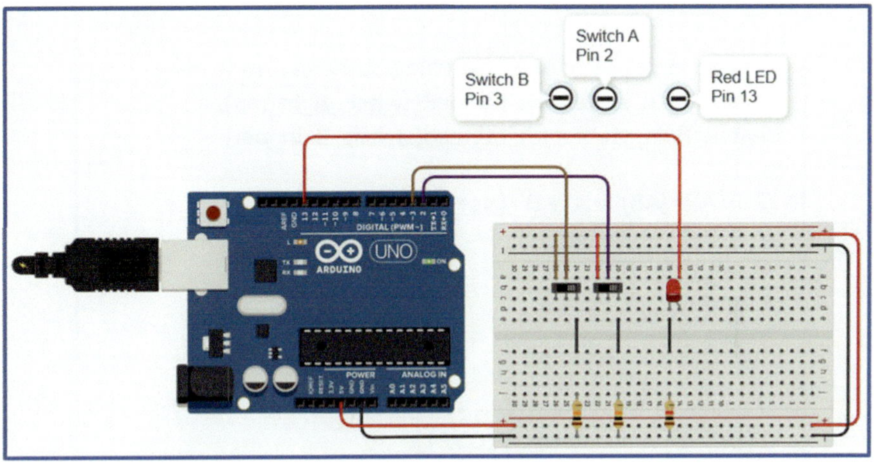


```
/*
Filename:
DigFun-12 NAND logic gate

Read in 2 digital (Boolean-valued) inputs (A and B) wired using
slide switches, and develop a digital output, Z=(AB)'=A'+B'
*/

// Assign user-understandable names to Arduino I/O pins
const int switch_A_in_pin = 2;
const int switch_B_in_pin = 3;
const int red_led_out_pin = 13;

// Status of Input pins is monitored by variables
boolean switch_A_in_state;
boolean switch_B_in_state;

void setup()
{
 // Assign user-understandable pin names as Inputs or
Outputs
  pinMode(switch_A_in_pin, INPUT);
  pinMode(switch_B_in_pin, INPUT);
  pinMode(red_led_out_pin, OUTPUT);
}
```

```
void loop()
{
 // Read in values of input switches, storing current state
 switch_A_in_state = digitalRead(switch_A_in_pin);
 switch_B_in_state = digitalRead(switch_B_in_pin);

 //  B  A Z=(AB)' = A' + B'  (Not AND = NAND)
 // 0 0 1
 // 0 1 1
 // 1 0 1
 // 1 1 0

 // Evaluate expression: Z=A'+B'
 if ( switch_A_in_state == LOW ||
      switch_B_in_state == LOW){
   // When either switch A is open or B is open or
   // both are open, turn output LED on
   digitalWrite(red_led_out_pin, HIGH);
 }
 else {
   // When both switches are closed, turn output LED off
    digitalWrite(red_led_out_pin, LOW);
 }

}
```

Lab-13 NOR logic gate

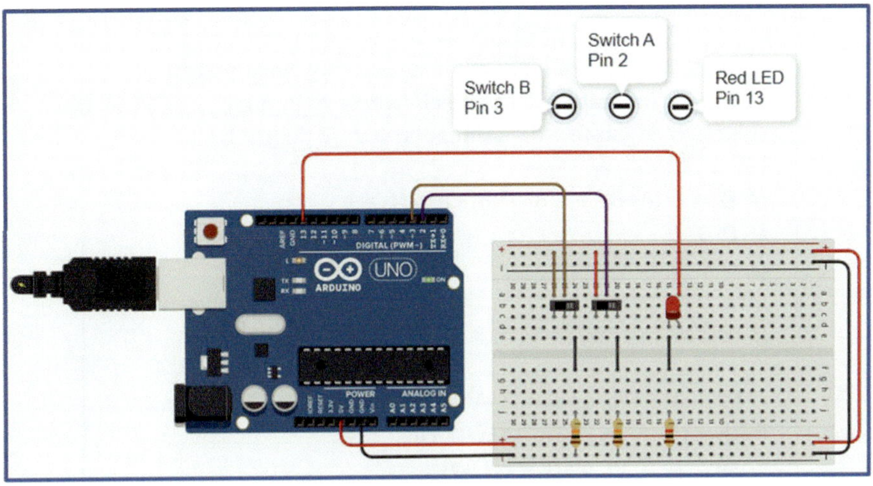


```
/*
Filename:
DigFun-13 NOR logic gate

Read in 2 digital (Boolean-valued) inputs (A and B) wired using
slide switches, and develop a digital output, Z=(A+B)'=A'B'
*/

// Assign user-understandable names to Arduino I/O pins
const int switch_A_in_pin = 2;
const int switch_B_in_pin = 3;
const int red_led_out_pin = 13;

// Status of Input pins is monitored by variables
boolean switch_A_in_state;
boolean switch_B_in_state;

void setup()
{
 // Assign user-understandable pin names as Inputs or
Outputs
 pinMode(switch_A_in_pin, INPUT);
 pinMode(switch_B_in_pin, INPUT);
 pinMode(red_led_out_pin, OUTPUT);
}
```

```
void loop()
{
 // Read in values of input switches, storing current state
 switch_A_in_state = digitalRead(switch_A_in_pin);
 switch_B_in_state = digitalRead(switch_B_in_pin);

 //  B  A Z=(A+B)' = A'B'  (Not OR = NOR)
 // 0 0 1
 // 0 1 0
 // 1 0 0
 // 1 1 0

 // Evaluate expression: Z=A'B'
 if ( switch_A_in_state == LOW &&
     switch_B_in_state == LOW){
   // When both switches A and B are open, turn output LED
on
   digitalWrite(red_led_out_pin, HIGH);
 }
 else {
   // When either or both switches are closed,
   // turn output LED off
   digitalWrite(red_led_out_pin, LOW);
 }

}
```

Lab-14 XOR logic gate

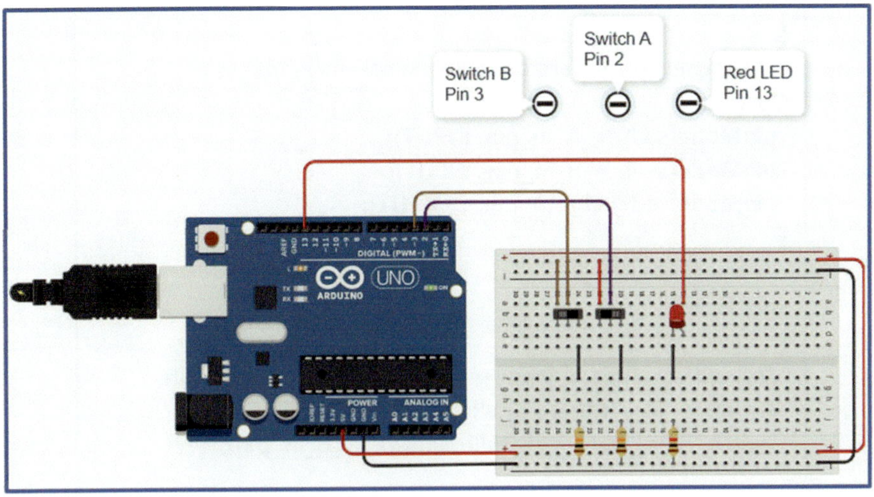


```
/*
Filename:
DigFun-14 XOR logic gate

Read in 2 digital (Boolean-valued) inputs (A and B) wired using
slide switches, and develop a digital output, Z=AB'+A'B
*/

// Assign user-understandable names to Arduino I/O pins
const int switch_A_in_pin = 2;
const int switch_B_in_pin = 3;
const int red_led_out_pin = 13;

// Status of Input pins is monitored by variables
boolean switch_A_in_state;
boolean switch_B_in_state;
```

```
void setup()
{
 // Assign user-understandable pin names as Inputs or
Outputs
 pinMode(switch_A_in_pin, INPUT);
 pinMode(switch_B_in_pin, INPUT);
 pinMode(red_led_out_pin, OUTPUT);
}

void loop()
{
 // Read in values of input switches, storing current state
 switch_A_in_state = digitalRead(switch_A_in_pin);
 switch_B_in_state = digitalRead(switch_B_in_pin);

 //  B  A Z=A xor B
 // 0 0 0
 // 0 1 1
 // 1 0 1
 // 1 1 0

 // Evaluate expression: Z=AB'+A'B
 if ( (switch_A_in_state == HIGH && // A closed and B open
      switch_B_in_state == LOW)
                          || // or
     (switch_A_in_state == LOW &&  // A open and B closed
      switch_B_in_state == HIGH) ){
   // When exactly one switch is closed, turn output LED on
   digitalWrite(red_led_out_pin, HIGH);
 }
 else {
   // When either both switches are closed or both are open,
   // turn output LED off
   digitalWrite(red_led_out_pin, LOW);
 }

}
```

Lab-15: XNOR logic gate

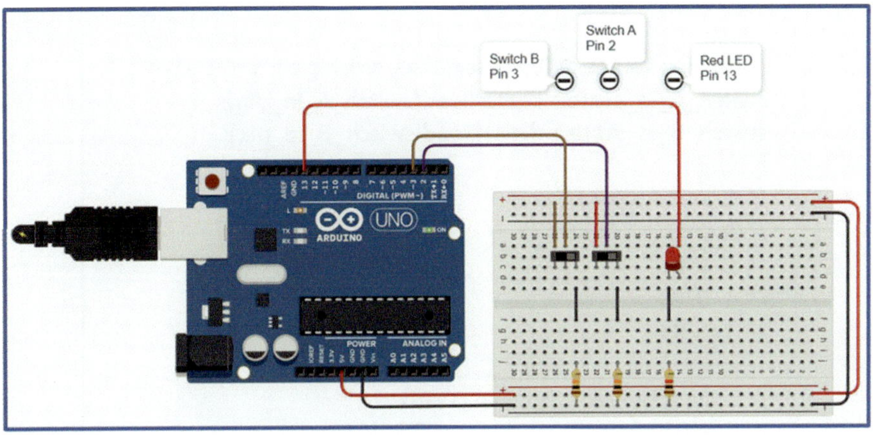


```
/*
Filename:
DigFun-15 XNOR logic gate

Read in 2 digital (Boolean-valued) inputs (A and B) wired using
slide switches, and develop a digital output, Z=A'B'+AB
*/

// Assign user-understandable names to Arduino I/O pins
const int switch_A_in_pin = 2;
const int switch_B_in_pin = 3;
const int red_led_out_pin = 13;

// Status of Input pins is monitored by variables
boolean switch_A_in_state;
boolean switch_B_in_state;

void setup()
{
 // Assign user-understandable pin names as Inputs or
Outputs
 pinMode(switch_A_in_pin, INPUT);
 pinMode(switch_B_in_pin, INPUT);
 pinMode(red_led_out_pin, OUTPUT);
}
```

```
void loop()
{
 // Read in values of input switches, storing current state
 switch_A_in_state = digitalRead(switch_A_in_pin);
 switch_B_in_state = digitalRead(switch_B_in_pin);

 //  B  A Z=A xnor B
 // 0 0 1
 // 0 1 0
 // 1 0 0
 // 1 1 1

 // Evaluate expression: Z=A'B'+AB
 if ( (switch_A_in_state == LOW &&  // A open and B open
      switch_B_in_state == LOW)
                              || // or
     (switch_A_in_state == HIGH && // A closed and B closed
      switch_B_in_state == HIGH) ){
  // When either both switches are closed or both are open,
  // turn output LED on
  digitalWrite(red_led_out_pin, HIGH);
 }
 else {
  // When exactly one switch is closed, turn output LED on
  digitalWrite(red_led_out_pin, LOW);
 }

}
```

DigFun-16a: Combinational Logic Circuits

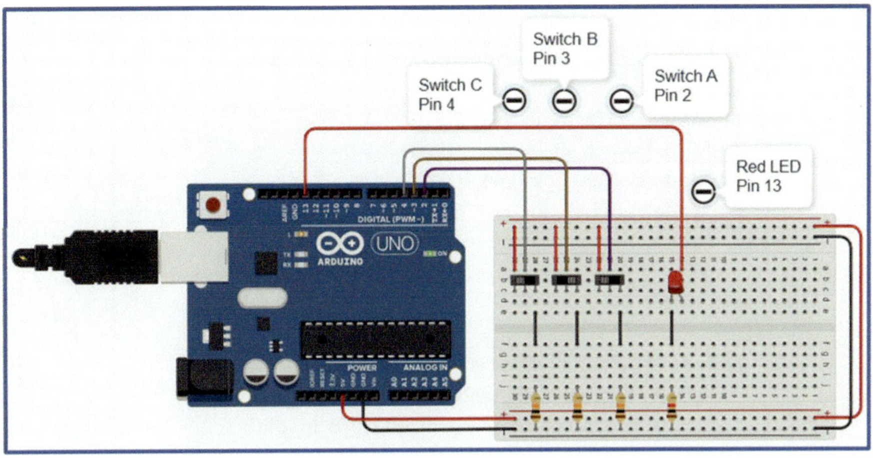


```
/*
Filename:
DigFun-16a Combinational logic gates

Read in 3 digital (Boolean-valued) inputs (A, B, C) wired
using slide switches, and develop a digital output, Z=A+BC',
simulating the operation of combinational logic gates, NOT,
AND, OR.
*/

// Assign user-understandable names to Arduino I/O pins
const int switch_A_in_pin = 2;
const int switch_B_in_pin = 3;
const int switch_C_in_pin = 4;
const int red_led_out_pin = 13;

// Status of I/O pins will be monitored by these variables
boolean switch_A_in_state;
boolean switch_B_in_state;
boolean switch_C_in_state;
```

```
void setup()
{
 // Assign user-understandable pin names as Inputs or
Outputs
 pinMode(switch_A_in_pin, INPUT);
 pinMode(switch_B_in_pin, INPUT);
 pinMode(switch_C_in_pin, INPUT);
 pinMode(red_led_out_pin, OUTPUT);
}

void loop()
{
 // Read in values of input switches, storing current state
 switch_A_in_state = digitalRead(switch_A_in_pin);
 switch_B_in_state = digitalRead(switch_B_in_pin);
 switch_C_in_state = digitalRead(switch_C_in_pin);

 // C B A  C' BC' Z=A+BC' Z'=A'(B'+C)
 // 0 0 0  1  0   0       1
 // 0 0 1  1  0   1       0
 // 0 1 0  1  1   1       0
 // 0 1 1  1  1   1       0
 // 1 0 0  0  0   0       1
 // 1 0 1  0  0   1       0
 // 1 1 0  0  0   0       1
 // 1 1 1  0  0   1       0

 // Evaluate expression: A+BC'
 if ( switch_A_in_state == HIGH ||
     ( switch_B_in_state == HIGH ) &&
     ( switch_C_in_state == LOW )
   ){
   // A+BC' is True, turn output LED on
   digitalWrite(red_led_out_pin, HIGH);
 }
 else {
  // A+BC' is False, turn output LED off
  digitalWrite(red_led_out_pin, LOW);
 }

}
```

Lab-16b: Combinational logic gates and status display

```
/*
Filename:
DigFun-16b Combinational logic gates and status display

Read in 3 digital (Boolean-valued) inputs (A, B, C) wired
using slide switches, and develop a digital output, Z=A+BC',
simulating the operation of combinational logic gates, NOT
AND, OR. I/O status is shown on screen using Serial Monitor.
*/


// Assign user-understandable names to Arduino I/O pins
const int switch_A_in_pin = 2;
const int switch_B_in_pin = 3;
const int switch_C_in_pin = 4;
const int red_led_out_pin = 13;

// Status of I/O pins will be monitored by these variables
boolean switch_A_in_state;
boolean switch_B_in_state;
boolean switch_C_in_state;
boolean red_led_out_state;

// 0.5s delay between successive runs of the loop() function
// for improving stability and easier viewing of Serial Monitor
const int loop_delay_time = 500;
```

```
void setup()
{
 // Assign user-understandable pin names as Inputs or
Outputs
 pinMode(switch_A_in_pin, INPUT);
 pinMode(switch_B_in_pin, INPUT);
 pinMode(switch_C_in_pin, INPUT);
 pinMode(red_led_out_pin, OUTPUT);

 // Initiate serial communication between Arduino and IDE
 // for displaying status of inputs and outputs
 Serial.begin(9600);
}

void loop()
{
 // Read in values of input switches, storing current state
 switch_A_in_state = digitalRead(switch_A_in_pin);
 switch_B_in_state = digitalRead(switch_B_in_pin);
 switch_C_in_state = digitalRead(switch_C_in_pin);

 Serial.print("switch_A_in_state is: ");
 Serial.println(switch_A_in_state); // print input A value
 Serial.print("switch_B_in_state is: ");
 Serial.println(switch_B_in_state); // print input B value
 Serial.print("switch_C_in_state is: ");
 Serial.println(switch_C_in_state); // print input C value


 // Given 3, boolean-valued inputs (A, B, C), the Truth Table
 // has 2^3=8 rows, count values from 0-7
 //
 // Count  C B A  C'  BC'  Z=A+BC'  Z'=A'(B'+C)
 //  0     0 0 0  1   0    0        1
 //  1     0 0 1  1   0    1        0
 //  2     0 1 0  1   1    1        0
 //  3     0 1 1  1   1    1        0
 //  4     1 0 0  0   0    0        1
 //  5     1 0 1  0   0    1        0
 //  6     1 1 0  0   0    0        1
 //  7     1 1 1  0   0    1        0
```

```
if ( switch_A_in_state == HIGH ||
     ( switch_B_in_state == HIGH ) &&
     ( switch_C_in_state == LOW )
   ){
  // A+BC' is True, turn output LED on
  digitalWrite(red_led_out_pin, HIGH);
}
else {
  // A+BC' is False, turn output LED off
  digitalWrite(red_led_out_pin, LOW);
}

// Read value of output Z, storing current state
red_led_out_state = digitalRead(red_led_out_pin);

Serial.print("Red LED output pin is: ");
// Serial.println(digitalRead(red_led_out_pin));
// print out variable Z value
Serial.println(red_led_out_state);
Serial.println(); // blank lines between output for clarity
delay(loop_delay_time); // delay between reads for stabilty
}
```

Lab-17a Rotary pot control of LED and Serial Monitor display

Interfacing the Arduino with analog inputs, changing the brightness of an output LED based on the value of a variable resistor (potentiometer)

```
/*
Filename:
DigFun-17a Rotary pot control of LED and Serial Monitor
display

Read in analog inputs (A0 of Arduino) using a potentiometer,
and when the voltage crosses a set threshold develop a digital
output, Z
*/

// Assign user-understandable names to Arduino I/O pins
// We connect a rotary potentiometer as analog input at pin
A0
// connect a red LED output at pin 8
// The #define statement, without semi-colon at the end
// creates a fixed the fixed value (label) when the Arduino
// sketch is compiled, replacing the defined word that follows
// (note, const int cannot be used, as A0 is not an integer)
#define rotary_pot_in_pin A0
const int red_led_out_pin = 13;

// Status of I/O pins will be monitored by these variables
// The value of the analog voltage read in on pin A0 is stored
// in an integer variable. This is a 10 bit value, which allows
// for up to 2^10=1024 values, ranging from 0 to 1023, or
// approximately 1000 different values.
```

```
// The variable type int can be used to store values in the
// range approximately -32K to +32K
int rotary_pot_in_value;
boolean red_led_out_state;

// Threshold beyond which output LED should be set
// Currently set to approximately a quarter of the range
// of values that can be read in on Analog pin A0
const int pot_threshold = 250;

// 0.5s delay between successive runs of the loop() function
// for easier viewing of the Serial Monitor display
const int loop_delay_time = 500;


void setup()
{
  // Analog pins A0-A5 are preassigned as analog INPUTs, so
  // these do not need to be assigned in the setup() as we
  // do for the digital Input pins
  pinMode(red_led_out_pin, OUTPUT);

  //Initialize serial communication at 9600 bits per second
  Serial.begin(9600);
}


void loop()
{
  // Use the inbuilt function
  // Arduino analogRead(Specify-Arduino-Analog-pin)
  // to record the current dial position and assign the value
  // to the rotaryPotIn_Value integer variable
  rotary_pot_in_value = analogRead(rotary_pot_in_pin);
  Serial.print("The analog value read in at pin A0 is = ");
  Serial.println(rotary_pot_in_value);

  // Evaluate whether the current pot value exceeds theshold
  if ( rotary_pot_in_value > pot_threshold ) {
    digitalWrite(red_led_out_pin, HIGH); // LED switched on
  }
  else {
    digitalWrite(red_led_out_pin, LOW);  // LED switched off
  }
```

```
  // Read value of output Z, storing current state
  red_led_out_state = digitalRead(red_led_out_pin);

  Serial.print("Red LED output pin is: ");
  // Serial.println(digitalRead(red_led_out_pin));
  // print out variable Z value
  Serial.println(red_led_out_state);
  Serial.println(); // blank lines between output for clarity

  delay(loop_delay_time);
}
```

Lab-17b: Rotary pot control of LED on-off duration and Serial

Interfacing the Arduino with analog inputs, changing the brightness of an output LED based on the value of a variable resistor (potentiometer)

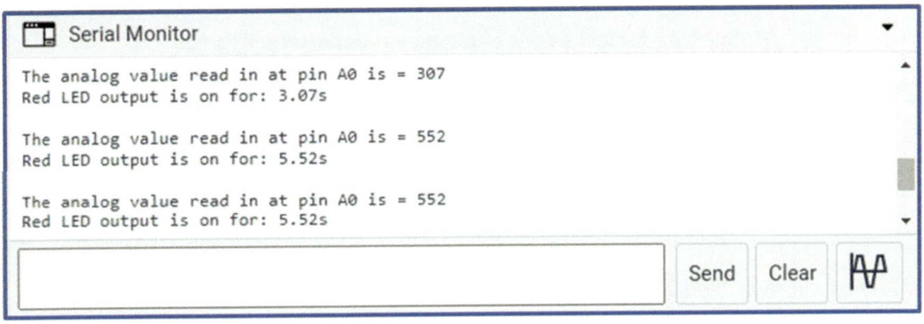


```
/*
Filename:
DigFun-17b Rotary pot control of LED on-off duration and
Serial

Read in analog input (A0 of Arduino) using a potentiometer,
and control the on-time for digital output, Z
*/

// Assign user-understandable names to Arduino I/O pins
// We connect a rotary potentiometer as analog input at pin
A0
// connect a red LED output at pin 8
// The #define statement, wihtout semi-colon at the end
// creates a fixed the fixed value (label) when the Arduino
// sketch is compiled, replacing the defined word that follows
// (note, const int cannot be used, as A0 is not an integer)
#define rotary_pot_in_pin A0
const int red_led_out_pin = 13;

// Status of Input pin will be monitored by the variable
// The value of the analog voltage read in on pin A0 is stored
// in an integer variable. This is a 10 bit value, which allows
// for up to 2^10=1024 values, ranging from 0 to 1023, or
// approximately 1000 different values.
// The variable type int can be used to store values in the
// range approximately -32K to +32K
int rotary_pot_in_value;
```

```
// Multiplier value of pot by 10 for converting it to seconds
// Pot values read in on the analog pin A0 vary between 0-
1023,
// otherwise for low values of A0, the on-duration will be too
// short to view clearly
const int pot_time_scaling = 10;

// 0.1s delay between successive runs of the loop() function
// for easier viewing of the Serial Monitor display
const int loop_delay_time = 100;

void setup()
{
 // Analog pins A0-A5 are preassigned as analog INPUTs, so
 // these do not need to be assigned in the setup() as we
 // do for the digital Input pins
 pinMode(red_led_out_pin, OUTPUT);

 //Initialize serial communication at 9600 bits per second(bps)
 Serial.begin(9600);
}

void loop()
{
 // Use the inbuilt function
 // Arduino analogRead(Specify-Arduino-Arduino-pin)
 // to record the current dial position and assign the value
 // to the rotaryPotIn_Value integer variable
 rotary_pot_in_value = analogRead(rotary_pot_in_pin);
 Serial.print("The analog value read in at pin A0 is = ");
 Serial.println(rotary_pot_in_value);

 // Switch the LED on and off for the duration read in by
 // the analogRead function. Since the delay (value-in-ms)
 // function reads in values between 0-1000 (approx.), the
 // corresponding on and off values would range from
 // 0-1000ms. To see a more sizable delay, we scale it by
 // 10 times. Now, with the dial turned to max,
 // it will be 10s.
 digitalWrite(red_led_out_pin, HIGH); // LED switched on
 // LED on duration is set by pot value read in
 delay(rotary_pot_in_value*pot_time_scaling);
```

```
 digitalWrite(red_led_out_pin, LOW);  // LED switched off
 // LED off duration is set by pot value read in
 delay(rotary_pot_in_value*pot_time_scaling);

 Serial.print("Red LED output is on for: ");
 Serial.print(rotary_pot_in_value*pot_time_scaling/1000.0);
 Serial.println("s");
 Serial.println(); // blank lines between output for clarity

 delay(loop_delay_time);
}
```

Lab-18a Photoresistor control of LED and Serial display

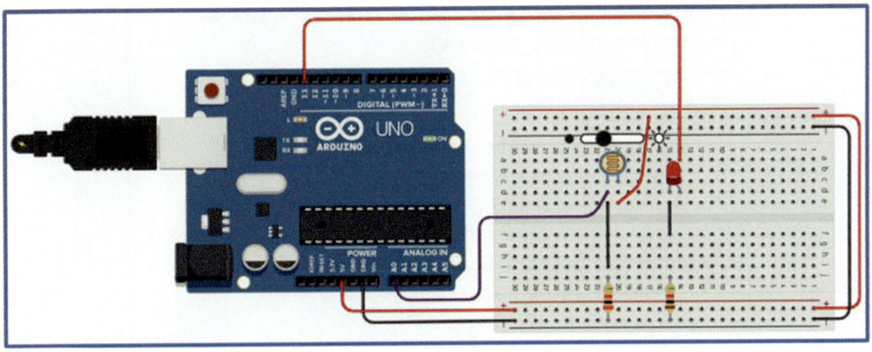

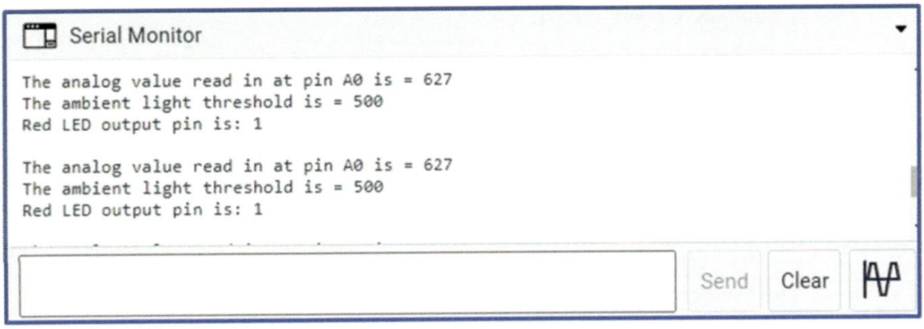

```
/*
Filename:
DigFun-18 Photoresistor control of LED and Serial display

Read in analog input (A0 of Arduino) using a photoresistor
(light dependent resistor), and when the ambient value
crosses
a set threshold develop a digital output, Z
*/

// Assign user-understandable names to Arduino I/O pins
// We connect a rotary potentiometer as analog input at pin
A0
// connect a red LED output at pin 8
// The #define statement, wihtout semi-colon at the end
// creates a fixed the fixed value (label) when the Arduino
// sketch is compiled, replacing the defined word that follows
// (note, const int cannot be used, as A0 is not an integer)
#define photoresistor_in_pin A0
const int red_led_out_pin = 13;


// Status of I/O pins will be monitored by these variables
// The value of the analog voltage read in on pin A0 is stored
// in an integer variable. This is a 10 bit value, which allows
// for up to 2^10=1024 values, ranging from 0 to 1023, or
// approximately 1000 different values.
// The variable type int can be used to store values in the
// range approximately -32K to +32K
int photoresistor_in_value;
boolean red_led_out_state;

// Threshold beyond which output LED should be set
// Currently set to approximately a half of the range
// of values that can be read in on Analog pin A0
const int ambient_light_threshold = 500;

// 0.5s delay between successive runs of the loop() function
// for easier viewing of the Serial Monitor display
const int loop_delay_time = 500;
```

```
void setup()
{
 // Analog pins A0-A5 are preassigned as analog INPUTs, so
 // these do not need to be assigned in the setup() as we
 // do for the digital Input pins
 pinMode(red_led_out_pin, OUTPUT);

 //Initialize serial communication at 9600 bits per second
(bps)
 Serial.begin(9600);
}

void loop()
{
 // Use the inbuilt function
 // Arduino analogRead(Specify-Arduino-Arduino-pin)
 // to record the current dial position and assign the value
 // to the rotaryPotIn_Value integer variable
 photoresistor_in_value = analogRead(photoresistor_in_pin);
 Serial.print("The analog value read in at pin A0 is = ");
 Serial.println(photoresistor_in_value);
```

```
 Serial.print("The ambient light threshold is = ");
 Serial.println(ambient_light_threshold);
 // Evaluate whether the current pot value exceeds theshold
 if ( photoresistor_in_value > ambient_light_threshold ) {
        digitalWrite(red_led_out_pin, HIGH); // LED switched
on
 }
 else {
   digitalWrite(red_led_out_pin, LOW);  // LED switched off
 }

 // Read value of output Z, storing current state
 red_led_out_state = digitalRead(red_led_out_pin);

 Serial.print("Red LED output pin is: ");
 // Serial.println(digitalRead(red_led_out_pin));
 // print out variable Z value
 Serial.println(red_led_out_state);
 Serial.println(); // blank lines between output for clarity
 delay(loop_delay_time);
}
```

Lab-18b Photoresistor control of LED using PWM

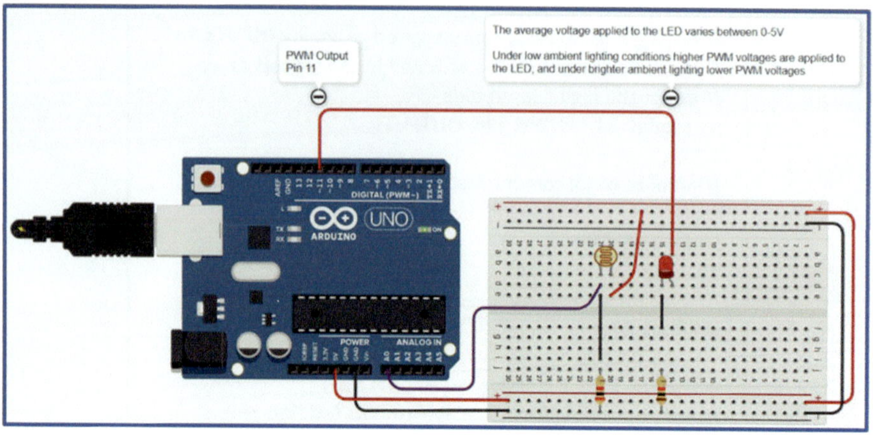

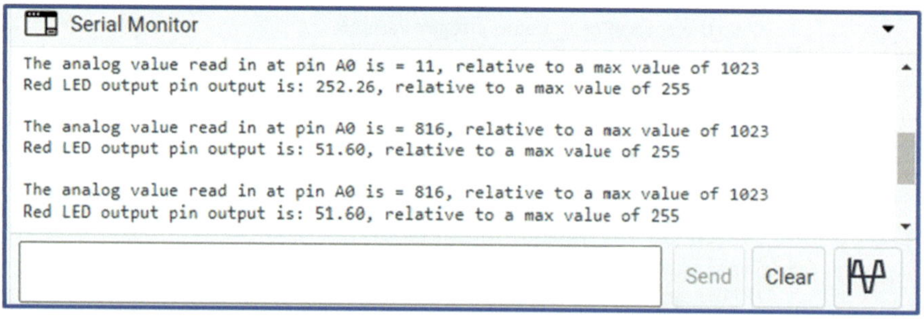


```
/*
Filename:
DigFun-18b Photoresistor control of LED using PWM

Read in analog input (A0 of Arduino) using a photoresistor
(light dependent resistor), and based on the ambient lighting
develop a Pulse Width Modulated (PWM) output for
brightening
or dimming the output of an LED
*/
```

```
// Assign user-understandable names to Arduino I/O pins
// We connect a rotary potentiometer as analog input at pin
A0
// connect a red LED output at pin 8
// The #define statement, wihtout semi-colon at the end
// creates a fixed the fixed value (label) when the Arduino
// sketch is compiled, replacing the defined word that follows
// (note, const int cannot be used, as A0 is not an integer)
#define photoresistor_in_pin A0
const int red_led_out_pin = 11;

// Status of I/O pins will be monitored by these variables
// The value of the analog voltage read in on pin A0 is stored
// in an integer variable. This is a 10 bit value, which allows
// for up to 2^10=1024 values, ranging from 0 to 1023, or
// approximately 1000 different values.
// The variable type int can be used to store values in the
// range approximately -32K to +32K
int photoresistor_in_value;

// Based on value of ambient light adjust Pulse Width
Modulated
// (PWM) output going to the LED; ranging between 0 to 255
float red_led_pwm_value;

// 0.5s delay between successive runs of the loop() function
// for easier viewing of the Serial Monitor display
const int loop_delay_time = 500;


void setup()
{
  // Analog pins A0-A5 are preassigned as analog INPUTs, so
  // these do not need to be assigned in the setup() as we
  // do for the digital Input pins
  pinMode(red_led_out_pin, OUTPUT);

  //Initialize serial communication at 9600 bits per second
(bps)
  Serial.begin(9600);
}
```

```
void loop()
{
 // Use the inbuilt function
 // Arduino analogRead(Specify-Arduino-Arduino-pin)
 // to record the current dial position and assign the value
 // to the rotaryPotIn_Value integer variable
 photoresistor_in_value = analogRead(photoresistor_in_pin);
 Serial.print("The analog value read in at pin A0 is = ");
 Serial.print(photoresistor_in_value);
 Serial.println(", relative to a max value of 1023");

 // Under low lighting conditions, increase output to LED
 // Under bright lighting conditions, decrease output to LED
 red_led_pwm_value = 255-
(photoresistor_in_value/1023.0)*255.0;

 // Only Arduino pins 3, 5, 6, 9, 10, or 11 permit PWM ops
 // analogWrite(Specify-pwm-pin, value-between-0-to-255)
 analogWrite(red_led_out_pin, red_led_pwm_value);

 Serial.print("Red LED output pin output is: ");
 Serial.print(red_led_pwm_value);
 Serial.println(", relative to a max value of 255");
 Serial.println(); // blank lines between output for clarity
 delay(loop_delay_time);
}
```

Lab-19: Switch control of 7-segment display

Controlling a 7-segment display using an Arduino

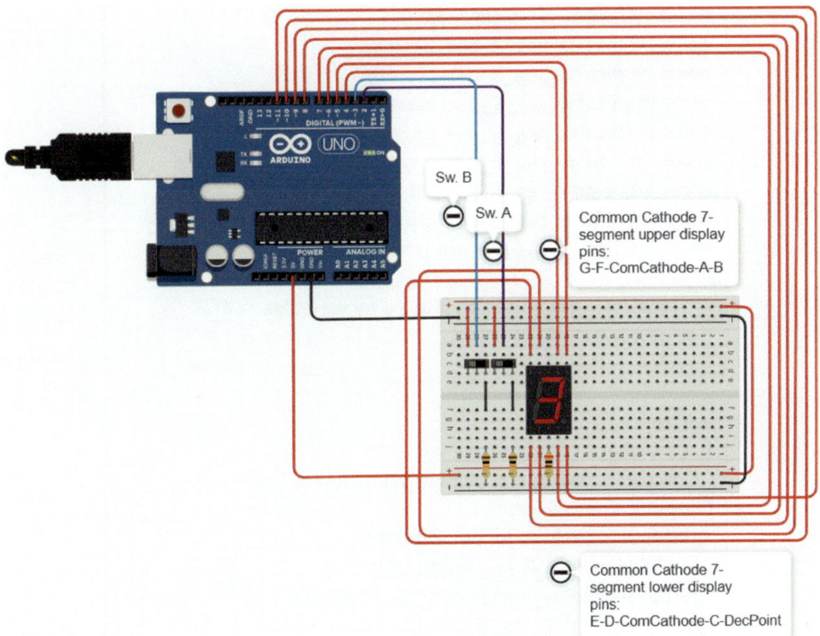


```
/*
Filename:
DigFun-19 Switch control of 7-segment display

Read in 2 digital (Boolean-valued) inputs (A and B) wired using
slide switches, and display the count on a 7-segment display
*/
```

```
// Assign user-understandable names to Arduino I/O pins
const int switch_A_in_pin = 2;
const int switch_B_in_pin = 3;
const int display_seg_A_out_pin = 4;
const int display_seg_B_out_pin = 5;
const int display_seg_C_out_pin = 6;
const int display_seg_D_out_pin = 7;
const int display_seg_E_out_pin = 8;
const int display_seg_F_out_pin = 9;
const int display_seg_G_out_pin = 10;
const int display_seg_DP_out_pin = 11;

// Status of Input pins is monitored by variables
boolean switch_A_in_state;
boolean switch_B_in_state;

void setup()
{
 // Assign user-understandable pin names as Inputs or
Outputs
 pinMode(switch_A_in_pin, INPUT);
 pinMode(switch_B_in_pin, INPUT);
 pinMode(display_seg_A_out_pin, OUTPUT);
 pinMode(display_seg_B_out_pin, OUTPUT);
 pinMode(display_seg_C_out_pin, OUTPUT);
 pinMode(display_seg_D_out_pin, OUTPUT);
 pinMode(display_seg_E_out_pin, OUTPUT);
 pinMode(display_seg_F_out_pin, OUTPUT);
 pinMode(display_seg_G_out_pin, OUTPUT);
 pinMode(display_seg_DP_out_pin, OUTPUT);
}

void loop()
{
 // Read in values of input switches, storing current state
 switch_A_in_state = digitalRead(switch_A_in_pin);
 switch_B_in_state = digitalRead(switch_B_in_pin);

 // B A Count
 // 0 0 0
 // 0 1 1
 // 1 0 2
 // 1 1 3
```

```
if ((switch_A_in_state == 0) &&
   (switch_B_in_state == 0)){
  // count at 0
  digitalWrite(display_seg_A_out_pin, HIGH);
  digitalWrite(display_seg_B_out_pin, HIGH);
  digitalWrite(display_seg_C_out_pin, HIGH);
  digitalWrite(display_seg_D_out_pin, HIGH);
  digitalWrite(display_seg_E_out_pin, HIGH);
  digitalWrite(display_seg_F_out_pin, HIGH);
  digitalWrite(display_seg_G_out_pin, LOW);
  // Also, put the decimal point on to indicate the start
  // of the counting sequence
  digitalWrite(display_seg_DP_out_pin, HIGH);
}
else if ((switch_A_in_state == 1) &&
       (switch_B_in_state == 0)){ // count at 1
  digitalWrite(display_seg_A_out_pin, LOW);
  digitalWrite(display_seg_B_out_pin, HIGH);
  digitalWrite(display_seg_C_out_pin, HIGH);
  digitalWrite(display_seg_D_out_pin, LOW);
  digitalWrite(display_seg_E_out_pin, LOW);
  digitalWrite(display_seg_F_out_pin, LOW);
  digitalWrite(display_seg_G_out_pin, LOW);
}
else if ((switch_A_in_state == 0) &&
       (switch_B_in_state == 1)){ // count at 2
  digitalWrite(display_seg_A_out_pin, HIGH);
  digitalWrite(display_seg_B_out_pin, HIGH);
  digitalWrite(display_seg_C_out_pin, LOW);
  digitalWrite(display_seg_D_out_pin, HIGH);
  digitalWrite(display_seg_E_out_pin, HIGH);
  digitalWrite(display_seg_F_out_pin, LOW);
  digitalWrite(display_seg_G_out_pin, HIGH);
}
else { // count at 3
  digitalWrite(display_seg_A_out_pin, HIGH);
  digitalWrite(display_seg_B_out_pin, HIGH);
  digitalWrite(display_seg_C_out_pin, HIGH);
  digitalWrite(display_seg_D_out_pin, HIGH);
  digitalWrite(display_seg_E_out_pin, LOW);
  digitalWrite(display_seg_F_out_pin, LOW);
  digitalWrite(display_seg_G_out_pin, HIGH);
}
}
```

Lab-20 Memory remembering status of inputs and Toggle mode

Adding memory (latch/flip-flop) functionality by remembering the status of an input, and also including "toggle mode" that turns an output on and off at regular intervals when both the inputs are activate

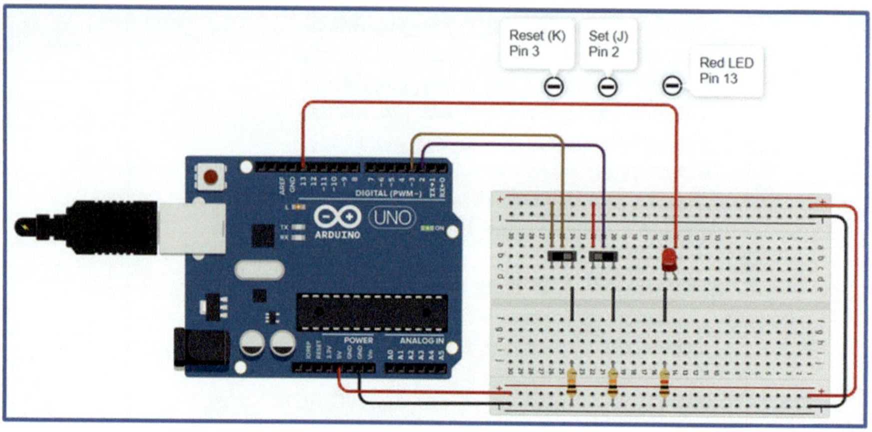

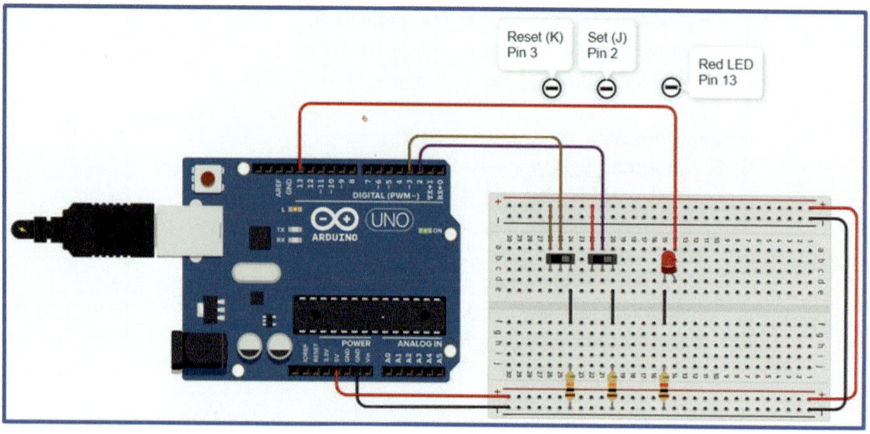


```
/*
Filename:
DigFun-20 Memory for remembering status of inputs and Toggle
mode

Read in 2 digital (Boolean-valued) inputs (Set/J and Reset/K)
wired using slide switches, and develop a digital memory output
that remembers the previous state of the output. When both
inputs are active, the output enters Toggle mode, turn the
output on and off at regular intervals.
*/
```

```
// Assign user-understandable names to Arduino I/O pins
const int switch_J_in_pin = 2;
const int switch_K_in_pin = 3;
const int red_led_out_pin = 13;

// Status of Input pins is monitored by variables
boolean switch_J_in_state;
boolean switch_K_in_state;

// When device enters Toggle mode, include a 0.5s delay
// between the on and off
const int toggle_delay_time = 500; // 500 ms on-off duration

// Memory variable for tracking status of output
// We will explicitly set it when J=1, K=0;
// explicity clear it when J=0, K=1;
// do not change it when J=0, K=0
boolean red_led_out_mem;


void setup()
{
 // Assign user-understandable pin names as Inputs or
Outputs
 pinMode(switch_J_in_pin, INPUT);
 pinMode(switch_K_in_pin, INPUT);
 pinMode(red_led_out_pin, OUTPUT);
}


void loop()
{
 // Read in values of input switches, storing current state
 switch_J_in_state = digitalRead(switch_J_in_pin);
 switch_K_in_state = digitalRead(switch_K_in_pin);

 // K J Q
 // 0 0 Mem
 // 0 1 1
 // 1 0 0
 // 1 1 Toggle
```

```
if ( switch_J_in_state == HIGH &&
   switch_K_in_state == LOW){ // J=1 and K=0
  // Set the output high when switch J is closed and K is open
  digitalWrite(red_led_out_pin, HIGH); // Output on
  red_led_out_mem = HIGH; // set the memory variable
}
else if ( switch_J_in_state == LOW &&
  switch_K_in_state == HIGH){ // S=1 and R=0
  // Set the output low when switch J is open and K is closed
  digitalWrite(red_led_out_pin, LOW); // Output on
  red_led_out_mem = LOW; // clear the memory variable
}

else if ( switch_J_in_state == LOW &&
  switch_K_in_state == LOW){ // J=0 and K=0
  // when both inputs J and K are open, value of the output
  // is determined by the status of switches previously;
  // do not change the value of  memory, simply use it here
  digitalWrite(red_led_out_pin, red_led_out_mem);
  // output remains at previously set value
}

else {
  //When both J and K are closed, initiate toggle mode
        digitalWrite(red_led_out_pin, HIGH);
        // Output on for specified time
  delay(toggle_delay_time);
        digitalWrite(red_led_out_pin, LOW);
  // Output off for specified time
  delay(toggle_delay_time);
}

}
```

Lab-21a Simple 2-bit up counter with reset

Multiple ways to program a 2-bit counter are shown in Lab-21a-c, showing that the same goal can be achieved in multiple way. Chose any of these methods, create your own, research online for alternative ways to program multi-bit counters on the Arduino.

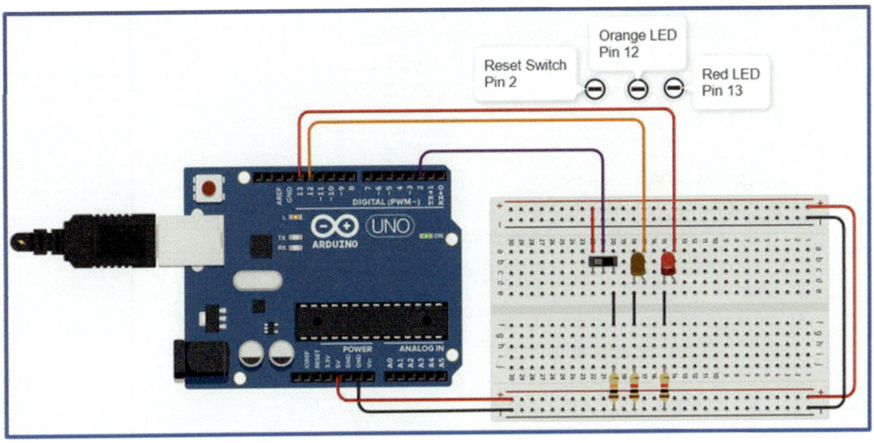


```
/*
Filename:
DigFun-21a Simple 2-bit up counter with reset

4 states (0 to 3) of a 2-bit up counter are implemented
using 2 LEDs, with a time delay between different states
*/


// Assign user-understandable names to Arduino I/O pins
const int switch_Reset_in_pin = 2;
const int red_led_out_pin = 13;
const int orange_led_out_pin = 12;

// 1s delay between successive counter values
const int count_delay_time = 1000;
```

```
// Status of reset Input pins is monitored by variable
boolean switch_Reset_in_state;
void setup()
{
 // Assign user-understandable pin names as Inputs or
Outputs
  pinMode(switch_Reset_in_pin, INPUT);
  pinMode(red_led_out_pin, OUTPUT);
  pinMode(orange_led_out_pin, OUTPUT);
}

void loop()
{
  // Read in values of input switch, storing current state
  switch_Reset_in_state = digitalRead(switch_Reset_in_pin);

  //  Pin 12    Pin 13  Count
  // Orange LED Red LED
  //   0       0     0
  //   0       1     1
  //   1       0     2
  //   1       1     3

  // Evaluate status of the reset input
  if ( switch_Reset_in_state == HIGH ){
    // When the reset switch is closed start the count
    // Count 0
    digitalWrite(red_led_out_pin, LOW);
    digitalWrite(orange_led_out_pin, LOW);
    delay(count_delay_time);

    // Count 1
    digitalWrite(red_led_out_pin, HIGH);
    digitalWrite(orange_led_out_pin, LOW);
    delay(count_delay_time);

    // Count 2
    digitalWrite(red_led_out_pin, LOW);
    digitalWrite(orange_led_out_pin, HIGH);
    delay(count_delay_time);
```

```
    // Count 3
    digitalWrite(red_led_out_pin, HIGH);
    digitalWrite(orange_led_out_pin, HIGH);
    delay(count_delay_time);
  }
  else {
    // When the reset switch is open reset after the
    // current counting sequence completes
    digitalWrite(red_led_out_pin, LOW);
    digitalWrite(orange_led_out_pin, LOW);
  }

}
```

Lab-21b 2-bit up counter with reset using for loop

```
/*
Filename:
DigFun-21b A 2-bit up counter with reset using for loop

4 states (0 to 3) of a 2-bit up counter are implemented
using 2 LEDs, with for loop moving between different states
*/


// Assign user-understandable names to Arduino I/O pins
const int switch_Reset_in_pin = 2;
const int red_led_out_pin = 13;
const int orange_led_out_pin = 12;

// 1s delay between successive counter values
const int count_delay_time = 1000;

// Status of reset Input pins is monitored by variable
boolean switch_Reset_in_state;

void setup()
{
 // Assign user-understandable pin names as Inputs or
Outputs
  pinMode(switch_Reset_in_pin, INPUT);
  pinMode(red_led_out_pin, OUTPUT);
  pinMode(orange_led_out_pin, OUTPUT);
}

void loop()
{
  // Read in values of input switch, storing current state
  switch_Reset_in_state = digitalRead(switch_Reset_in_pin);

  //  Pin 12    Pin 13  Count
  // Orange LED Red LED
  //   0         0      0
  //   0         1      1
  //   1         0      2
  //   1         1      3
```

```
 // Evaluate status of the reset input
 if ( switch_Reset_in_state == HIGH ){
   // When the reset switch is closed start the count
   for (int i=0; i<4; i++){
     if ((i==1)||(i==3)) {
       //Red Led on when count, i, is 1 or 3
       digitalWrite(red_led_out_pin, HIGH);
     }
     else {
       digitalWrite(red_led_out_pin, LOW);
     }

     if (i>1){
       // Orange on when count, i, is 2 or 3
       digitalWrite(orange_led_out_pin, HIGH);
     }
     else {
       digitalWrite(orange_led_out_pin, LOW);
     }
     delay(count_delay_time);
   }
 }
 else {
   // When the reset switch is open reset after the
   // current counting sequence completes
   digitalWrite(red_led_out_pin, LOW);
   digitalWrite(orange_led_out_pin, LOW);
 }

}
```

Lab-21c Alt 2-bit up counter with reset using for loop

```
/*
Filename:
DigFun-21c Alt 2-bit up counter with reset using for loop

4 states (0 to 3) of a 2-bit up counter are implemented
using 2 LEDs, with for loop moving between different states
*/


// Assign user-understandable names to Arduino I/O pins
const int switch_Reset_in_pin = 2;
const int red_led_out_pin = 13;
const int orange_led_out_pin = 12;

// 1s delay between successive counter values
const int count_delay_time = 1000;

// Status of reset Input pins is monitored by variable
boolean switch_Reset_in_state;

void setup()
{
 // Assign user-understandable pin names as Inputs or
Outputs
 pinMode(switch_Reset_in_pin, INPUT);
 pinMode(red_led_out_pin, OUTPUT);
 pinMode(orange_led_out_pin, OUTPUT);
}

void loop()
{
 // Read in values of input switch, storing current state
 switch_Reset_in_state = digitalRead(switch_Reset_in_pin);

 // Pin 12   Pin 13  Count   Count%2=Remainder
 // Orange LED Red LED      (1 for odd numbers)
 //  0       0     0        0
 //  0       1     1        1
 //  1       0     2        0
 //  1       1     3        1
```

```
// Evaluate status of the reset input
if ( switch_Reset_in_state == HIGH ){

  for (int i=0; i<4; i++){
   if ( i%2==1 ){
    // remainder of i/2 is 1 for odd numbers:1,3
    digitalWrite(red_led_out_pin, HIGH);
   }
   else {
    digitalWrite(red_led_out_pin, LOW);
   }

   if ( i>1 ){
    // 2nd Led on when count is 2 or 3, i.e. i>i
    digitalWrite(orange_led_out_pin, HIGH);
   }
   else {
    digitalWrite(orange_led_out_pin, LOW);
   }
   delay(count_delay_time);
  }

 }
 else {
  // When the reset switch is open reset after the
  // current counting sequence completes
   digitalWrite(red_led_out_pin, LOW);
   digitalWrite(orange_led_out_pin, LOW);
 }

}
```

Lab-22: Combining sequential and combinational functions

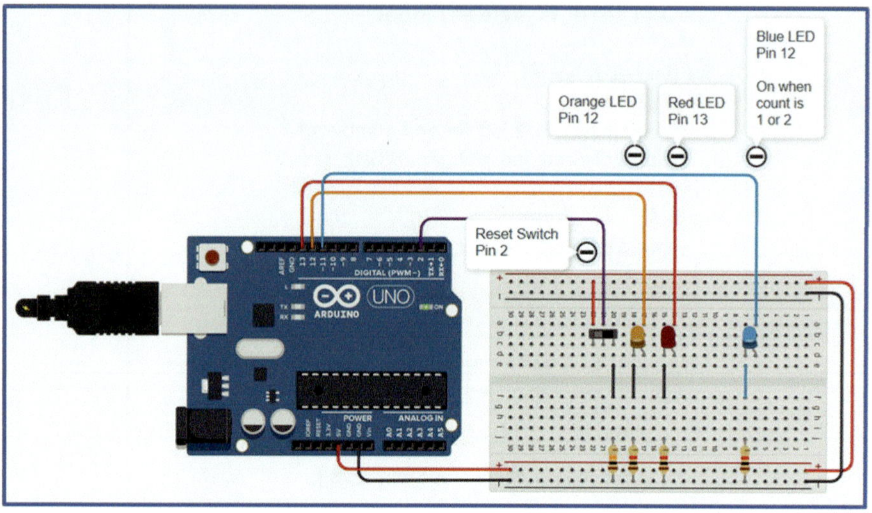


```
/*
Filename:
DigFun-22 Combining sequential and combinational logic

Decode specific states of an up-counter for turning on
an output LED
*/

// Assign user-understandable names to Arduino I/O pins
const int switch_Reset_in_pin = 2;
const int red_led_out_pin = 13;
const int orange_led_out_pin = 12;
const int blue_led_out_pin = 11;

// 3s delay between count values
const int count_delay_time = 3000;

// Status of reset Input pins is monitored by variable
boolean switch_Reset_in_state;
```

```
void setup()
{
 // Assign user-understandable pin names as Inputs or
Outputs
 pinMode(switch_Reset_in_pin, INPUT);
 pinMode(red_led_out_pin, OUTPUT);
 pinMode(orange_led_out_pin, OUTPUT);
 pinMode(blue_led_out_pin, OUTPUT);
}


void loop()
{
 // Read in values of input switch, storing current state
 switch_Reset_in_state = digitalRead(switch_Reset_in_pin);

 // LED-12 LED-13 Count
 //  0      0      0
 //  0      1      1
 //  1      0      2
 //  1      1      3

 // Evaluate status of the reset input
 if ( switch_Reset_in_state == HIGH ){

   for (int i=0; i<4; i++){
    if ((i==1)||(i==3)) {
     //Red Led on when count, i, is 1 or 3
     digitalWrite(red_led_out_pin, HIGH);
    }
    else {
     digitalWrite(red_led_out_pin, LOW);

     if (i>1){
      // Orange Led on when count, i, is 2 or 3
       digitalWrite(orange_led_out_pin, HIGH);
     }
     else {
       digitalWrite(orange_led_out_pin, LOW);
     }
    }
```

```
    // Turn the blue LED on when the count is either 1 or 2
    if ((i==1)||(i==2) ) {
          digitalWrite(blue_led_out_pin, HIGH);
    }
    else {
      digitalWrite(blue_led_out_pin, LOW);
      }

     delay(count_delay);

   }
  }
  else {
   digitalWrite(red_led_out_pin, LOW);
   digitalWrite(orange_led_out_pin, LOW);
  }

}
```

Lab-23a Control Piezo buzzer frequency and duration

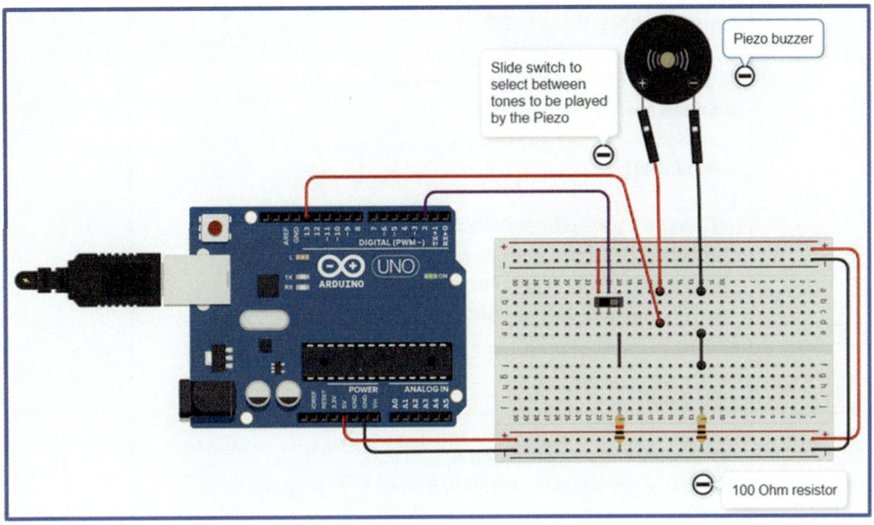


```
/*
Filename:
DigFun-23a Control Piezo buzzer frequency and duration

Read in 1 digital (Boolean-valued) input (A) wired using a
slide switch, and based on the value play either one tone
or another for different durations
*/

// Assign user-understandable names to Arduino I/O pins
const int switch_A_in_pin = 2;
const int piezo_buzzer_out_pin = 13;

// Play higher frequency, 1500 Hz, when switch open
const int high_frequency = 500;

// Play low frequency, 500 Hz, when switch open
const int low_frequency = 500;

// 1s between successive loops
const int inter_loop_time = 1000;

// Longer duration sound
const int long_time = 800;
```

```
// Shorter duration sound
const int short_time = 100;

// Status of Input pin is monitored by a variable
boolean switch_A_in_state;

void setup()
{
 // Assign user-understandable pin names as Inputs or
Outputs
 pinMode(switch_A_in_pin, INPUT);
 pinMode(piezo_buzzer_out_pin, OUTPUT);
}

void loop()
{
 // Evaluate status of switch A
 switch_A_in_state = digitalRead(switch_A_in_pin);

 if ( switch_A_in_state == HIGH ){
  // When switch A closed, play higher freq for longer
  // tone(pin13, 1500Hz, 800ms)
  tone(piezo_buzzer_out_pin, high_frequency, long_time);
  // tone (pin, frequency, on duration in ms)
  delay(inter_loop_time);
  // Results in a gap of 1000ms-800ms=200ms since the
  // tone and the the next commands, in this case delay
  // start simultaneously, and the tone stops after 800ms
  // so there is a 200ms=.2s gap with no sound before the'
  // loop repeats
 }
 else {
  // switch A open, play lower freq for shorter duration
  // tone(pin13, 500Hz, 100ms)
  tone(piezo_buzzer_out_pin, low_frequency, short_time);
  delay(inter_loop_time); // Gap of 900ms
 }

 /* A piezoelectric material changes its shape when a voltage
 is applied to it, converting a voltage applied to it into
 a mechancial vibrarion, which in air can produces an audible
 sound when it is placed in a casing or by using a diaphram.
 */
}
```

Lab-23b Play musical notes or other sound on Piezo buzzer

```
/*
Filename:
DigFun-23b Play musical notes or other sound on Piezo buzzer

Read in 1 digital (Boolean-valued) input (A) wired using a
slide switch, and based on the value play either do-re-mi
or a low frequency tone
*/


// Assign user-understandable names to Arduino I/O pins
const int switch_A_in_pin = 2;
const int piezo_buzzer_out_pin = 13;

// Play low frequency, 500 Hz, when switch open
const int low_frequency = 500;

// Frequencies corresponding to different musical notes
// stored in an integer array
const int tones[8]= {262,  //c
              294,  //d
              330,  //e
              349,  //f
              392,  //g
              440,  //a
              494,  //b
              523   //c, next octave
             };

// duration for playing each note
const int sound_time = 500;

// 1s between successive loops
const int inter_loop_time = 1000;

// Status of Input pin is monitored by a variable
boolean switch_A_in_pin_state;
```

```
void setup()
{
 // Assign user-understandable pin names as Inputs or
Outputs
 pinMode(switch_A_in_pin, INPUT);
 pinMode(piezo_buzzer_out_pin, OUTPUT);
}


void loop()
{
 // Evaluate status of switch A
 switch_A_in_pin_state = digitalRead(switch_A_in_pin);

 if ( switch_A_in_pin_state == HIGH ){
  // When switch A closed, play do-re-mi
   for (int tone_count=0; tone_count<8; tone_count++){
    //tone(pin, frequency, duration)
    tone(piezo_buzzer_out_pin, tones[tone_count],
sound_time);
         delay (inter_loop_time);

   // tone(piezo_buzzer_out_pin, 1000, 800);
   // tone (pin, frequency, on duration in ms)
   // delay(inter_loop_time)
   // Results in a gap of 1000ms-800ms=200ms since the
   // tone and the next commands, in this case delay
   // start simultaneously, and the tone stops after 800ms
   // so there is a 200ms=.2s gap with no sound before the'
   // loop repeats
   }
 }
 else { // switch A open, play low freq
  tone(piezo_buzzer_out_pin, low_frequency, sound_time);
  delay(inter_loop_time); // Gap of 800ms
 }

}
```

Lab-24a Control a higher voltage DC motor

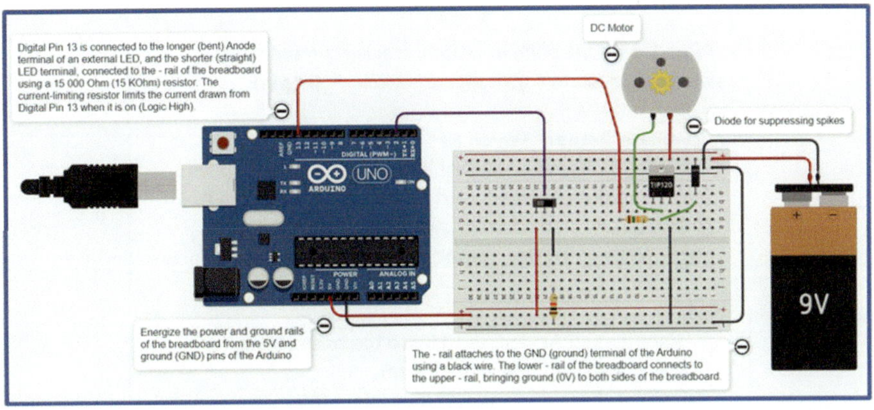


```
/*
Filename:
DigFun-24a Control a higher voltage DC motor

Read in 1 digital (Boolean-valued) input (A) wired using a
slide switch, and if closed turn a higher voltage DC motor on
using a transistor
*/

// Assign user-understandable names to Arduino I/O pins
const int switch_A_in_pin = 2;
const int transistor_motor_out_pin = 13;

// Status of Input pin is monitored by a variable
boolean switch_A_in_state;

void setup()
{
 // Assign user-understandable pin names as Inputs or
Outputs
 pinMode(switch_A_in_pin, INPUT);
 pinMode(transistor_motor_out_pin, OUTPUT);
}
```

```
void loop()
{
  // Read in values of input switch, storing current state
  switch_A_in_state = digitalRead(switch_A_in_pin);

  // Evaluate status of switch A
  if ( switch_A_in_state == HIGH ){
    // When switch A closed send 5V to base of the TIP 120
    // NPN power transistor for switching it on
    digitalWrite(transistor_motor_out_pin, HIGH);
  }
  else {
    // When switch A open, send 0V to the base of the
    // NPN transistor for switching it off
    digitalWrite(transistor_motor_out_pin, LOW);
  }

  /* The TIP 120 power transistor has 3 terminals:
  Emitter, which should be connected to the common ground,
  Base, which should be connected to the output of the
Arduino
  through a suitable resistor,
  Collector, which should be connected to one end of the
motor,
  and the other end of the motor should be connected to the
  higher DC power terminal, in this case the top of the
  breadboard which is connected to +9V DC.

  In order to suppress voltage spike when the motor is
switched
  off, a "freewheeling" or "flyback" diode is connected for
  dissipating the energy.

  Note that the black ground wire from the Arduino and of
black
  ground wire from the 9V DC supply should be connected for
  providing a common reference for the control circuit.
  */
}
```

Lab-24b Control the position of a servo motor

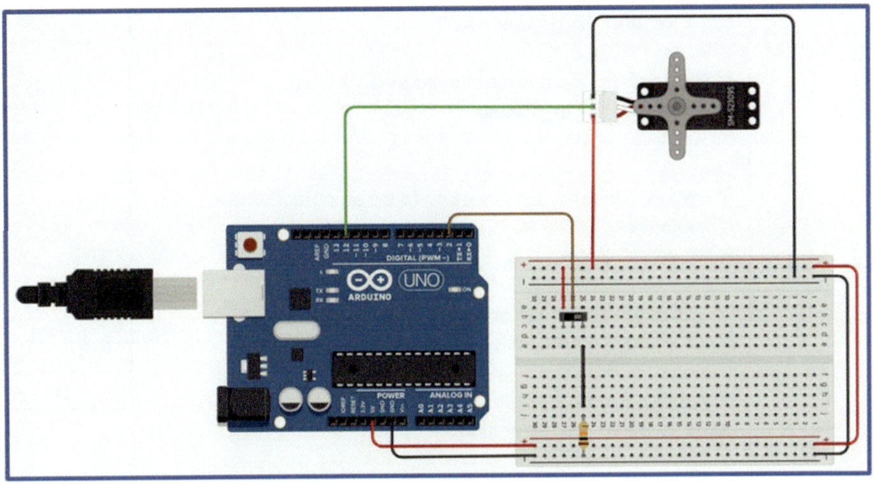


```
/*
Filename:
DigFun-24b Control the position of a servo motor

Read in 1 digital (Boolean-valued) input (A) wired using a
slide switch, and if closed rotate a servo motor first one way
by a specified number of degrees, and then the other way;
otherwise return to home position
*/


// Include the specialized Servo.h library at the start of in
// this sketch. Servo.h includes the read() function which
// allows us to determine the current angle it is at; the
// write() function for setting to move to a specified angle;
// and following creation of a "Servo" type object,
// attaching it to a specific pin of the Arduino. Doing so
// connects the signal/command pin of the servo to the
// specified Arduino output in the software.
#include <Servo.h>

// Assign user-understandable names to Arduino I/O pins
const int slider_switch_in_pin = 2;
const int servo_motor_out_pin = 12;
```

```
// time delay between successive loops
const int inter_loop_time = 100;

// Status of Input pin is monitored by a variable
int slider_switch_in_status;


// Declare the servo we are using to be an object of the
// Servo class
Servo servo_motor;

// 0.5s time steps while moving the servo
const int unit_time = 500;

// position 0 degrees home
const int servo_home_position = 0;

void setup()
{
 // Assign user-understandable pin names as Inputs or
Outputs
 pinMode(slider_switch_in_pin, INPUT);
 //Attach the servo motor to the Arduino pin
 servo_motor.attach(servo_motor_out_pin);
}

void loop()
{
 // If slide-switch is on turn the servo one way and back
 // and if the slide-switch is off lock it at home position
 slider_switch_in_status = digitalRead(slider_switch_in_pin);
 if (slider_switch_in_status == HIGH){

   for (int position=0; position<15; position++){
     servo_motor.write(position);
     delay (unit_time);
   }

   for (int position=15; position>0; position--){
     servo_motor.write(position);
     delay (unit_time);
   }
 }
```

```
else {
  // After reset or reload of program, if sliderswitch is off
  // the servo performs a homing operation, then moves to
the
  // 0 degree position
  servo_motor.write(servo_home_position);
 }

 delay(inter_loop_time);

 /*
 Servo motors can be configured to move to a specific
 position (in degrees) or by specific amounts. Internal
 sensors provide feedback about the motors position and
 velocity. In addition to a power and ground pin, the servo
 has a signal/command pin. Servos are continuously powered,
 often with a 50Hz square wave which has a 5-10% duty cycle
 (on time/total time), with the a control pulse applied
 every 20ms. Typically, servos have a range of 180 degrees.
 A pulse of duration 1ms (within the 20ms window) applied to
 its signal line will move the servo to the 0 degree position;
 while one of 2 me will rotate it to other end (180 degrees),
 with 1.5ms positioning it 90 degrees, and so on. A type of
 software Pulse Width Modulation (PWM) is used to control
the
 position of the servo. The servo can be connected to any of
 the Digital pins on the Arduino board.
 */

}
```

Lab-25 Message on LCD

Displaying a customized message (Lab it to learn! Time Flies: ###s, where # represents number of seconds since the program is first run or since a rest) on a LCD using inbuilt libraries and sample public domain code available for free through Tinkercad.

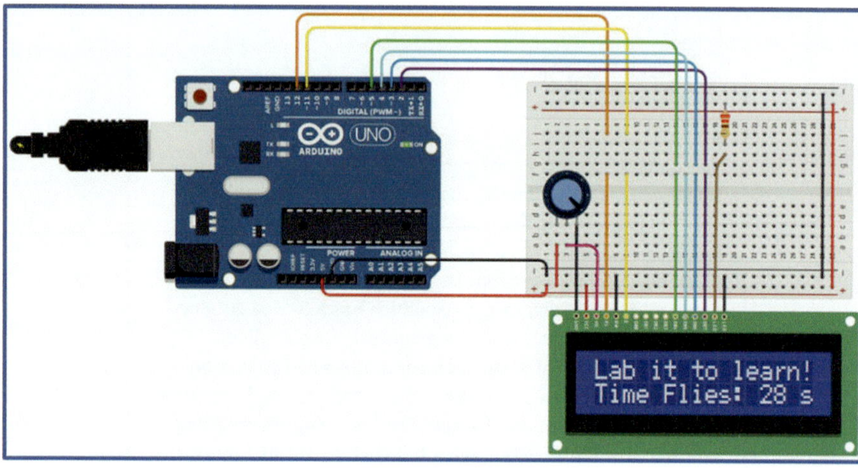


```
/*
Filename:
DigFun-25 Message on LCD

This sketch initially printed "Hello World!" to the LCD and
showed the time in seconds since the program started. It is
now modified to print:
  "Lab it to Learn!"
  "Time Flies: ###s
; where # represents the time elapsed in seconds since the
program was first run or reset.

Our colleague and friend Dale Patrick was a proponent of
hands-on learning in an interactive lab environment, enjoying
staying immersed in active experimentation. You, too, may find
that the lightness which hobby-learning the things you are
interested in, such as electronics, offers is one of the most
durable types of learning, allowing you to be present here and
now, aware of the task at hand, of your lab environment,
joyfully immersed, tinkering.

Modified 31 Jan. 2022 by Vigs Chandra, Steve Fardo and  Brian Fardo.
*/
```

```
//
// C++ code
//
/*
 LiquidCrystal Library - Hello World

 Demonstrates the use of a 16x2 LCD display.
 The LiquidCrystal library works with all LCD
 displays that are compatible with the Hitachi
 HD44780 driver. There are many of them out
 there, and you  can usually tell them by the
 16-pin interface.

 The circuit:
 * LCD RS pin to digital pin 12
 * LCD Enable pin to digital pin 11
 * LCD D4 pin to digital pin 5
 * LCD D5 pin to digital pin 4
 * LCD D6 pin to digital pin 3
 * LCD D7 pin to digital pin 2
 * LCD R/W pin to ground
 * LCD VSS pin to ground
 * LCD VCC pin to 5V
 * 10K resistor:
 * ends to +5V and ground
 * wiper to LCD VO pin (pin 3)

 Library originally added 18 Apr 2008  by David
 A. Mellis
 library modified 5 Jul 2009  by Limor Fried
 (http://www.ladyada.net)
 example added 9 Jul 2009  by Tom Igoe
 modified 22 Nov 2010  by Tom Igoe

 This example code is in the public domain.
 http://www.arduino.cc/en/Tutorial/LiquidCrystal
*/

#include <LiquidCrystal.h>

int seconds = 0;

LiquidCrystal lcd_1(12, 11, 5, 4, 3, 2);
```

```
void setup()
{
 lcd_1.begin(16, 2); // Set up the number of columns and
             // rows on the LCD.

 // Print a message to the LCD.
 lcd_1.print("Lab it to learn!");

 // set the cursor to column 0, line 1
 // (note: line 1 is the second row, since counting
 // begins with 0):
 lcd_1.setCursor(0, 1);
 lcd_1.print("Time Flies:    s");
}

void loop()
{
 // set the cursor to column 0, line 1
 // (note: line 1 is the second row, since counting
 // begins with 0, reposition cursor at column location after
 // the phrase "Time Flies: x   s"), at the x, at column 12.
 lcd_1.setCursor(12, 1);
 // print the number of seconds since reset:
 lcd_1.print(seconds);
 delay(1000); // Wait for 1000 millisecond(s)
 seconds += 1;
}
```

List of Programs

Program 23: **Lab-23a:** Control Piezo buzzer frequency and duration
Lab-23b: Play musical notes or other sound on Piezo buzzer

Program 24: **Lab-24a:** Control a higher voltage DC motor
Lab-24b: Control the position of a servo motor

Program 25: **Lab-25:** Message on LCD

This sketch initially printed "Hello World!" to the LCD and showed the time in seconds since the program started. It is now modified to print:

"Lab it to Learn!"

"Time Flies: ###s"

; where # represents the time elapsed in seconds since the program was first run or reset.

About the Authors

Dale R. Patrick was a Professor in the Department of Applied Engineering and Technology at Eastern Kentucky University. His experience includes teaching and organizing laboratory classes in electrical engineering/electronics technology for many years. He completed Bachelor of Science and Graduate Degrees at Indiana State University. His experience also includes research projects, technical teacher training, and energy consulting for business and industry. He and Dr. Fardo have co-authored over 30 textbooks and laboratory manuals.

Stephen W. Fardo is currently a Foundation Professor Emeritus in the Department of Applied Engineering and Technology at Eastern Kentucky University. His experience includes teaching and organizing laboratory classes in Electrical Engineering/ Electronics Technology for many years and coordinating / advising technical teacher programs for the State of Kentucky. He completed Bachelor of Science, Master of Science and Doctoral Degrees at Eastern Kentucky University and University of Kentucky. His experience also includes research projects, technical teacher education, and energy consulting for business and industry. He and Dale Patrick have co-authored over 30 textbooks and laboratory manuals.

Vigyan (Vigs) Chandra is a professor and coordinator of the Cyber Systems Technology-related undergraduate and graduate degree programs offered by the Departments of Computer Science and Information Technology (CSIT) and Engineering Technology at Eastern Kentucky University. He earned a doctoral degree from the University of Kentucky in Electrical and Computer Engineering and a master's degree in Career and Technical Education from Eastern Kentucky University. He holds certifications in various computer networking areas and teaches computer systems and applications, network hardware, communication systems, and digital, analog, and machine-control electronics. He is the recipient of the 2020–2021 Critical Thinking and Reading Teacher Award and the 2013 Golden Apple Award for Teaching Excellence at Eastern. His professional interests include implementing active teaching and learning strategies, metacognition, integrating open-source software/hardware with online control and computer networking technologies.

Brian W. Fardo is an Adjunct Professor in the Department of Applied Engineering and Technology and the Noel Studio for Academic Creativity at Eastern Kentucky University. His experience includes teaching classes in applied creative thinking, electrical systems, materials and processes, transportation, manufacturing, construction, engineering graphics, and quality control. He has a B.Sc. Degree in Industrial Distribution and Logistics, an M.Sc. in Technology Systems from East Carolina University and an M.Sc. in Career and Technical Education Eastern Kentucky University. He has completed coursework for a Doctoral Degree in Engineering/Technology Education at North Carolina State University. He currently owns and operates a telecommunications business.